Shaligram Pilgrimage
in the Nepal Himalayas

Shaligram Pilgrimage in the Nepal Himalayas

Holly Walters

Routledge
Taylor & Francis Group
LONDON AND NEW YORK

First published in 2020 by Amsterdam University Press Ltd.

Published 2025 by Routledge
4 Park Square, Milton Park, Abingdon, Oxon OX14 4RN
605 Third Avenue, New York, NY 10158

Routledge is an imprint of the Taylor & Francis Group, an informa business

ISBN: 9789463721721 (hbk)
ISBN: 9781041185994 (pbk)
ISBN: 9781003703594 (ebk)
NUR 740

Cover illustration: Mustang, Nepal, 2016
Photo taken by the author

Cover design: Coördesign, Leiden

DOI: 10.5117/9789463721721

For Product Safety Concerns and Information please contact our EU representative:
GPSR@taylorandfrancis.com
Taylor & Francis Verlag GmbH, Kaufingerstraße 24, 80331 München, Germany

For my husband, Chris, the incomparable householder

"Om Namo Bhagavathey Vishnavey Sri Salagrama
Nivasiney – Sarva Bheesta Bhalapradhaya
Sakala Thuridha Nivarine Salagrama Swahah!"
– *Salagrama mula mantra* (*From Shaligram Mahimai by Murali Battar*)

(I pray that the LORD Sriman Mahavishnu –
who is residing inside the Salagrama, which provides all wishes,
fulfills all desires – quickly answer all our prayers)

Table of Contents

Acknowledgements

My research for this book began in 2012, when I first arrived in India on a midnight flight to Kolkata with little more than a small grant and a bus ticket to Nabadwip. Since then, I have accrued many debts. First, my thanks go to the many pilgrims and devotees who took the time and interest to involve themselves in this work. Without their patience, careful correction, and suggestions for other avenues of inquiry, my research in both India and Nepal would not have been possible. In India, I owe a debt of gratitude to Rama Vigraha Das and his wife Muralipriya, without whom I would never have been introduced to Shaligram stones or their meanings. Those afternoons of fresh coconuts and conversation formed the very first foundations of what is written here, and their continued help from afar has only made the narrative and experiences richer. Additionally, I wish to express my gratitude to Mahalakshmi devadasi and Krishnalaulya devadasi, who took me under their care as I wandered the villages of West Bengal and saw to it that I became more family member than researcher.

In Nepal, I offer sincere thanks to Kul Bahadur Gurung, whose knowledge of and connections among travel companies and local travel guides ensured that an ethnography of mobility remained feasible, even when monsoons, landslides, and high altitudes seemed to suggest otherwise. While my research may have informed him as much as it informed me when it came to religious practices in Nepal, there is simply no substitute for the in-depth knowledge of mobility and concern for successful travel out in the field that he brought to this project. Additionally, no project such as this would have succeeded if not for the guidance and assistance of Dil Gurung, who brought me to Mustang and back again more than once and each time in one piece. Without his assistance and guidance in and around the villages of the Muktinath Valley and high above in the fossil beds of the Annapurna mountain range, this research would never have benefitted from the perspectives of Hindu, Buddhist, and Bonpo peoples in the way that it has. Finally, my thanks go to Dinesh and Renuka Thapa, who took me into their home for the better part of a year, sat me at their dinner table each night, and called me sister.

Deep thanks are also due to my three most dedicated mentors; Janet McIntosh, Ellen Schattschneider, and Sarah Lamb, who guided this work both with care and with keen minds in the theoretical and practical approaches of anthropology and ethnographic work in South Asia. Professor Frank Salomon of the University of Wisconsin-Madison was the first to begin my training in the anthropology of religion so many years ago, but it has been

Dr. McIntosh, Dr. Schattschneider, and Dr. Lamb whose pointed questions and profound theoretical mastery that has shaped the contours of this work.

Because this project crossed a number of disciplinary boundaries, I must also express my thanks to both Jen Bauer of the University of Tennessee and to Christian Klug of the University of Zurich in Switzerland for their kind assistance in fielding all of my paleontological questions. Without their patience and assistance in identifying the ammonite species represented in this project as well as taking the time to explain many of the important geological processes necessary for their formation, I would not have been able to join the discourses of science and the discourses of religion together in the ways that I have. I may have been a dinosaur enthusiast as a child, but their expertise has had value beyond measure.

My gratitude also goes to the Mellon-Sachar Foundation for the initial funding that took this project from India to the high Himalayas of Nepal in 2015 and also to the Fulbright Commission, who funded the entirety of the final year of research both in Kathmandu and all across Mustang from 2016 to 2017. Yamal Rajbhandary and Mily Pradhan provided invaluable support throughout the process, and I thank them for their faith in this research and genuine interest in its outcomes.

Finally, my thanks to the many loved ones who supported the research, writing, and completion of this book. Without their patience and sacrifice, such an endeavor would never have been possible. Thank you to my husband, Christopher, for always believing I could do it even when I doubted it over late night tea and endless revisions. Thank you to my parents and extended family, Patricia Buske, "Porky" Buske, Tom and Shari Harsdorf who followed my travels and tribulations as closely as distance would allow. And for inspiring me to great learning as a child and for continuing to celebrate my successes with exuberance usually reserved for visiting dignitaries, my last and sincerest thank you to my grandmother, Frieda Wiech, who passed away just one week after I arrived in Nepal in 2016. Though she knew her final days had come and I had offered to turn around and return home immediately, she would have none of it. I was exactly where she wanted me to be.

And last, but certainly not least, my thanks to the Kali Gandaki and the deities of Shaligram; for your appearances, for allowing me to follow along in your journeys, and for your kind guidance during *puja* and *darshan* in the many homes of welcoming devotees. I could only go where you would lead.

Note on Transliteration

While many books and articles transcribe words from Hindi, Nepali, Tibetan, and Sanskrit using standard diacritical conventions (*darśan, purāṇa*), I have chosen to transliterate personal names (Naga Baba, Shankaracarya), referents (Sri, Mataji), deity names (Shiva, Vishnu), place names (Muktinath, Pashupatinath, Kathmandu), and the names of scriptural texts along with Sanskrit, Hindi, or Nepali language source materials (Skanda Purana, Devibhagavata) into standard English. Due to both their multiple spelling variations across a number of linguistic fields and their inconsistent representation in quoted texts, my choice to render these words using standard English conventions is intended for both consistency and reader clarity.

Additionally, words that have become incorporated into standard English usage (Shaligram, ashram) have been neither italicized nor diacriticized, except when their use in the original language may differ slightly in meaning or context from the English usage (*śālagrāma*). Finally, my overall choice to use "Shaligram" in general throughout this work is also due to the fact that "Shaligram" itself has a variety of different spellings and pronunciations in different areas of South Asia. For example, *sāligrāma* (dental) is the typical pronunciation of the term among devotees throughout South India and to some degree in North India and Nepal, while other sources insist on the pronunciation *śālagrāma* (palatal) as more correct and referential to the original Sanskrit pronunciation. But since "Shaligram" is generally recognizable by all individuals referenced in this work, it will be the standard term used herein.

1 Living Fossils

Impressions of a Once and Future World

Abstract
For more than two thousand years, the veneration of sacred fossil ammonites, called Shaligrams, has been an integral part of ritual practice throughout South Asia. Originating from a single remote region of Himalayan Nepal, in the Kali Gandaki River Valley of Mustang, the ritual use of these stones today has become a significant focus of pilgrimage and exchange in India and Nepal and among the global diaspora. But Shaligrams are also deeply intertwined with divine movement, and the challenges of travel to Mustang have resulted in restrictions on a ritual practice that depends upon the mobility of people and stones for its continuation. As a result, many practitioners now believe that the worship of Shaligrams may be in danger of disappearing.

Keywords: Shaligram, pilgrimage, Hinduism, fossil, ammonite

Full fathom five thy father lies:
Of his bones are coral made:
Those are pearls that were his eyes:
Nothing of him that doth fade,
But doth suffer a sea-change
Into something rich and strange.
– *Shakespeare, The Tempest*

A Hindu pilgrim, recently arrived from South India, stood anxiously next to a bus stand in Mustang, Nepal. "I'm going to burn my passport," he said. "I'm going to destroy all my documents and go to Damodar. I came here (on pilgrimage) to find Shaligram and I will find Shaligram. You can't put borders on sacred land." I was taken aback. The Damodar Kund, a glacial lake several

Walters, Holly, *Shaligram Pilgrimage in the Nepal Himalayas*. Taylor & Francis Group, 2020
DOI: 10.5117/9789463721721_CH01

days' walk far to the north, lay beyond the boundary between Upper Mustang and Lower Mustang, and without special permits and astronomical fees, foreigners were not allowed to cross into the politically contentious zone between Chinese-occupied Tibet and Himalayan Nepal. But this was not the first time I would encounter these sentiments. More than once, a Hindu or Buddhist pilgrim would explain how they might hide their passports in a mountain crevasse, strip off their clothes and travel as mute hermits (so that their accents would not give them away) and steal across the border late at night or in an area where there were no roads for government jeeps to travel. In every case, the reasoning was the same: they had come in search of sacred stones, and there was no border that could stop them. This was Shaligram pilgrimage, and where the Shaligram goes, so do the people.

For more than two thousand years, the veneration of sacred fossil ammonite stones, called Shaligram Shila (or alternatively, *sāligrāma śila* or *śālagrāma śila*),[1] has been an integral part of Hindu ritual practice throughout Nepal and the Indian subcontinent. While ammonite fossils are common throughout the world, these unique types of black shale river fossils originate from a single remote region of Himalayan Nepal, in the Kali Gandaki River Valley of Mustang District. Today, the ritual use of these stones has become a significant focus of pilgrimage, religious co-participation, and exchange between Nepal and India as well as among the global Hindu diaspora. Their characteristic ridged spirals and ebon-black coloration readily reveal their presence in the silty waters of the river as pilgrims and devotees step carefully through powerful currents to reach Shaligrams just beginning to appear out of the eroding riverbanks. Each Shaligram is one-of-a-kind where the forces that formed it have left behind distinctive combinations of characteristics: spiral shell reliefs, white quartz lines, and perfectly rounded black shale nodules sometimes paired with small holes or intricate internal impressions where the original fossil mold has long since worn completely away. But these characteristics are not only geological in nature, they must also be religiously interpreted, to determine precisely which deity has made their presence available within the stone.

Viewed primarily as natural manifestations of the Hindu god Vishnu, Shaligrams are called *svarupa* ("natural forms") and are therefore inherently

1 "Shaligram" has a variety of different spellings and pronunciations in different areas of South Asia. For example, *sāligrāma* (dental) is the typical pronunciation of the term among devotees throughout South India and to some degree in North India and Nepal, while other sources insist on the pronunciation *śālagrāma* (palatal) as more correct and referential to the original Sanskrit pronunciation.

sacred. The meaning of "natural", then, is often locally articulated as something whose formation lies distinctly outside of human agency (meaning, for example, through the will of the gods or with the landscape). Shaligrams are natural in that they are not human-made and ultimately demonstrate their own agency. For this reason, they do not require any rites of consecration or invocation, such as the *prana pratiṣṭha* (lit. establishing of breath/life force),[2] when brought into homes or temples. Shaligrams are also highly valued as symbolic manifestations of divine movement, either through a geologically and mythologically formative journey down the sacred river (which runs from the Southern Tibetan plateau down through central Nepal and into Northern India) or transnationally in the hands of devout pilgrims. Pouring out into the river each year following the summer melt high in the mountains, Shaligrams are gathered up by pilgrims, tourists, and merchants alike. On their way out of the mountains, they travel through forests and cities, into temples and homes, across great expanses of time and space, carried by the indescribable forces of nature or the complex networks of pilgrimage and exchange that underlie their vital mobility. As divine forms, Shaligram stones are representative of power expressed as a journey through a sacred landscape, and in the high Himalayas, religion is constantly on the move.

The ramshackle bus finally trundled its way up the steep road and groaned to a stop among the crowd of waiting devotees. As destination requests for far-flung villages or sacred sites were shouted between drivers and potential passengers, the pilgrim went on to remark, "I've known many who have destroyed their passports to get to Kali Gandaki. They say that I am foreign so that's why I need permits for Damodar (the source of the river in nearby Upper Mustang). But I am Hindu, and this land is *Shaligramam* (a reference to Vishnu). I could never be foreign here."

In recent decades, the mobility of Shaligrams has also come to represent the mobility of pilgrims and the fluidity of ritual practices themselves. Given Mustang's long-standing status as a travel-restricted political red zone, Shaligrams are fast becoming metonymic for sacred landscapes that are continuously coming into conflict with political landscapes. Through competing claims to Tibetan, Nepali, and indigenous origins, the national

2 *Prana pratistha* refers to the ritual by which a *murti* (image of a god) is consecrated in a Hindu temple. Hymns and mantras are recited to invite the deity to be a resident guest, and the murti's eyes are opened for the first time. Practiced in the temples of Hinduism and Jainism, the ritual is considered to infuse life into the image and bring to it the numinous presence of divinity.

identity of Mustang is currently framed by the region's perilous political position near the borders of Tibet, where it acts as a buffer region between China, India, and central Nepal. As a result, many Hindu and Buddhist pilgrims have come to treat these national borders and Mustang's political isolation as affronts to religious identities that depend upon individual mobility and the movement of sacred stones to extend family and community belonging beyond the boundaries of nation, ethnicity, or caste.

It is then this combination of movement through vast expanses of geological time, across historical and mythological landscapes, and into the daily lives of families and communities at the conclusion of pilgrimage that defines what it is to be Shaligram and therefore for devotees in turn, what it is to be Hindu or Buddhist. For Shaligram practitioners, to be in the presence of a Shaligram is to be in the constant presence of the gods themselves. To ritually worship a Shaligram is to accept the deities as members of one's own family, and to begin their veneration properly one must go to the places where Shaligram appears.[3] The aniconic character of Shaligrams and their natural formation within the Kali Gandaki comprise the first part of their journey into sacrality, and in this way, their geological formations as well as their geographical migrations from mountain to lake to river, which gives them their characteristic appearance, are just as much a part of their religious narrative as their legends and stories. Framing Shaligram practices through the themes of movement and time along with divine personhood and multispecies will then begin to reveal what Shaligrams are as well as how Hindu and Buddhist devotees in South Asia have come to experience them.

Moving in Time with Life

At its core, a Shaligram is symbolic movement made physically manifest. The journey begins with the stones' geological and mythological travels

3 Unfortunately, there are no current studies which indicate approximately what percentage of Hindus or Buddhists are also Shaligram practitioners. Regardless, such a study would find defining Shaligram practice in this way difficult given that there are no specific standards of practice related to the worship of sacred stones in South Asia. In other words, some Hindus may use Shaligrams along with a wide variety of other deity icons in their worship while others might worship only Shaligrams alone. Some may also only worship occasionally (such as at a temple which houses a Shaligram) while others practice Shaligram rituals daily. Some may keep Shaligrams in their homes, others may not. In any case, the wide variety and extension of Shaligram traditions in South Asia imply that Shaligram worship is relatively common on the subcontinent and likely has been for some time.

down the sacred river and includes their equally divine transnational mobility in the hands of devout pilgrims returning to homes in regions and countries throughout the world. But to complicate matters, because the national identity of Mustang remains ambiguous and contested, the Nepali government has continued to make travel and access difficult for pilgrims as it attempts to control and homogenize the territory's identity under the central purview of Kathmandu. As a result of various restrictions on pilgrimage, the Shaligrams' own mobility has thus become weighted with special meaning: for example, a Shaligram's natural movement in eroding out of the mountains and tumbling down into the river to flow outwards into the landscape becomes both an analogue and an alter for pilgrims' own ideal mobility – a parallelism that is only made possible by the kinship between stones and people.

Throughout this book, I will use these and other examples to demonstrate how the co-mobility of stones and pilgrims is a kind of political practice that facilitates formations of national community and religious identity. Emerging from particular geographic and historical settings, Shaligram practices treat sacred stones as divine persons who are both linked to places and transcendent of them. As a result, both pilgrims and regional peoples also come to experience themselves as being "from" a particular sacred place even though they encounter pressures from outside entities – such as local or foreign governments, geological and political research studies, and national resistance movements – that mark the landscape as unstable and contested. As time goes on, it is then the divine personhood of the Shaligrams themselves, as deities endowed with consciousness and intent and as kith and kin to ritual practitioners, that continues to connect community and kinship networks across extraordinary expanses of time and space and beyond the boundaries of nation-state, gender, ethnicity, or caste.

With a theoretical grounding in the anthropological literatures of space and place-making (i.e., Gupta and Ferguson 1992; Hirsch and O'Hanlon 1995; and Basso 1996) in articulation with ethnographies of personhood and kinship in South Asia (Lamb 2000; Carsten 2000; Franklin and McKinnon 2001; and Uberoi 1993), this work links together current theories of cultural time and material temporality (Mieu 2015; Hodges 2008; Munn 1992; Hanks 1990; Parmentier 1987; Geertz 1975; and Evans-Pritchard 1940) with critical mobility studies (Ingold 2011; Hausner 2007; Urry 2002; Fisher 2001; and Graburn 1989). While the object-agency and object-personhood (Gentry 2016; Geismar 2011; Hoskins 2006; and Schattschneider 2003) of Shaligrams themselves situates this research within broader theories of cultural linking (de Bruijn and van Dijk 2012; Salazar 2010; Horst and Miller 2005; Castells

2004; and Appadurai 1986), I contend throughout this work that it is the life-long and generational relationships between Shaligrams (as both divine persons and objects), their practitioners, pilgrimage, and the landscape that generates new kinds of histories, transactions, and social belonging. As such, this work also resonates with many of the current threads of inquiry now gaining ground in the study of religion and materiality. Religious scholars in this area focus specifically on how religion happens through material culture, which may include everything from images, ritual objects, architecture and sacred space, art and archaeology, and religious objects produced for decoration or mass consumption. In addition to material forms, this growing sub-discipline also addresses the role of different practices that engage material religion in spiritual action, such as how various types of ritual language and performance, teaching and instruction, pilgrimage, magic and spiritual medicine, or liturgy and exegesis constitute and maintain religious worlds (Morgan 1999, 2010). As a departure from Thing Theory (Brown 2004), however, I am less attentive to "thingness" here as an aspect of human-object interactions and more concerned, for reasons that will become obvious, with the ways in which practitioners contend with the "entity-like presence" (Fowles 2010) of Shaligrams and the agency of the deities manifest within them.

Issues of temporality will also come up repeatedly in the formulation of Shaligrams as kin, where they link ancestors with descendants in such a way as to construct families and communities of the living, the dead, and the divine. Time is a valuable tool in the ethnographer's toolkit. This is not only because anthropologists should attend to their own and their participant's temporal views in the construction of intersubjective fieldwork (such as Fabian 1983) but, as in the case of Shaligram ritual practices, it is the layering of different kinds of cyclical and linear events in the lives of both individuals and communities that reifies these complex identities and relationships in the present day.

Because the theme of boundaries is so pervasive in my findings, I must also attend to what it means to traverse those boundaries. In previous studies of mobility in Africa and in the Middle East, roads, electricity and infrastructure, mobile phones, and the internet have all become objects of study in the ways that they remake political and economic power relations and introduce new spaces of peril and precarity (Bishara 2015; Dilger et al. 2012; McIntosh 2009; and Comaroff and Comaroff 1993). But instead of focusing on the rise of specific technologies in the growing concerns about the global "field", I offer a new view on an old topic: sacred stones as mobile techne. In this mobility paradigm, interrelationships between place and culture are reframed by the interrelationships between object and personhood/kinship so that the ways in which (im)mobility and (in) flexibility shape relationships are foregrounded. Mobile techne, a concrete and context-dependent method of making something mobile, will then also involve physical movement through landscapes as well as symbolic and spiritual movement through the *dham*, the immaterial landscapes and dwelling places of the deities.

In this ethnography, mobility itself is the locus of transformation. But rather than a transient space of liminal *communitas* (in the manner of Victor Turner) that lies between categories of social belonging, mobility is taken here as a permanent potentiality that is quite literally set in stone. The mobility of Shaligrams thus provides an ambiguous, shifting space of change and renewal continuously available for renegotiating one's place in an unstable world. Along the pilgrimage route required to obtain a Shaligram, mobility includes the political practice of recreating and resisting ideologies of national and ethnic belonging in Nepal and India, where political restrictions on the movements of both people and Shaligrams becomes a contested point for the realization of national and religious identities. Like the *bhola*, the "gullible fools", of the Kanwar pilgrimage in India (Singh 2017), Shaligram pilgrims also use the mobile spaces of pilgrimage to reassert the power and sovereignty of individual lives regardless of economic, caste, or national status.

As practitioners and pilgrims move outwards and return to their places of residence, the mobility of Shaligrams is then translated into ritual and divine personhood through their intimate ties to community and kinship networks of reciprocity and exchange; a transformation that will also position them within community and familial relationships in a time of great social upheaval and out-migration. This places this particular ethnography of mobility into a somewhat odd space in relation to current anthropological literatures that use mobility as a framework for ethnographic study. Given that these literatures tend to focus primarily on the movement of groups of people in the context of social mobility[4] (the symbolic movement of

4 Grusky, David B. and Erin Cumberworth (February 2010). "A National Protocol for Measuring Intergenerational Mobility" (PDF). Workshop on Advancing Social Science Theory: The Importance of Common Metrics. Washington, D.C.: National Academy of Science. Retrieved 15 July 2014.Lopreato, Joseph & Hazelrigg, Lawrence E. (December 1970). "Intragenerational versus Intergenerational Mobility in Relation to Sociopolitical Attitudes." *Social Forces*. University of North Carolina Press. 49 (2): 200-210. JSTOR 2576520. doi: 10.2307/2576520.
Weber, Max (1946). "Class, Status, Party." In H.H. Girth and C. Wright Mills. From *Max Weber: Essays in Sociology*. New York: Oxford University. pp. 180-195.

individuals and groups between social strata), physical population mobility[5] (such as migrant labor, tourism and travel, pilgrimage, and diaspora studies), or ascetism and concepts of freedom,[6] this work draws together frameworks that merge physical and symbolic movement as conducted through object-person relationships, in the form of Shaligram stones, as producers of meaning and as positions for the negotiation of identity.

In this work, I am interested in how mobility stops being a means to an end (or a place) but instead becomes a vital practice in its own right. In the spirit of multi-sited ethnography (Marcus 1995), Shaligram practice can move out of the more traditional frameworks of single-site analysis to reveal the macro-constructions of a larger social order that cross-cuts common dichotomies such as "local" and "global" or "lifeworlds" and "systems." This is why I will use the term "mobility" with a view towards several equally important valences: to mean physical movement across space, to mean the potential for movement among both people and their deities, to mean exchange within kinship networks or as commodities, and to mean the capacity for a Shaligram to exist in both individual life times and in historical, generational, or geological time. Whenever necessary, I point out which meanings are most salient to the argument at hand, but it is important to realize that people often use the term "mobility" to leverage multiple meanings at once and are not especially troubled by the apparently contradictory pivots between one meaning or another. Rather, it becomes mobility itself that remains the primary concern.

Collins, Patricia Hill (1998). "Toward a new vision: race, class and gender as categories of analysis and connection. "*Social Class and Stratification*: Classic Statements and Theoretical Debates. Boston: Rowman & Littlefield. pp. 231-247.
Wilkinson, Richard & Kate Pickett (2009). "The Spirit Level: Why Greater Equality Makes Societies Stronger." Bloomsbury Press.

5 Ramesh, S.S.; Ganju, D.D.; Mahapatra, B.B.; Mishra, R.M.; Saggurti, N.N. (2012). *Relationship between mobility, violence and HIV/STI among female sex workers in Andhra Pradesh, India*. BMC Public Health. 12 (1): 764.
Nancy, L.; Kaivan, M. "Women as agents of change: Female income and mobility in India." *Journal of Development Economics*. 94 (1): 941-17.
Kim, Y.Y. (2010). "Female individualization? Transnational mobility and media consumption of Asian women." *Media, Culture and Society*. 32 (1): 25-43.
Schoenbaum, N (2012). "Mobility Measures." *Brigham Young University Law Review*. 2012 (4): 1169.
Blossfeld, Hans-Peter (2009). Life Course Inequalities in the Globalisation Process. Burlington, VT: Ashgate. p. 59

6 Hausner, Sondra L. 2007. *Wandering with Sadhus: Ascetics in the Hindu Himalayas*. Indiana University Press.
Salazar, Noel B. Salazar, and Kiran Jayaram (eds.) 2017. *Keywords of Mobility: Critical Engagements*. Berghahn Books.

A Lifetime of Movement

Shaligram pilgrimage is the first vital preceding step to Shaligram practice, which transitions movement in space (pilgrimage and exchange) to movement in time (birth, life, and death alongside the generations of families and communities). Because of this, Shaligrams add a fascinating new dimension to recent studies that use temporality to capture ever more dynamic ways of "being in time". Because temporality is not merely a product of structural circumstances but is produced through concrete practices where people come to actively construct and embody time (Bourdieu 1977; Munn 1992), the links between spatial movement and temporal movement become clearer where practitioners continuously reinterpret pilgrimage both as a physical journey to specific places for the purposes of personal transformation and as an ongoing, lifelong process of aging and achieving milestones. In fact, "pilgrimage as life" was such a common metaphor among Shaligram practitioners that it often became unclear as to whether they were referring to actual plans for an upcoming pilgrimage or were commenting more generally on the transient state of existence.

The layering of time, mobility, and space is also especially important when the meanings of Shaligram ritual practices are expanded outward into the global South Asian diaspora, who often view Shaligrams as vital links anchoring them back to family members back home, to ancestors, to presiding household deities, and to the sacred lands of pilgrimage. As kin, Shaligrams then articulate with reflexive models of temporal kinship that permit people to establish kin relations with persons they meet, persons they are biologically related to, or with otherwise unrelated persons elsewhere, as well as with ancestors and deities largely distant from this particular moment in time. Therefore, as a kind of composite "kinship chronotype" (Ball et al. 2015), Shaligram relationships enable persons and communities to co-locate themselves and their kin or kin-like others in a wide variety of places and times and to use these ties to participate in collective belonging outside of typical boundaries of caste, religion, or nationality. Taking temporality to be an embodied symbolic process unfolding in practice (Mieu 2015) through which people imagine themselves to inhabit a present in relation to various kinds of pasts and futures (Munn 1992: 115-116), we can then begin to understand how a variety of large-scale political and historical processes affect practitioners' views about national mobility and belonging which they may then articulate through communal ritual practices using sacred stones.

This view of time and nationalism is the most initially productive relative to Akhil Gupta and James Ferguson's notion of "space": a critique of location, displacement, and community that unmoors naturalized notions of essential places from their representations in geographical locations (1992). By attending to spaces and places as continually constituted and renegotiated over time, this project leverages mobility as an analytical viewpoint from which to re-theorize anthropological notions of contact, contradiction, and integration. For example, like other ethnographies of space and pilgrimage, mobility constitutes an intriguing ethnographic point of intervention into how the conceptual and material dimensions of places and landscapes are central to the production of social life.

But unlike ethnographies that tend to focus more on issues of global economic restructuring, urban displacement, migratory flows, or deterritorialization (Olwig and Hastrup 1997; Low 1999; Dawson et al. 2014; and Haenn and Wilk 2016) as they have undermined assumptions about the fixity of people, this work demonstrates how objects (Shaligrams) – which are distinctly from a specific place, transcend that place, and yet carry notions of space and place with them – upend assumptions about the fixity of place itself. As a result, Shaligram mobility demonstrates how identity and belonging are perceived differently among a community of religious practitioners who often describe themselves as being "from" a sacred place of pilgrimage that they have generally only ever visited (or have never seen at all, in some cases). What is more, by attending to the movability of placeness itself, the mobility of Shaligrams can then take on even more complex meanings of flexibility and ambiguity.

As one elderly Tibetan woman living in Pokhara, a town roughly 200 km west of Kathmandu, once described it, "I have lived in Nepal since the '60s but I am not Nepali. I was born in Lokha (a city in southeastern Tibet) but I have lived on the other side of the Himalayas for most of my life, so I am not Tibetan really either. But I carry Shaligram with me. Do you see it in my *chuba*? It also comes from Tibet, but it was found at the refugee camp in Mustang where my family settled. It stays with me always. It can't be taken away. So, you see, that is where I am from."[7] Whether she meant from Mustang or from "Shaligram", she would not say.

Deities as Multispecies

The presence of a Shaligram stone within a Hindu household often marks the family as especially pious, and many devotees consider the worship

7 Conversation in English and Nepali. Transcribed from recorded conversation.

of Shaligrams as a way to link themselves, their ancestors, and their descendants with ancient Hindu stories and traditions and with sacred places extending thousands of years into the past. As is often the case in Shaligram worship, specific stones are also associated with the specific ancestors who acquired them on pilgrimages decades or even centuries previously. These Shaligrams are typically passed down from generation to generation of first and second sons, and many Shaligram devotees can recount the long genealogical histories of both their families and their Shaligrams accordingly. Ritual stones of this magnitude are usually venerated with daily *pujas* (ritual worship) and offerings of water, tulsi (holy basil) leaves, flowers, food, sandalwood paste, and turmeric and vermillion (*kumkum*) powder. The presence of Shaligrams is also indispensable to the performance of yearly festivals and important ceremonies (the Marriage of Tulsi and Shaligram festival being one especially salient example). Shaligrams often make appearances at weddings, funerals, house-warming functions (such as *grhapravesha* and *vastu-puja*), during pacificatory rites of various types (*shanti*), and at any point where the welfare of the household and family may be at stake.

In temples, Shaligrams play important roles in the construction of deity altars. The famous image of Vishnu at Badrinath is reported to be carved out of a Shaligram, as is the image of Krishna at Uḍupi in Karnataka and the Shiva Linga within the main temple of Pashupatinath in Kathmandu. Also, the present deity altar in Sri Padmanabhaswamy temple located in Thiruvananthapuram in Kerala is said to be constructed of more than 12,000 Shaligram stones arranged to form the icons of Padmanabha (Vishnu) who is reclining on the serpent Ananta. The serpent has five hoods facing inwards which signify contemplation while the deity's right hand has then been placed over a Shiva Lingam. Lakshmi, the Goddess of Prosperity, and Bhudevi the Goddess of Earth (two consorts of Vishnu) are arranged on either side of the main icon while Brahma emerges on a lotus emanating from Vishnu's navel. Similarly, the reclining Vishnu deity residing in another temple in Thiruvattar, about 50-60 kilometers from Trivandrum, is reported to be made out of 16,000 Shaligram stones.

In other cases, Shaligram stones are said to reside inside deity icons (*murti* – meaning divine image), such as the wooden icon of Lord Jagannath in Puri, at the Venkateshwar Temple in Tirupati, at the Dwakadheesh Temple in Dwarka, and in the Krishna Rukmini Temple at Bhet Dwarka in Gujarat where they act as *padartha* (literally "object" or "category" but used here to mean "essence of existence") within the images. According to many pilgrims to Muktinath Temple in Mustang, the icon of the Buddha of Compassion, Avalokiteshvara (Tibetan: *Chenrezig*),[8] who also serves simultaneously as the icon of Vishnu, sits on the very first Shaligram to have ever been discovered in Nepal and who therefore holds within him the divine essence of all Shaligrams as a whole. In his seminal work on Shaligram mythography, S.K. Ramachandra Rao notes that it is a Shaligram stone that officiates as the *snapana-murti*, or the icon for bathing, at the shrine of Natha-dvura in Rajasthan and that a group of Shaligrams remains the principal focus of daily worship at the temple of Vengaḍam at Tirupati Tirumalai (1996: 2-4). Lastly, the largest and heaviest temple Shaligram known currently resides in the Jagannath Temple at Puri in Orissa, while the largest collection of Shaligrams outside of India is at ISKCON's Karuna Bhavan temple in Scotland.

For those deity altars that cannot accommodate Shaligrams inside the *murti*, it is not uncommon to see small collections of Shaligrams resting at the deity's feet, where they are the subject of daily *pujas* in connection with the iconic deities associated with them, or in the case of large temple collections, arranged on a side altar in full view of all ritual activities. In other cases, garlands (*mala*) of Shaligrams are used to decorate deities at particularly auspicious times, and many of the Shaligrams comprising these garlands are said to have been collected by successive groups of pilgrims over spans of hundreds or even thousands of years. One such garland, for example, occasionally used at the temple of Mayapur in Northern India is made of silver strands attaching one hundred and eight small Shaligrams together in a manner similar to a *japa mala* (Hindu rosary). And finally, though Shaligram practices are most common among the Vaishnava (Vishnu worshippers) traditions of Hinduism, among Shaivas (Shiva worshippers), Shaligram stones may also be used in homes and in temples as forms of the Shiva Lingam.[9]

Though Shaligram stones might initially appear to be good candidates for broader "object ethnographies" that track the movement of commodities

8 Throughout this ethnography, I will often refer to Avalokiteshvara as a buddha rather than a bodhisattva, which is more common in Tibetan religious texts and among religious scholars and specialists. This is because the people of Mustang, as well as elsewhere in Nepal, specifically used "buddha" when referring to Avalokiteshvara/Chenrezig and did not typically use the word "bodhisattva", if at all.

9 Shiva linga, the ostensibly phallic representation of Shiva (which is more often interpreted as "the pillar of fire" rather than as a castrated phallus), can come in either natural formations (like the white quartz bāṇa-liṅgas found in the river Narmada) which are referred to as "svayambhū liṅgas" or as man-made, called "mānuṣa liṅgas". There is also a reasonably common practice of making temporary lingas out of clay, cow-dung, flowers, or grain which are typically consumed or destroyed following the ritual worship.

through globalized free markets, resource distribution, and the commodification of religious objects for tourism, Shaligram practitioners actively resist this paradigm of cultural exchange in a variety of ways. Drawing on networks of kinship exchange, they routinely point out that no Shaligram should ever be traded for money but should, rather, only move from person to person through inheritance, through marriage or family transfer, or gifting. Furthermore, they have also begun to leverage digital technologies, such as internet communication (see Chapter 8) in order to "hide Shaligrams from money" and to facilitate the exchange of stones among people who are otherwise unable to travel to Mustang on pilgrimage. As a result, one of the central arguments of this work, Shaligrams as divine persons and kin, is more closely engaged with the mode of research termed multi-species ethnography.

In previous anthropological studies, non-humans (animals, plants, mountains, deities, etc.) were often relegated to the margins of discussion and framed as either symbols, food, part of the landscape, or otherwise peripheral and supplementary to human action. In recent years, however, ethnographies of animals (Ingold 1994; Sanders 1999; and Irvine 2004), insects (Raffles 2010), plants (Tsing 2005), fungi (Tsing 2016), microbes (Latour 1988; Paxson 2008; and Helmreich 2009), and even "earth beings" such as mountains (de la Cadena 2010) have shifted non-human agents and entities out of the realm of Agamben's *zoe* or "bare life," that which is acultural and killable, and into the purview of *bios*, being possessed of legibly biographical and political lives (Agamben 1998). As once living (fossilized ammonites) and now alive again (deities), Shaligrams present something of a conundrum to the study of organisms whose lives and deaths are so intimately linked to human social worlds. Indeed, they challenge even the very notion of what it means to "be alive" in the first place. While unkillable in the conventional sense, Shaligrams are nevertheless viewed as distinctly and actively living, carrying communal and family activities into what Eduardo Kohn might call a human "entanglement with other kinds of lived selves" (2007: 4). In other words, a Shaligram, while no longer an organism in the biological sense (though it once was), is reborn out of the landscape into a new kind of life whose livelihood is shaped by the religious, political, economic, and cultural forces that surround it (Kirksey and Helmreich 2010). Like many other multispecies ethnographies, this is an account of where nature and culture break down.

This view of Shaligrams through the lens of "multispecies" proceeds to complicate a number of conversations in the anthropology of religion.

Firstly, Shaligram practices do not fit into any popular dichotomy that views science/religion or objectivism/relativism as theoretically opposed. More importantly, Shaligrams as divine persons and kin also continue to question whether there is such a thing that we might call "religion" universally (Lambek 2008; Bubandt et al. 2012). Secondly, considerations of Shaligram ritual and pilgrimage complicate the conversation as to whether religion constitutes an experience (associated with particular psychological and phenomenological ideas of the sacred; see Eliade 1959; Otto 1958; and Van der Leeuw 1938) or a presence, i.e., a distinct social and material reality (Taves 2009; Engelke 2007). This is because the presence of Shaligrams facilitates relationships with the divine through multiple competing authorities, texts, actions, and objects. Furthermore, the discourses of science and religion, even among Shaligram practitioners, tend to be blended together as two related (and not mutually exclusive) "mythologies" that work together to explain the continued importance of Shaligram veneration in South Asia and elsewhere.[10]

The incorporation of Shaligrams in family life also challenges the distinction between "sacred" and "secular" categories that serve to elevate religion out of the context of everyday life. For the majority of Shaligram practitioners, the gods participate just as much in the mundanities of cooking and eating, work and rest, household maintenance, gardening and animal care, and child-rearing as they do in the interactions between people, spirits, and religious ideals. This is how Shaligram practices pull religion out of the medical, phenomenological, and naturalistic models of "spirit possession" and "mediumship" especially popular in modern anthropological accounts of religion as well as categories of "folk" or "primitive" religion versus "high" religion more common in theological studies. For these reasons, I pay particular attention to the multiple roles that Shaligrams play in the day to day, in wider political and social concerns, and in related conceptions of geological and mythological time. These distinctions between object, person, deity, and fossil will then help me to demonstrate complementary ways of studying religion that interrogates and describes the things that people may or may not define as "religious" versus systems that anthropologists might delineate as "religion".

10 As a bridge between concepts of Nature and Culture, then, Shaligram ritual practices serve to highlight a methodological split between critical theorists in the humanities and social scientists more focused on the natural sciences (often as played out in multi-species ethnography and Science and Technology Studies).

Precious Persons of Stone

It may sound paradoxical to link objects with persons in this manner, but this work is indeed about an alternative view of the boundaries between "human" and "non-human" as it relates to shifting boundaries between the "sacred" and the "everyday". While scholars have already described the ways in which personhood in South Asia is often constructed through external relationships with other persons, places, objects, and ideas (Deleuze 1992; Lamb 2004), this research furthers those descriptions by demonstrating how divine persons (particularly divine objects as persons) are similarly constituted. Shaligram practitioners do not refer to Shaligrams as "stones", and the paleontological term "fossil" is unsurprisingly contentious. Rather, Shaligrams are typically referred to simply as "bodies" or otherwise given nominal distinctions using gendered pronouns (His/Her) depending on which deity is materially manifest. More importantly, Shaligram practitioners themselves do not generally use representational language when referring to Shaligrams. In other words, Shaligrams do not symbolize or "stand in" for deities, they *are* deities. The use of representational terms was more common on my part than on the part of my research participants, who were often quick to point out any misconceptions regarding who and what might be present at a given moment. My analysis of symbolic meanings is therefore largely my own and intended to clarify the relationships between broader cultural systems of mobility and religious practice than on the manifest nature of Shaligrams themselves. As a result, personhood here is therefore repositioned as a process that includes bodies that "are" present and bodies "as if" they are present so as to blur the distinctions between reality and its representations.

What is important to stress here once again is that Shaligrams are a part of the broader, everyday interactions between Hindus, Buddhists, Bonpos, and the divine – where the gods are immanent in the world, simultaneously transcendent, and embodied in multiple different kinds of earthly forms. Offerings and gift-giving to these forms are then meant to draw the deity's favor and to nurture good relationships with them through physical exchanges. The simultaneous presence of the divine in material bodies (such as stones, trees, elephants, rivers, etc.) then helps to mediate needs, problems, and conflicts in people's everyday lives by creating connections between an individual, a family, or a community's present circumstances and the actions or desires of the gods. There are also no specific standards of practice related to either Shaligram stones or to Hindu deity images (*murti*). This means that some Hindus use Shaligrams in association with a wide variety of other ritual objects, altar objects, sacred places, and deity icons, while others use Shaligrams alone. Some devotees might also only worship occasionally (such as at a temple that houses a Shaligram or on festival days), while others practice Shaligram rituals daily. Some may keep Shaligrams in their homes; others may prefer to keep them in places of community worship. In any case, I do not in any way imply here that Shaligram veneration stands uniquely separate from the routine and familiar interactions that the vast majority of devotees have with the daily appearances of the divine.

Shaligrams are generally contextualized within larger ritual systems that venerate naturally occurring objects interpreted as divine persons or divine person-like beings (such as Shiva Linga stones, Dwarka Shilas,[11] rudraksha seeds, mountains and rivers, trees and forests, stars and celestial bodies, and certain animals[12]), but they are also commonly associated with specific deity *murti* (especially statues) with whom they share household and altar space. Therefore it is, again, important to note here that "person" does not specifically refer to a "human" but to a being that can have agency, speak, engage in social relationships and exchanges with other people, be cared for and care for others in return, have a life course and go through life-cycle rituals, such as a marriage or a funeral.

Shaligrams are also situated within cultural systems of reincarnation, within the concept of karmic life cycles, that view birthmarks, congenital abnormalities, and other notable characteristics on human bodies as clues to a person's past life experiences. For example, psychologist Ian Stevenson's *Reincarnation and Biology*[13] contains ten such examples of children in India with various birthmarks or birth defects that were said to correspond to places where their previous personalities were shot, injured, or otherwise fatally wounded. As persons, then, Shaligrams are equally integral to pilgrimage circuits as humans are, especially in terms of landscapes and practices

11 A type of coral stone obtained from the Gomati river (Gomti River) in Dvaraka. Often worshipped as manifestations of Vishnu along with Shaligrams.

12 "The Death of a Hungry God: The electrocution of a wild elephant in a village in northeast India illustrates how these formidable beings are experienced as both animal and deity" – https://www.sapiens.org/culture/elephants-india-religion/?utm_source=SAPIENS.org+Subscribers&utm_campaign=117193ae4e-Email+Blast+12.22.2017&utm_medium=email&utm_term=0_18b7e41cd8-117193ae4e-216302925 Accessed 1 February 2018.

13 Ian Stevenson, 1997. *Reincarnation and Biology: A Contribution to the Etiology of Birthmark and Birth Defects*. Praeger. Reactions to Stevenson's work are highly mixed given his belief that birthmarks and "maternal impressions" were, in fact, evidence of previous lives and, specifically, indicated the manner in which the previous personality had died. In his *New York Times* obituary, Margalit Fox wrote that while Stevenson's supporters saw him as a misunderstood genius, mainstream peer review simply ignored his research as earnest but gullible.

that span hundreds of miles and reach across the national and geological borders of multiple countries and sociopolitical identities. Their "birth" in the Kali Gandaki River indexes the beginning of a new kind of belonging where landscapes of pilgrimage are repositioned as "homelands" in a different sense; as the birthplaces of family members, deities, and ancestors manifest in naturally occurring stone.

What is a Shaligram?

From the viewpoints of both religion and science, there remains a fair question as to what one might mean when one says "Shaligram". The ontology of Shaligrams is, therefore, a theme central to much of this work. Briefly, ontology refers to the nature of being, the nature of reality, or theories of being. This means engaging with Shaligrams as inhabitants of a different "world" and not merely as objects in a particular "worldview" (Kohn 2013: 9-10). In a sense, this entire book is an exploration of the ontology or nature of being of Shaligrams and will encompass a variety of analytical categories related to place, divine person, and kin.

In the discourses of geological and paleontological science, Shaligrams are ammonite fossils. Ammonites are the common name given to the subclass Ammonoidea, an extinct order of cephalopod that, despite their outward similarities to the modern-day chambered nautilus, are more closely related to other living coleoids like squid and cuttlefish. The first occurrence of ammonites dates back to the Devonian period around 400 million years ago. The last surviving lineages disappeared, along with the dinosaurs, around 65 million years ago following the Cretaceous-Paleogene extinction event. According to the Geological Survey of India, printed in 1904, Shaligram ammonites date specifically from the Early Oxfordian to the Late Tithonian Age near the end of the Jurassic Period some 165-140 million years ago (1904: 46). Up to around 40 million years ago, the land that is now Mustang, Nepal was submerged beneath a shallow ocean called the Tethys Sea located at the southern edge of one of two continents called Laurasia. As the Indian subcontinent broke away from the east coast of the continent of Gondwanaland somewhere around 80 million years ago, it moved northwards, eventually crashing into the south coastal regions of Laurasia and resulting in the massive geological uplift that created the Himalayan mountains. But as soon as the mountains were born, they were destined to die by erosion.

After the Tethys Sea was completely drained, the fossilized remains of its seafloor were left to slowly wear out of the slopes of the rising mountainsides.

Consequently, the ammonites would tumble out of the mud shales and slate beds and into the rivers to churn their way smooth – a vital part of the movement that eventually transforms them into Shaligrams. In almost every respect, Shaligrams (and the ammonites that precede them) symbolize a crossing of lived culture with tectonic history – where each stone acknowledges the vast span of Deep Geological Time compared to a human lifetime. This discourse is, however, extremely contentious within the religious discourse of Hindu pilgrimage, and many pilgrims who journey to Mustang to obtain the stones express significant ambiguity in reading Shaligrams through the lens of paleontology. In many ways, their ambiguity recalls the conundrum of the Shakespearean lines with which I opened this chapter, where the pearl opacity of the subject's eyes hint of a corpse transformed: dead yet seeing, completely still but quite alive.

Since antiquity, ammonites have been associated with religion or with religious histories. Part of the challenge of writing about Shaligrams comes from the many layers of time and levels of antiquity that must be sorted through. For example, many Hindu and Buddhist pilgrims still refer to the scientific classification of ammonites as its own kind of "mythology": as a series of stories about events, persons, and arbitrary categories that took place in the past and explain specific phenomena in the present. This is a "mythology" they will then eventually link with their own for determining the answer to the ultimate question: "What is a Shaligram?" As origin stories, the fossil history of the Himalayas and the tales of world creation as relayed in the Puranas are often taken to be equally authoritative though value-laden in different ways. For the former, the taxonomical units of geology and paleontology are viewed as ways in which new forms of life are brought into being and described so that non-practitioners (i.e., Westerners and "modernized" South Asians) might be able to understand the significance of Shaligrams in space and time, couched in the language of logic and biology. For the latter, the progression of events within sacred texts renders Shaligrams' kinship and descent from gods to men meaningful and relevant to issues in the present day. Or, the fossil taxa of ammonites are made comparable – are made the same way, for the same reasons, of the same elements – as the religious taxa of deities and ancestors.

There is a primordial layer, the ancient times of ammonites themselves. Then there are their Greek and Latin source names along with their Vedic categorizations and Indian descriptions; another time of the "ancients". There is also their history of research within the rise of both Eastern and Western sciences, followed by their personal histories, which animates all earlier times in the shape of the Shaligram in hand. In ancient Rome,[14] ammonites were known as *"Cormu Ammonis"*, *"Corni de Ammone"*, or *"Cornamone"* because their shapes were thought to resemble the tightly coiled ram's horns used to represent the Egyptian god Ammon. Pliny the Elder (AD 23-AD 79) even referred to them in the 37th volume of his work *Naturalis Historia*. In it, he writes: "The *Hammonis cornu* is among the holiest gems of Ethiopia, it is golden in colour and shows the shape of a ram's horn; one assures that it causes fortune-telling dreams" (see also Nelson 1968). The "golden colour" he refers to is a likely reference to the fact that many ammonite fossils, including Shaligrams, are often covered in iron pyrites, which give them a sparkling golden appearance. Georgius Agricola, sometimes referred to as "the father of mineralogy" and the author of *De Re Metallica*, a work based on Pliny's *Naturalis Historia*, also referred to ammonites as *Ammonis Cornu*. Even today, ammonite genus names often end with *-ceras*, the Greek word (κέρας) for "horn".

The Swiss naturalist Conrad Gessner included some ammonite illustrations in his work *De rerum fossilium* (1565), but even toward the end of seventeenth century, it is especially interesting to note that the organic nature of ammonites remained under debate (a debate that takes places in the Hindu Scriptures as well). Robert Hooke, the famed experimental scientist and nemesis of Sir Isaac Newton, was fascinated by the logarithmic coil of ammonite shells and their regularly arranged septa (recall the classic image of the golden ratio). It was he who reached the conclusion that ammonites were not only of organic origin but also widely resembled the nautilus and may therefore be related. However, it wasn't until 1716 that ammonites would finally join scientific taxonomy with a classification scheme first recorded by another Swiss naturalist, Johann Jacob Scheuchzer. The modern form of the word ammonite was coined by the French zoologist Jean Guillaume Bruguière in 1790, but it wasn't until 1884 that the subclass Ammonoidea was finally formalized in modern zoological taxonomy (Romano 2014).

14 The *Shaligram Kosh* makes the claim that early Greek geographers referred to the river where Shaligrams were found as "Kondochetts". This is, however, highly unlikely given the history of the region and the extent of Greek influences up to and including the time of Alexander the Great. While Kondochetts was apparently a reference to a river on the Indian subcontinent in Greek geography, it is unclear as to whether or not it is a reference to the Kali Gandaki, and the claim is not otherwise verifiable.
'Kondochates'/'Condochates', Arrian, _Indica_ 4.4 translation: https://sourcebooks.fordham.edu/ancient/arrian-bookVIII-India.asp. Attested as Κονδοχάτην in the accusative, so Κονδοχάτης nominative. Accessed 6 September 2019.

In China, ammonites were called horn stones (*jiao-shih*) and were typically used in traditional medicine. Japanese texts, on the other hand, refer to them as chrysanthemum stones (*kiku-ishi*), and Buddhists interpreted their clockwise spirals (a representation of the direction in which the universe rotates) as a focus for meditation or as symbols of the eight-spoked dharma wheel (an interpretation currently shared by many Buddhist pilgrims to Mustang as well). Additionally, among ancient Celts, these fossils have been interpreted as a kind of petrified venomous snake (*ophites*) and referred to as "serpent stones". In medieval England, ammonites (along with various other types of fossils) were taken as evidence for the actions of Biblical saints such St. Patrick, St. Keyne Wyry of Wiltshire (c. 461-505), or St. Hilda of Whitby (c. 614-680). According to Sir Walter Scott's *Marmion*,[15] fossil ammonites were serpents that infested the region of Whitby before the coming of St. Hilda, who subsequently defeated the serpents and turned them to stone on the site where she intended to build an abbey (see also Skeat 1912).[16] In the Americas, Cretaceous baclitid ammonites were also once collected by the indigenous peoples as "buffalo stones" and were kept in medicine pouches as aids in corralling bison (Mayor 2005). Called *Iniskim*, members of the Blackfoot First Nations continue to harvest bright opalescent ammonites for ceremonial purposes even today.[17] Furthermore, aside from their role as Shaligrams, ammonites also have a long and storied history more broadly in what Van Der Geer refers to as the fossil folklore of South Asia. She relates in detail, for example, entire regions of fossil beds containing not only ammonites but ancient giraffes, elephants, and tortoises near the Siwalik Hills of the

15 Lovett, Edward (September 1905). "The Whitby Snake-Ammonite Myth." *Folk-Lore*. 16 (3): 333-334.

16 Skeat, W.W., 1912. ""Snakestones" and stone thunderbolts as subjects for systematic investigation." *Folk-lore*, 23: 45-80. Additionally, during the 19th century, it was not uncommon for people to carve images of snake's heads around the bottom aperture of the ammonite shell.

17 See also: Rainbow Ammonites and Bison Stones available at https://albertashistoricplaces.wordpress.com/2018/01/10/rainbow-fossils-and-bison-calling/ Accessed 10 September 2018.
Further Reading:
Etter, W. 2015. Early Ideas about Fossil Cephalopods. *Swiss Journal of Palaeontology* 134: 177-186.
Monks, N. and P. Palmer. 2002. *Ammonites*. Natural History Museum, London, London, England.
Mychaluk, K.A., A.A. Levinson, and R.L. Hall. 2001. Ammolite: Iridescent Fossilized Ammonite from Southern Alberta, Canada. *Gems & Gemology* 37: 4-25.
Peck, T.R. 2002. Archaeological Recovered Ammonites: Evidence for Long-Term Continuity in Nitsitapii Ritual. *Plains Anthropologist* 47: 147-164.
Reeves, B.O.K. 1993. Iniskim: A Sacred Nisitapii Religious Tradition. In *Kunaitupii: Coming Together on Native Sacred Sites, Their Sacredness, Conservation, and Interpretation*, edited by B.O.K. Reeves and M.A. Kennedy, pp. 194-259.

Himalayas in India, which are used as evidence of the great cosmic battle of Kurukshetra as described in the Mahabharata epic and which are also visited by religious pilgrims from all over the world (2008).

According to the *Epigraphia Indica* (vol. 2, p. 204), the earliest evidence of Shaligram worship in India dates back to the second century BCE, with an inscription near Mewar in Rajasthan that mentions a shrine for the twin gods Vasudeva and Samkarsana as being made out of Shaligram stones. There are additional inscriptions, one dating back to the 1st century BCE in Madhyapradesh, for example, that also describe the worship of Vishnu in the form of Shaligram, along with the well-known *Mora* inscription near Mathura, dating to roughly the same period, which mentions the "*five worshipful heroes of the Vṛṣṇi dynasty in their luminous stone forms: śālagrāmas, bhagavatām, vṛhṣṇīnām, pañ ca-vīrāṇāṃ, pratimāḥ* [...]" (*Epigraphia Indica*, vol. 24. 194 ff.). Hindu scholars interpret this inscription as a likely reference to the five vyuha forms (incarnations of a divine attribute and not full deity incarnations) of Vishnu: Vasudeva, Samkarsana, Pradyumna, Anirudda, and Samba (son of Pradyumna) (Rao 1996: 4).

In the first millennium AD, Shaligram practices were finally written down in the Puranic scriptures and commentaries, though an effort to standardize their interpretations and rituals wouldn't come about until much later with many of the *bhakti* (devotional) reforms of the late fifteenth century. In South India, the Hindu saints Ramanujacharya (around AD 1017-1137) in Tamil Nadu and Madhvacharya (AD 1238-1317) in Karnataka would also set forth ritual proscriptions still followed by the Hindu Vaishnava and Smarta traditions today. In North India, the traditions of Gaudiya Vaishnavism (in West Bengal) and Sri Vaishnavism (Vishishtadvaita) as well as the Hare Krishna sect have maintained their own Shaligram practices, many of which have now been exported to the West, leading to new demands for stones in places far beyond South Asia. While few Vedic texts mention Shaligrams specifically, the Brahmavaivarta Purana, Garuda Purana, and Skanda Purana, as well as the commentaries of the eighth-century philosopher Shankaracharya,[18] are currently considered the most authoritative. Despite this, the majority of Shaligram practices remain at their heart largely composed of oral traditions, regional variations, tradition and sect specifics, and individual preferences.

18 While the use of Shaligrams in worship can be traced back before the time of the poet-saint Adi Shankara, Shankara's commentary of verse 1.6.1 of the Taittiriya Upanishad and his commentary of verse 1.3.14 of the Brahma Sutras demonstrate that the use of Shaligram stones was a well-established Hindu practice by the time of his composition

The presence of Shaligrams in collections of archaeological artifacts excavated from earthen mounds inside caves occupied by the very first inhabitants of Mustang indicate that Shaligram practices likely predate the arrival of Hinduism in Nepal by several centuries and may have begun as a localized shamanic practice later adopted and disseminated by the spread of Vedic religion in the late centuries BCE. This is not surprising given the commonality of aniconic imagery in the early religions of South Asia. As Diana Eck writes: "the most ancient non-Vedic cultus of India was almost certainly aniconic" (here referring to a lack of anthropomorphic characteristics). Stones, natural symbols, and earthen mounds signified the presence of the deity long before the iconic images of the great gods came to occupy the sancta of temples and shrines" (1986: 44). Even fewer modern books and manuscripts discuss Shaligram pilgrimage or ritual practices in any depth, usually relegating them to a passing mention in the context of other cultural concerns or political issues. As of this writing, no detailed ethnographic descriptions of Shaligram pilgrimage exist in the academic literature, and almost no accounts of Shaligram practices have been analyzed at length in the corpus of the social sciences.

While Vedic and Puranic texts are often consulted as foundational authorities for Shaligram ritual practices, they do not encompass the depth and breadth of Shaligram pilgrimage and ritual practice among Hindus, Buddhists, and Bonpos (Himalayan shamans) today. Therefore, any account of Shaligrams must be attentive to both change and continuity: to sort out the processes and influences of various cultural contexts, cultural exchanges over time, and the issues of great distances between the mobility of pilgrimage and the spaces of veneration. To encounter a Shaligram at any one given point is to experience its significance particularly for that context, a kind of localization which, though enlightening, is potentially unrevealing of its broader meanings, substance, and connections. Privileging any single historical moment, including this one, at which the scholar might enter the scene does not help us to understand "why a Shaligram?". Undoubtedly, any one of these moments would be informative as to that particular Shaligram's use and importance in that context, but without a more expansive view that includes the movement of person, object, and narrative together beginning with pilgrimage to the high Himalayas of Mustang, to destination temples throughout South Asia and elsewhere, to the homes and communities of devotees the world over, a greater understanding of the profound nature of Shaligrams will remain obscured.

The following chapters explore the cultural meanings of the material world in motion for the religious communities of South Asia who venerate

Shaligram stones. They describe how space, time, and boundaries, especially the fluidity of political, geographical, and material boundaries, are constructed and experienced by Shaligram devotees in contemporary Nepal and India. In the place where both immigrant and indigenous Hindus, Buddhists, local Bonpos, and their deities converge, I found that religious, ethnic, and political identities became fluid and unstable, deities became manifest in the objects of the natural world, and people began to speak of a fossil that was not just a fossil but a living member of the family and of the community. Through this ethnographic exploration, this research then shows how these particular aspects of material religious practice are used to create and reproduce personal and familial identities as well as community belonging and cohesion among members of various, outwardly disparate South Asian religious traditions. It also discusses how these attachments through mobility are translated into anti-nationalist and boundary-rejecting political practices by allowing for the agency of stones who have become bodies and divine persons in their own right.

Writing an Inconstant World

While discussions of scriptural traditions will be important, this book discusses Shaligram pilgrimage and veneration from an ethnographic perspective, not a textual one. This is partly because texts (including reading Shaligrams themselves as texts) play only a partial role in the overall complexity of Shaligram practices as a whole and partly because actual ethnographic accounts of people who use Shaligrams in their daily lives are almost non-existent. An essential task of ethnography is to convey a sense of the lived experiences and practices of people – in this case to demonstrate how devotees, landscapes, and Shaligrams actually interact – rather than reproduce textual ideals or religious ideologies that are never quite truly realized in the day to day. In truth, most of the complex intricacies of actual Shaligram practice bear little superficial resemblance to their descriptions in religious texts. Yet these systems are connected, not only in how people view their own positions within the greater context of Hindu and Buddhist traditions but also in the way in which devotees reconcile various contradictions that arise between day-to-day practices and sacred ideals. To some, the myths and stories contained in the Puranic texts are taken as a kind of practical instruction; a set of divinely inspired guidelines for conducting oneself properly in the presence of the sacred and through life in general. To others, such esoteric reading is irrelevant (thought to be mainly abstract and symbolic) to the kinds of intimate and direct sense experience required to truly apprehend the material world and the divinity within and beyond it. For my part, it was vital to understand how Shaligrams fit into people's lives and experiences – as mediums of exchange, as symbols of religious affiliation, as material manifestations of the divine, as objects of cultural or political communication and organization, and as members of the family and community.

Instead of focusing on the textual histories of Shaligrams, then, I use real-life stories, quotations, conversations, and observations from the periods I spent working with Hindu and Buddhist devotees to convey the complex dimensions of Shaligram pilgrimage and practice such as it was in the first decades following the re-opening of Mustang, Nepal to foreign travel in 1992. In the sense that I am using it, ethnography is the study of communal meaning-making, the description of material practices and experiences as they appear in particular places at particular times. Shaligram devotees' descriptions of space, place, object, and movement therefore constitute the basis for my arguments. Through the use of Shaligram stones by adherents in multiple religious traditions (including co-participatory and hybrid forms), this work shows how mobility and transiency itself become the basis by which power and sovereignty are reclaimed and expressed. Through thinking about multiple different communities as they are unified by the movement of a particular object that is both rooted in a place and transcendent of all places, we will arrive at new ways of understanding mobility as a factor of collective identity and new ways of imagining how persons are embodied in objects and how objects therefore become persons.

To that end, there is also a fair amount of information that I had no choice but to leave out. In the future, there could be any number of books written on specific Shaligram practices in specific places, including the use of the courts in Kolkata, India to determine the "paternity" of Shaligrams for the purposes of inheritance, or the Shaligram festival traditions of Tamil Nadu, or even the ritual interpretations of particular Shaligrams specific to the Brahmin castes of Western India. But alas, only so much can be included in any one work, and I have chosen to begin this one where the Shaligrams themselves begin – in Mustang, Nepal.

Into the Foothills

For more than two thousand years, the veneration of sacred fossil ammonites, called Shaligrams, has been an integral part of ritual practice throughout

Nepal and India. Originating from a single remote region of Himalayan Nepal in the Kali Gandaki River Valley of Mustang District, ritual use of these stones today has become a significant focus of pilgrimage, religious co-participation, and exchange throughout South Asia and among the global Hindu and Buddhist diaspora.

Viewed primarily as natural manifestations of the Hindu god Vishnu, Shaligrams are considered to be inherently sacred not only because they are not man-made but because the workings of the landscape (i.e., the processes of geological formation) has imbued them with a living essence and agency of their own. For this reason, Shaligrams require no rites of consecration or invocation when brought into homes or temples as presiding deities over the family and the community. In other words, the gods do not come to inhabit them nor are Shaligrams strictly symbolically representative of deities. They *are* them.

Shaligrams are also deeply intertwined with understandings of divine movement, either through a geologically and mythologically formative journey down the sacred river or transnationally in the hands of devout pilgrims. Pouring out into the river each year following the summer melt high in the mountains, Shaligrams are gathered up by pilgrims, tourists, and merchants alike. On their way out of the mountains, they travel through forests and cities, into temples and homes, across great expanses of time and space, carried by the indescribable forces of nature or the complex networks of pilgrimage and kinship exchange that eventually come to define their "lives" as gods and as family members. Shaligrams are therefore described as a kind of divine person, and one who is often in the habit of making pilgrimages throughout Nepal, into India, and across the world.

In Shaligrams, the discourses of science and religion meet, blend, and become comparable methods designed for a singular purpose: to narrate the past so that it explains the present. For this reason, many Shaligram practitioners often refer to both science and their own religious stories as "mythology" – or conversely, to both as factual truth if only from slightly different perspectives. Shaligrams as both fossils and deities then challenge us to question many of the taken-for-granted ways we connect the past and the present and what we think of as knowing through science versus knowledge in religion. In the end, these questions might help us to come to a better understanding of our own and other ways of being in the world, both physically and spiritually, or to understand how "fossil", "person", and "deity" are not necessarily mutually exclusive categories of existence but rather indications of a variety of experiences, relationships, and narratives that center on the veneration of a particularly unique kind of ammonite.

How, then, might an ethnography of sacred fossils be possible? In short, the answer is: a new methodology, one that takes into account both human and non-human actors as well as a community of practitioners who are not otherwise linked by nationality, ethnic identity, or even religion. This methodology must then also attend to mobility in the comings and goings of pilgrims along with the movement of sacred stones from family to family, community to community, and even from individual to individual over time. Without it, the larger systems at stake in the world of Shaligram veneration may not become adequately visible or might be lost in the particulars of time and place.

Structure of This Book

This ethnography is divided into eight chapters and a conclusion, each of which is meant to "follow" the Shaligram stones outwards from their origins in Mustang, Nepal to their eventual destinations throughout South Asia and the world. Chapters 1 and 2 lay out the theoretical and anthropological groundwork of ritual and material practice for the ethnography of Shaligram religious and social worlds that follows. These chapters also act as the theoretical pivot of the work, joining the discourses of religion, science, personhood, and place-making together to demonstrate the ways in which Shaligrams are constituted as divine persons and act as agents in their own right. Contrary to popular Western viewpoints, geological processes (including fossilization) and social processes (such as ritual) in the formation of Shaligrams are not mutually exclusive and do not necessarily constitute two opposing versions of the creation of a single entity. This argument then demonstrates that both the physical and cultural constitutions of Shaligrams are consistent with general understandings of personhood in South Asia and are not, in fact, incongruous with formations of families and communities involving human persons.

Chapters 3 and 4 detail the histories and ritual landscapes surrounding the region of Mustang, Nepal and of the temple site of Muktinath. By addressing issues of political conflict, religious fluidity, and the corpus of Shaligram creation stories specifically tied to the Kali Gandaki River Valley, I demonstrate how the changing paradigm of mobility and nationalism along the pilgrimage routes has contextualized and influenced modern Shaligram practice as well as how government and scholarly narratives of the region have deeply influenced the ways in which pilgrims and local peoples speak about their own understandings of the world. These chapters present an anthropological overview of the political and cultural issues currently facing Mustang, describing its history of conflict, migration, and religious blending as a way to address the conflicts between sacred and political landscapes from which Shaligrams are produced, collected, and exchanged.

Chapters 5, 6, and 7 constitute the core of the ethnography by addressing issues of mobility and pilgrimage at the point of Shaligram origin and the subsequent social and ritual life of the stones once they return "home". These chapters are much more empirical than theoretical and are based directly on the stories, experiences, and narratives of Shaligram pilgrims and devotees themselves. By taking up topics of physical space and movement as they transition into symbolic space and movement, I demonstrate how devotees construct and reproduce meaning out of the material world and then leverage those meanings in political practice.

Chapter 7, "River Roads," is set in Mustang, Nepal between 2015 and 2017. In this chapter I show how a shared mythic view of the landscape constitutes the first linkage by which Shaligram devotees create a shared identity, despite differences in almost all other aspects of their lives. I contrast this with the experiences of resident peoples, many of whom rely on pilgrimage and tourism to support themselves economically. I argue that it is then mobility itself that becomes the ultimate expression of power and autonomy on the margins of a developing state. Where the sacred and the everyday become fluid, both pilgrims and residents continuously re-instantiate a sacred landscape over a political one, favoring religious affiliations over national identities in a space of relative political disorder.

In Chapter 8, "Ashes and Immortality", I address the nature of Shaligram "death" and the problem of Shaligram commodification in a digital world. As an intervention in the growth of virtual religious practices online, I note how the rising issue of global markets for selling stones in South Asia (and now increasingly abroad) has resulted in a kind of decolonized Shaligram practice that uses internet platforms as a method for concealing Shaligram mobility from monetary exchanges. As globally mobile religious commodities, however, Shaligrams are not diminished as agents in the eyes of devotees but rather begin to take on even greater symbolic meaning as representatives of the plight of human bodies caught in webs of marketing and global capitalism.

Apart from contextualizing events, I have kept detailed historical discussions of Mustang, Nepal (and of the complex national relations between Nepal and India generally) to a minimum. More in-depth histories and ethnographies of Mustang and its peoples are available elsewhere (Fisher 2001; Craig 2008; Ramble 1983, 2002; Snellgrove 1961; Dhungel 2002; and

Messerschmidt and Gurung 1974) and should be consulted by any student or academic wishing to learn more about the region and the more expansive ethnographic underpinnings of my analysis of Mustang. I include historical discussions here as a way of contextualizing the origins of Shaligram pilgrimage as well as the modern political and economic concerns facing the mobility of Shaligrams today.

Finally, while certain aspects of Shaligram practice are relatively consistent from one circumstance to another, most others are framed by time, place, religious affiliation, and history. This means that, should one encounter a Shaligram or Shaligram devotee at any given point, their specific ideologies and practices might not be rendered here exactly as one encounters them at that moment. This book is written in the hope that the reader might gain a larger, more overall sense of what Shaligrams are and what their meanings and practices entail for Hindu and Buddhist devotees in South Asia and, to some degree, among the South Asian diaspora. But also, perhaps more significantly, by delving into the ways in which one diverse and disparate group practices their faith and forges connections between persons and ideas at particular moments in time, we might come to better understand our own ways of being in the world, both physically and spiritually. In a world where "living fossil" no longer simply refers to the living and breathing simulacra of a more ancient creature petrified in stone, it may be possible to imagine, for a time, a stone that has lived, died, and once again come alive.

Bibliography

Agamben, Giorgio. 1998. *Homo Sacer: Sovereign Power and Bare Life*. Translated by Daniel Heller-Roazen. Stanford: Stanford University Press.

Appadurai, Arjun. 1986. *The Social Life of Things, Commodities in Cultural Perspectives*. Cambridge: Cambridge University Press.

Ball, Christopher, and Nicholas Harkness, ed. 2015. "Kinship Chronotopes: A special issue," *Anthropological Quarterly* 88 (2).

Basso, Keith H. 1979. *Portraits of 'The Whiteman': Linguistic Play and Cultural Symbols Among the Western Apache*. Cambridge: Cambridge University Press.

—. 1996. "Wisdom Sits in Places: Landscape and Language Among the Western Apache." In *Wisdom Sits in Places: Landscape and Language among the Western Apache*, pp. 105-149. Albuquerque: University of New Mexico Press.

Bishara, Amahl. 2015. "Driving while Palestinian in Israel and the West Bank: The politics of disorientation and the routes of a subaltern knowledge," *American Ethnologist* 42: 33-54.

Bourdieu, Pierre. 1977. *Outline of a Theory of Practice*. Cambridge: Cambridge University Press.

Brown, Bill. 2004. *A Sense of Things*. Chicago: University of Chicago Press.

Bubandt, Nils, and Martijn van Beek. 2012. *Varieties of Secularism in Asia: Anthropological Explorations of Religion, Politics and the Spiritual*. New York: Routledge.

Carsten, Janet. 2000. *Cultures of Relatedness: New Approaches to the Study of Kinship*. Cambridge: Cambridge University Press.

Castells, Manuel, ed. 2004. *Network Society: A Cross Cultural Perspective*. Cheltenham: Edward Elgar Publishing.

Comaroff, Jean, and John L. Comaroff. 1993. "Introduction." In *Modernity and its Malcontents: Ritual and Power in Post-Colonial Africa*, edited by John L. Comaroff and Jean Comaroff. Chicago: University of Chicago Press.

Craig, Sienna. 2001. "A Tale of Two Temples: Culture, Capital, and Community in Mustang, Nepal." Paper Presented at the New York Conference on Asian Studies in Ithaca, NY.

—. 2008. *Horses Like Lightening: A Story of Passage Through the Himalayas*. Somerville: Wisdom Publications.

Dawson, Allan Charles, Laura Zanotti, and Ismael Vaccaro, eds. 2014. *Negotiating Territoriality: Spatial Dialogues Between State and Tradition*. New York: Routledge.

de Bruijn, Mirjam, and Rijk van Dijk. 2012. "Connecting and Change in African Societies: Examples of 'Ethnographies of Linking' in Anthropology," *Anthropologica* vol. 54, no. 1: 45-59.

De la Cadena, Marisol. 2010. "Indigenous Cosmopolitics In the Andes: Conceptual Reflections beyond 'Politics'," *Cultural Anthropology* 25: 334-370.

Deleuze, Gilles. 1992. "Postscript on the societies of Control." Article available at *libcom.org*.

—. 1994. *Difference and Repetition*. London: Athlone.

Dhungel, Ramesh K. 2002. *The Kingdom of Lo (Mustang): A Historical Study*. Kathmandu: Tashi Gephel Foundation.

Dilger, Hansjörg, Abdoulaye Kane, and Stacey A. Langwick, eds. 2012. *Medicine, Mobility, and Power in Global Africa: Transnational Health and Healing*. Bloomington: Indiana University Press.

Eck, Diana. 1986. "Darshan of the Image," *India International Centre Quarterly* vol. 13, no. 1, IMAGES (March): 43-53.

—. 1998. *Darśan: Seeing the Divine Image in India*. New York: Columbia University Press.

—. 2012. *India: A Sacred Geography*. New York: Three Rivers Press.

Eliade, Mircea, and Trask, Willard R. 1959. *The sacred and the profane: The nature of religion*. New York: Harcourt; Chicago: Brace & World.

Engelke, Matthew. 2007. *A Problem of Presence: Beyond Scripture in an African Church* (*The Anthropology of Christianity*). Berkeley: University of California Press.

Evans-Pritchard, E. 1940. *The Nuer*. Oxford: Clarendon.

Fabian, Johannes. 1983. *Time and the other: How anthropology makes its object*. New York: Columbia University Press.

Fisher, William F. 1987. *The Re-Creation of Tradition: Ethnicity, Migration, and Social Change Among the Thakali of Central Nepal*. PhD diss., Columbia University.

—. 2001. *Fluid Boundaries: Forming and Transforming Identity in Nepal*. New York: Columbia University Press.

Fowles, Severin. 2010. “People Without Things.” Springer Science and Business Media. http://www.campusincamps.ps/wp-content/uploads/2014/04/People-Without-Things.pdf. Retrieved 19 September 2019.

Franklin, Sarah, and Susan McKinnon. 2001. *Relative Values: Reconfiguring Kinship Studies*. Durham: Duke University Press.

Geertz, Clifford. 1973. *The Interpretation of Cultures*. New York: Basic Books Classics.

—. 1973 [1966]. “Person, time and conduct in Bali.” In *The Interpretation of Cultures*, pp. 360-411. New York: Basic Books.

—. 1975a. “Common Sense as a Cultural System,” *The Antioch Review* 33 (1): 5-26.

—. 1975b. *Kinship in Bali*. Chicago: University of Chicago Press.

Geismar, Haidy. 2011. “Material Culture Studies and other Ways to Theorize Objects: A Primer to a Regional Debate,” *Comparative Studies in Society and History* 53 (1): 210-218.

Gentry, James Duncan. 2016. *Power Objects in Tibetan Buddhism*. (Brill’s Tibetan Studies Library). Boston: Brill Publishers.

Geological Survey of India. 1904. *Geological Survey of India, Vol. XXI, Part 1*. Government of India. London and Calcutta.

Graburn, Nelson. [1977] 1989. “Tourism: The Sacred Journey.” In *Hosts and Guests: The Anthropology of Tourism*, edited by Valene L. Smith. Philadelphia: University of Pennsylvania Press.

Gupta, Akhil, and James Ferguson. 1992. “Beyond ‘Culture’: Space, Identity, and the Politics of Difference,” *Cultural Anthropology* 7 (1): 6-23; reprinted in *Culture, Power, Place* (edited by Gupta and Ferguson, 1997, pp. 33-51) and in *Anthropology of Globalization* (edited by Inda and Rosaldo, 2002, pp. 65-80).

Haenn, Nora, and Richard Wilk, eds. 2016. *The environment in anthropology: A reader in ecology, culture and sustainable living*. 2d ed. New York: New York University Press.

Hanks, William F. 1996. “Exorcism and the Description of Participant Roles.” In *Natural Histories of Discourse*, edited by Michael Silverstein and Greg Urban, pp. 160-202. Chicago: University of Chicago Press.

—. 1990. *Referential Practice: Language and Lived Space Among the Maya*. Chicago: University of Chicago Press.

Hausner, Sondra L. 2007. *Wandering with Sadhus: Ascetics in the Hindu Himalayas.* Bloomington: Indiana University Press.

Helmreich. Stefan. 2009. *Alien Ocean: Anthropological Voyages in Microbial Seas.* Berkeley/Los Angeles: University of California Press.

Hirsch, Eric, and Michael O'Hanlon. 1995. *Anthropology of Landscape: Perspectives on Place and Space.* Clarendon: Oxford.

Hodges, Matt. 2008. "Rethinking Time's Arrow: Bergson, Deleuze and the Anthropology of Time," *Anthropological Theory,* 8 (4): 399-429.

Horst, Heather A., and Daniel Miller. 2005. "From Kinship to Link-up. Cell Phones and Social Networking in Jamaica," *Current Anthropology* 46 (5): 755-778.

Hoskins, Janet. 2006. "Agency, biography. And objects." In *Handbook of Material Culture,* edited by C. Tiller et al., pp. 74-84. London: Sage Publications.

Ingold, Tim. 1994. *What is an animal.* London: Routledge.

—. 2000. *The Perception of the Environment: Essays in Livelihood, Dwelling, and Skill.* London: Routledge.

—. 2011. *Redrawing Anthropology: Materials, Movements, Lines.* Ashgate: Aldershot.

Irvine, Leslie. 2004. *If You Tame Me: Understanding Our Connection with Animals (Animals Culture and Society).* Philadelphia: Temple University Press.

Kirksey, S. Eben, and Stefan Helmreich. 2010. "The Emergence of Multispecies Ethnography," *Cultural Anthropology* 25 (4): 545-576.

Kohn, Eduardo. 2007. "How Dogs Dream: Amazonian Natures and the Politics of Transspecies Engagement," *American Ethnologist* 34 (1): 3-24.

—. 2013. *How Forests Think: Toward an Anthropology Beyond the Human.* Berkeley: University of California Press.

—. 2014. "What an Ontological Anthropology Might Mean." Theorizing the Contemporary, Cultural Anthropology website. https://culanth.org/fieldsights/463-what-an-ontological-anthropology-might-mean. Accessed 13 January 2014.

Lamb, Sarah. 1997. "The Making and Unmaking of Persons: Notes on Aging and Gender in North India," *Ethos* 25 (3): 279-302.

—. 2000. *White Saris and Sweet Mangoes: Aging, Gender, and Body in North India.* Berkeley: University of California Press.

—. 2004. "Making and Unmaking of Persons: Gender and Body in Northeast India." In *Gender in Cross-Cultural Perspective,* 4th ed., edited by Caroline Brettell and Carolyn Sargent, pp. 230-240. Upper Saddle River, NJ: Prentice Hall.

Lambek, Michael. 2008. "Provincializing God? Provocations from an Anthropology of Religion." In *Religion: Beyond a Concept,* edited by Hent de Vries, pp. 120-138. New York: Fordham University Press.

Latour, Bruno. 1988. *The Pasteurization of France.* Translated by Alan Sheridan and John Law. Cambridge, MA: Harvard University Press.

—. 1993. *We have never been modern.* Cambridge, MA: Harvard University Press.

—. 1999. "On Recalling ANT." In *Actor Network Theory and After*, edited by Law and Hassard, pp. 15-26. Oxford: Blackwell Publishers.

Leeuw, Gerardus van der. 1986. *Religion in Essence and Manifestation*. Translated by J.E. Turner. London: G. Allen & Unwin.

Low, Setha, ed. 1999. *Theorizing the city: The new urban anthropology reader*. New Brunswick, NJ: Rutgers University Press.

Marcus, George E. 1995. "Ethnography in/of the World System: The Emergence of Multi-Sited Ethnography," *Annual Review of Anthropology* vol. 24 (October): 95-117.

Mayor, Adrienne. 2005. *Fossil legends of the first Americans*. Princeton: Princeton University Press.

McIntosh, Janet. 2005. "Liminal Meanings: Sexually Charged Giriama Funerary Ritual and Unsettled Participant Frameworks," *Language and Communication* 25: 39-60.

—. 2009. *The Edge of Islam: Power, Personhood, and Ethnoreligious Boundaries on the Kenya Coast*. Durham: Duke University Press.

Messerschmidt, Donald A., and Nareshwar Jang Gurung. 1974. "Parallel Trade and Innovation in Central Nepal: The Cases of the Gurung and Thakali Subbas Compared." In *Contributions to the Anthropology of Nepal*, edited by C. von Fürer-Haimendorf, pp. 197-221. Warminster, England: Aris and Phillips.

Mieu, George, Paul. 2015. "'Beach-Boy Elders' and 'Young Big-Men': Subverting the Temporalities of Ageing in Kenya's Ethno-Erotic Economies," *Ethnos* vol. 80 (4): 472-496.

Morgan, David. 1999. *Visual Piety: A History and Theory of Popular Religious Images*. Berkeley: University of California Press.

—, ed. 2010. *Religion and Material Culture: The Matter of Belief*. London: Routledge.

Munn, Nancy D. 1970. "The Transformation of Subjects into Objects in Walbin and Pitjantjatjara Myth." In *Australian Aboriginal Anthropology: Modern Studies in the Social Anthropology of the Australian Aborigines*, edited by Ronald M. Berndt. Nedlands, W.A.: University of Western Australia Press.

—. 1992. "The Cultural Anthropology of Time: A Critical Essay," *Annual Review of Anthropology* 21: 93-123.

Nelson, C.M. 1968. "Ammonites: Ammon's Horns into Cephalopods," *Journal of the Society for the Bibliography of Natural History* 5: 1-18.

Olwig, Karen Fog, and Kirsten Hastrup, eds. 1997. *Siting culture: The shifting anthropological object*. New York: Routledge.

Otto, Rudolf. 1958. *The Idea of the Holy*. Oxford: Oxford University Press.

Parmentier, Richard. 1987. *The Sacred Remains: Myth, History and Polity in Belau*. Chicago: University of Chicago Press.

Paxson, Heather. 2008. "Post-Pasteurian Cultures: The Microbiopolitics of Raw-Milk Cheese in the United States," *Cultural Anthropology* 23 (1): 15-47.

Raffles, Hugh. 2010. *Insectopedia*. 1st ed. New York: Pantheon Books.

Ramble, Charles. 1983. "The Founding of a Tibetan Village: The Popular Transformation of History," *Kailash* 10 (1): 267-290.

—. 1990. "How Buddhist are Buddhist Communities? The Construction of Tradition in Two Lamaist Villages," *Journal of the Anthropological Society of Oxford* 21 (2): 185-197.

—. 1992. The Archive of Baragaon. http://www.tibetanhistory.net/documents-from-mustang/baragaou/. Accessed 10 October 2016.

—. 2002. "Temporal disjunction and collectivity in Mustang, Nepal," *Current Anthropology* 43, supplement, August-October.

—. 2008a. *Tibetan Sources for a Social History of Mustang, Nepal. Volume 1: The Archive of Te*. Halle: International Institute for Tibetan and Buddhist Studies.

—. 2008b. *The Navel of the Demoness: Tibetan Buddhism and Civil Religion in the Highlands of Nepal*. Oxford: Oxford University Press.

—. 2014. "The Complexity of Tibetan Pilgrimage." In *Searching for the Dharma, Finding Salvation – Buddhist Pilgrimage in Time and Space*, Proceedings of the Workshop "Buddhist Pilgrimage in History and Present Times" at the Lumbini International Research Institute (LIRI), Lumbini, 11-13 January 2010. Edited by Christoph Cueppers and Max Deeg. Lumbini International Research Institute 2014.

—, and C. Seeber. 1995. "Dead and Living Settlements in the Shoyul of Mustang," *Ancient Nepal* no. 138: 107-130.

—, and Michael Vinding. 1987. "The Bem-chag Village Record and the Early History of Mustang District," *Kailash* 13 (1-2): 5-47.

Rao, S.K. Ramachandra. 1996. *Shaligram Kosh* [*Śālagrāma – Kosha*]. Sri Satguru Publications. Delhi: Indian Books Center.

Romano, Marco. 2014. "From petrified snakes, through giant 'foraminifers', to extinct cephalopods: the early history of ammonite studies in the Italian peninsula," *Historical Biology* 27 (2): 214-235.

Salazar, Noel B. 2010. "Towards an Anthropology of Cultural Mobilities," *Crossings: Journal of Migration and Culture* 1 (1): 53-68.

Sanders, Clint. 1999. *Understanding dogs*. Philadelphia: Temple University Press.

Schattschneider, Ellen. 2003. *Immortal Wishes: Labor and Transcendence on a Japanese Sacred Mountain*. Durham: Duke University Press Books.

Singh, Vikash. 2017. *Uprising of the Fools: Pilgrimage as Moral Protest in Contemporary India*. Stanford: Stanford University Press.

Snellgrove, David. 1958. *Buddhist Himalaya*. Oxford: Bruno Cassirer.

—. 1961. *Himalayan Pilgrimage: A Study of Tibetan Religion by a Traveller Through Western Nepal*. Oxford. Cassirer.

—. 1965. "Review of the *Sherpas of Nepal* by C. Furer-Haimendorf," *Asian Review of Art and Letters* New Series 2.1.

Taves, Ann. 2009. *Religious Experience Reconsidered: A Building-Block Approach to the Study of Religion and Other Special Things*. Princeton: Princeton University Press.

Tsing, Anna Lowenhaupt. 1993. *In the Realm of the Diamond Queen*. Princeton: Princeton University Press.

—. 2005. *Friction: An Ethnography of Global Connection*. Princeton and Oxford: Princeton University Press.

—. 2016. *The Mushroom at the End of the World: On the Possibility of Life in Capitalist Ruins*. Princeton: Princeton University Press.

Uberoi, Patricia. 1993. *Family, Kinship, and Marriage in India*. Delhi: Oxford University Press.

Urry, John. 2000. *Sociology Beyond Societies: Mobilities for the Twenty-First Century*. London: Psychology Press.

—. 2002. "Mobility and Proximity," *Sociology* 36 (2): 255-273.

Van der Geer, Alexandra, Michael Dermitzakis, and John De Vos. 2008. "Fossil Folklore from India: The Siwalik Hills and the Mahabharata," *Folkore* 119: 71-92.

Waterhouse, E. 1940. *Religion in Essence and Manifestation*. By G. Van Der Leeuw. Translated by J. E. Turner. (London: George Allen & Unwin.1938, pp. 709).

2 Spiral Notebooks

A Multi-Local Shaligram Ethnography

Abstract
Shaligrams are both fossils and living deities, born of the sacred landscape of Mustang. For pilgrims, Mustang is home to multiple sacred sites belonging to Hinduism, Buddhism, and Bon. The land is also conceptualized as a body, wherein the Kali Gandaki River Valley is simultaneously the location where the Hindu deity Vishnu manifests himself as a sacred stone as well as the place where the corpse of a great Buddhist/Bon demoness (*sinmo*) is continuously subdued through ritual and sacred architecture. Any ethnography of Shaligrams must therefore account for intersections of mobility, time, place, and access. This is because the consolidation of movement and ritual is what enlivens Shaligrams and begins the process wherein they become living members of a community.

Keywords: ethnography, Hinduism, Buddhism, Bon, Himalayas

"Wherever you go becomes a part of you somehow."
– *Anita Desai*

If ethnography is the "writing of culture", then it must also include the perspectives of the researcher, embedded in specific cultural contexts and working to make some narrative sense out of many bits and pieces of confusing and sometimes contradictory information. When I began this research in India in 2012, I had initially set out to study the construction of deity altars in Hindu homes contrasted with deity care in temples. I knew that this kind of work would require a fair degree of long-distance traveling as well as what Clifford Geertz refers to as "deep hanging-out" when it came time to visit individual homes and spend time with local families. Fortunately, attempting something as vague and imprecise as hanging out

Walters, Holly, *Shaligram Pilgrimage in the Nepal Himalayas*. Taylor & Francis Group, 2020
DOI: 10.5117/9789463721721_CH02

was quickly expedited by the amount of work that always needs to be done in households full of children and extended relatives. As I helped cook evening meals, offered to assist with the care of the household deities, and did my best to distract rambunctious children or feed hungry animals, I quickly found warm welcomes in many different places. It was then, in a village in West Bengal, that I first encountered Shaligrams.

The first time I heard the term "Shaligram", I was almost dismayed that, despite several years of studying religion in South Asia, I had never heard of such a thing before. As I learned about the collections of strange black stones resting on their silver *puja* trays or tiny brass chairs, hearing tales of pilgrimage and inheritance from elderly men, attentive sons, devout mothers, and ascetic widows, I began to get a sense that something much more was at stake. But it also fueled a growing intellectual concern. How was I to assess a system of cultural values and meanings when the object of focus was located in temples and homes stretching all the way across Nepal, down the entire Indian subcontinent, and now little by little into Europe, Australia, and the USA? How was I going to "arrive on the scene", so to speak, when Shaligrams were expected to move from sites of ritual practice, to be kept in temples, and then to be distributed to devotees who then continued to distribute them even further; to their families and friends and sometimes to other temples of pilgrimage and veneration?

In many instances, Shaligrams are the consummate gift, the nature of their exchange and reciprocity clearly acting to enmesh both giver and receiver in webs of meaning, relations, and obligations (Mauss 1954). Therefore, I knew that any choice of place that I might make could only be temporary and that any Shaligrams in question could simply end up passing me by on the way there, moving right along with the people dedicated to caring for them. The only possibility then would be to follow the Shaligrams, and therefore the people, wherever they might go.

Shortly after I began my work in Northern India, with much urging from my friends and research participants, it soon became clear that I would have to go to Nepal, specifically to Mustang, the spiritual and geological origin point of all Shaligrams. I first arrived in Kathmandu in June of 2015, only five short weeks after the massive Gorkha Earthquake that devastated large parts of the country. My work was to begin at Pashupatinath, one of Hinduism's most sacred temple locations, which contains two well-known Shaligram veneration sites: a small temple-house near the shores of the Bagmati river and a deep well that sits at the head of a large stone slab used to wash the dead before they are brought to their funeral pyres on the cremation grounds just below. The neighborhood surrounding Pashupatinath also serves as the home of a number of Shaligram devotee families and more than one active family of Shaligram sellers, who have been making the pilgrimage and venerating sacred stones from the Kali Gandaki for nearly four generations. From there, I would head to Mustang.

This ethnography is my interpretation of the years I spent among Shaligram practitioners in South Asia, rooted in the theoretical frameworks of social science. This research took place between 2012 and 2017 with Hindu (Vaishnava, Shaiva, and Smarta) and Buddhist Shaligram devotees in Nepal and India. Many of these devotees were also pilgrims to the Kali Gandaki River and to the temple site of Muktinath in Mustang, where I was able to not only participate in pilgrimage and the ritual acquisition of sacred stones but in a few cases, was able to follow the pilgrims and stones outwards to destination temples and homes in the Kathmandu Valley and in Northern India.

Among those traveling to the Kali Gandaki, I spent time with *pujari* (temple priests) and other ritual specialists, a few of whom had journeyed to Nepal to write their own pilgrimage pamphlets on Shaligrams and many of whom had studied Shaligram interpretations in their home temples from New Delhi to Mumbai, to Hyderabad, to Chennai, to Sri Lanka. Months-long sojourns in Mustang allowed me to spend time with locals as well. In some cases, they identified themselves as Bon adherents, sometimes Hindu or Buddhist, and at other times claimed no religious affiliation at all. In other cases, this time allowed me to work specifically with those who spend their spring and summer months catering to the needs of pilgrims or collecting Shaligrams themselves, oftentimes to sell.

Bodies and Landscapes

My first journey to Mustang was, in a word, breathtaking. I arrived at the dusty airport in Jomsom (at roughly 3,100 meters), a moderately sized town that serves as the district's administrative headquarters, on an early morning flight from Pokhara, a foothills tourist town just below the Annapurna massif to the southeast. Stepping off the tiny twin-engine Otter aircraft and into the thin Himalayan air was made only more staggering by the surrounding 8000+ meter peaks of the region's two most prominent mountains, Nilgiri and Dhaulagiri. Located in Nepal's Dhaulagiri Zone, Jomsom is only some 15 minutes by mountain flight but nearly 180 km and more than a five-day walk from Pokhara. Even today, many native Mustangis as well as Shaligram pilgrims still traverse the treacherous distance on foot or by horseback. For centuries, this region had been one of the primary thoroughfares for trans-Himalayan trade. Yak and mule caravans once traveled through the Kali Gandaki River valley exchanging highland salt for lowland grains, braving the extreme high-altitude winds to trade goods between the Gangetic plains and the Tibetan plateau.

Mustang is currently divided into upper (northern) and lower (southern) regions: distinctions that are both locally and nationally relevant and that have had profound economic, social, political, and cultural ramifications. Following the Chinese invasion of Tibet in 1950, both Upper and Lower Mustang were closed to travel, and even today Upper Mustang remains highly restricted to foreigners. The village of Kagbeni marks the first boundary between the two divisions. Foreigners are not allowed to travel north beyond Kagbeni without special permissions, expensive trekking permits, and the services of a guide – an issue that has become especially relevant to Shaligram pilgrimage in the area, since the lake that principally produces Shaligrams, the Damodar Kund, lies in Upper Mustang. Kagbeni is also one of the main stops on the Shaligram pilgrimage route. It sits directly on the banks of the Kali Gandaki, and it is the furthest south that Hindu pilgrims can reliably find significant numbers of Shaligrams by wading through the river themselves.

Mustang is also home to a number of Tibetan Buddhist traditions, to the Bon religion (Tibetan: བོན – also spelled Bön), a pre-Buddhist Tibetan tradition sometimes described as shamanic or animist, and to various shamanistic traditions and practices dating back to the first occupation of the valley's cave systems sometime around 800 BCE (Chetri et Al. 2004: 15).[1] But as is typical throughout much of Nepal, distinctions within and between religious traditions are often vague. In what William F. Fisher refers to as the "river metaphor of culture", relationships between narrative, person, and object do not readily conform to notions of "pure" culture, despite the continual marketing of Mustang to trekkers as a kind of "lost kingdom of Tibet". This means that no particular religious shrine, sacred object, or even ritual practice can be said to "belong" specifically to one tradition or another. In fact, sacred sites are often shared by various traditions simultaneously. Practitioners at these sites then routinely blend rituals, celebrations, and deities together in myriad ways.

Just as the peoples of Mustang shape and are shaped by changing conditions, constraints, historical contexts, and reinterpretations, so too are Shaligrams and the pilgrimage that devotees depend on to obtain them.

1 See also *Mustang: The Land of Fascination*. 2004. King Mahendra Trust for Nature Conservation, p. 15.

Mustang is, in fact, a focal point of several kinds of religious pilgrimage and contains a high number of famous sites, including my primary goal at the time, the combined Hindu/Buddhist/Bon shrine at Muktinath (4,100 meters), the final pilgrimage destination where Shaligrams are principally venerated. Given the confluence of cultural and religious exchange in this region, it is not surprising then that many of these pilgrimage sites and circuits are overlaid with extensive mythologies tied not only to place but to movement across the landscape as a whole.

For Hindus, the linking of sacred spaces (and pilgrimage locations) with the presence of divine bodies is relatively common. For example, the pilgrimage site of Amarnath contains a column of ice interpreted as a Shiva Lingam (phallus), a mountain shrine in Garhwal (Kashmir) is said to be a part of the body of a buffalo briefly incarnated as Shiva, and even Mount Kailash, the Tibetan mountain home of Shiva and his consort Parvati, is often described in terms of anthropomorphic forms. The landscape of Mustang is no different. Consequently, there are a number of variations of the origin story of the Kali Gandaki River and of Shaligrams in Mustang. The most common (and in some sense most popular) version of the story involves the tale of the demon Jalandhar and his virtuous wife Brinda/ Vrinda. There are, in fact, two versions of this story recorded in the Puranas: the Padma Purana (*kriya-yoga-sara* section) and in the Prakrtikhanda chapter of the Brahmavaivarta Purana, where Jalandhar and Brinda are substituted with the prince Shaṃkhacuḍa and his wife Tulasi (the events of the story are, however, much the same). As I will discuss in greater detail later on, a hybrid version of the tale involving Jalandhar and Tulasi is the most commonly related version of this story in the context of Shaligram origins and pilgrimage (see Chapter 6).

In some Puranic accounts, women are also warned that they will accumulate various karmic sins by touching or worshipping Shaligrams and should, if necessary, only worship them from afar. Such restrictions are, however, typically only found in texts that date to the late medieval period (Rao 1996: 40-41) and today, a large majority of Shaligram practitioners take these passages as later superstitions that were added to the texts due to prevailing attitudes about gender at the time. In practice, Shaligram pilgrims and practitioners are often women, especially if they are the wives of high-caste men and have the responsibilities of caring for household altars. Subsequently, many Shaligram devotees treat these restrictions in much the same way that they also view the rumors that Shaligrams will produce daily quantities of gold (an idea stemming from the presence of iron pyrites in many Shaligrams): as irrational.

Despite this, some Hindu traditions still maintain strictures on the participation of women in Shaligram worship. This became, on occasion, a barrier to my own research, particularly in circumstances where I, as a female researcher, was not permitted into certain spaces or could not gain the satisfactory confidence of ritual specialists in order to discuss their perspectives at length. Over the course of my travels, it was one of the few times that my gender, more than my foreignness, shaped my interactions with pilgrims and devotees in the interests of my research. It was not a barrier to Shaligram pilgrimage, however, because the landscape of Mustang (and of the temple complex of Muktinath) is considered by many religious traditions, including Hindus and Buddhists, to be an especially potent locus of female divine power. This is one reason why many of the Buddhist and Hindu holy sites in Mustang are tended by women (particularly Buddhist nuns), despite the fact that Shaligram ritual specialization is almost entirely under the purview of elder men.

For Buddhists, the story of Mustang's landscape is equally complex and includes a wide pantheon of deities, many of whom predate the arrival of orthodox Buddhism to the region. I had heard a number of Buddhist stories in the early months of my research relating tales of demonesses (*sinmo*) specifically tied to landscapes.[2] For example, a Buddhist friend of mine in Kathmandu once explained that the seventh-century monastery of Katsel (*ska tshal*) in Tibet was built at the bottom of the Kyichu Valley in order to pin down the right shoulder of a powerful *sinmo* who routinely caused natural disasters. Afterwards, in order to fully subdue the *sinmo*, the then king of Tibet, Songtsen Gampo, had four more temples and monasteries built on geomantically significant places so that the *sinmo*'s body would be permanently bound to the earth using the holy places of the Buddha to force her into the topography of the ground. Another story involving Songtsen Gampo explains that even the very founding of Tibet itself was the result of the king's construction of twelve temples that "tamed" the demoness (a *rakshasi*), whose supine body now symbolizes the entire country.[3]

Later, I heard a similar story regarding the monastery of Simtokha Dzong in the Thimpu Valley of Bhutan. Built in 1629, the site of the current monastic

2 In Sanskritized literatures, *sinmo* are often referred to as a type of *rakshasa*, a kind of earth-demon common in South Asian mythologies.

3 Four of the temples are referred to as "Tadul" (mtha 'dul – The Border Taming Temples), while the remaining eight are referred to as the "Yangdul" (yang 'dul; Further-Taming Temples). The monastery of Thradug stands on the sinmo's left shoulder, Uru Katshal on her right shoulder, Jokhang on her heart, on her left breast Samye, on her right breast the rock monastery Yerpa, Taktshang on her left leg, and Ganden on her mouth.

school was said to have been chosen because the carefully planned geometry of the sacred spaces was needed to guard over a demon that had vanished into a rock nearby; hence the name Simtokha, from *sinmo* (demoness) and *do* (stone). In Mustang, several local villagers were also happy to explain that, each year in Lo Monthang in Upper Mustang, they held a festival to commemorate the deity Dorje Jono's defeat of his demon father Tenchi. This is because Dorje Jono was responsible for sending the yearly rains to refresh the high mountain water pools. This is why it was necessary, they explained, to re-enact the story by engaging in vigorous dancing and horn blowing in order to banish the demons so that the water would come on time. Finally, as I was preparing to leave Mustang for the second time, a visiting Buddhist monk from India on his way to one of the regional *gompas* (Buddhist temple/ecclesiastical building) explained that even the great Siddhartha Gautama, the historical Buddha, had once tangled with demons. After all, it was the great demon lord Mara who had attempted to disrupt the Buddha's meditative journey into enlightenment by summoning nine great storms. It wasn't until the Buddha then called upon the earth itself to witness and support him that the demon was cast out.

Prior to my third trip to Mustang in 2017, I had read Sienna Craig's account of another such story in her work *Horses Like Lightening*, which recalled a *sinmo* who once terrorized the land of Mustang before the coming of Guru Rinpoche (also called Padmasambhava, the eighth-century Indian Buddhist master who helped construct the first Buddhist monastery in Tibet and who is also revered among New Bon lineages for his tantric cycles). With this in mind, I began to ask the local peoples near the village of Ranipauwa (the village next to Muktinath temple) about the *sinmo* of Mustang. On my way up from the jeep-stand a few hundred feet from the village gate, I finally encountered a Mustangi Buddhist and a village resident who was keen to relate the tale. At my continued questioning, he explained that there was indeed once a *sinmo* who had inhabited the land, but she was subdued by Guru Rinpoche as he traveled through Mustang on his way to Tibet, spreading the teachings of the Buddha as he went.

"There was once a *sinmo* in Lo," he began. "She was never still and traveled all over the Himalayas, sometimes hiding in caves. Someone once told me that she was fire-like, and if you saw her, she could burn you up by looking at her. She was also very old and sometimes would possess animals until they died of fright or she would summon great storms that would destroy all the fields and gardens or trick herds into wandering away. She caused all kinds of problems, and people said that she could appear in dreams and would bring sickness. She also traveled on the wind, and it was her who made all the wars happen between the people in the mountains. One day, the *sinmo* came to Mustang and decided to stay here permanently. She had been told by other *sinmo* that a great teacher was coming from India and would soon go to Tibet, so maybe she thought that in Mustang he wouldn't find her. But this was Guru Rinpoche." He pointed to the massive golden statue of the guru overlooking the village a short distance away. "But the *sinmo* was not afraid. Like the land, there was no compassion. She made it so the people became angry and started fighting with one another over stupid things. Everyone wanted to be rich and become important men, so they went after money and things like that. Sometimes I think that is why there is still so much trouble here. But Guru Rinpoche was a very powerful teacher, and he came to Mustang looking for the *sinmo*. He came to Mustang flying on a tiger skin and carrying a great *dorje* (a thunderbolt scepter). When the *sinmo* saw his *dorje*, she was finally afraid. But she also did not want to leave, you know?" He patted his horse contentedly.

"It was because of the Buddha's teachings that he could do this, and because of that the *sinmo* ran away and hid. Mustang has so many caves and mountains and rivers, she thought she could get away. But then Guru Rinpoche took out his bag of Tibetan salt and he started to leave the salt wherever he traveled [I recalled from earlier that this salt was obtained from other earth spirits he had tamed in other places]. Even though the *sinmo* also tried to disguise herself as an old woman, Guru Rinpoche could still see her, and he continued to follow her all over the land. The *sinmo* thought that the land would protect her, but it didn't, and Guru Rinpoche could see her no matter where she went. Finally, he struck her with his *dorje* and tore her open. He poured out her blood and that is why the dirt is red in some places and pulled out her intestines to cover over the plains. He threw her liver to Ghami, but now there are only ruins there. Then he destroyed her heart by cutting it up into a hundred and eight pieces, you know like the one hundred and eight[4] water fountains at Muktinath, and then he buried them under the *chorten* (Tibetan: "religious construction" – a reference to the many large monument shrines of *mani* stones found throughout Mustang). That is why we must maintain these *chorten* carefully because as long as they hold down the *sinmo's* body, she can't get up and terrorize us again."[5]

4 The significance of the 108 number is open to interpretation. But 108 has long been considered a sacred number in Hinduism and Buddhism. Traditionally, malas, or garlands of prayer beads, come as a string of 108 beads. A mala is used for counting as you repeat a mantra – much like the Catholic rosary. Mathematicians of Vedic culture viewed 108 as a number of the wholeness of existence.

5 Another version of this story, which also recounts the nature of sacred landscapes in Mustang, is detailed in Sienna Craig's book "Horses Like Lightening." (pp. 217-220). This conversation was

As is not uncommon in Tibetan mythologies, the land of Mustang is an animate one that requires "taming" – a world of uncontrolled natural forces that are often interpreted as stand-ins for the conversion of local peoples to Buddhism. Within these ancient landscapes, wrathful gods guard their earthly domains, but they also protect and preserve their peoples within it. In many of these tales, *sinmo* are especially difficult to deal with, and stories of their violent encounters with famous Buddhist figures abound within both Tibetan and Mustangi traditions. As in Tibet, Mustang's *sinmo* are both literally and figuratively tied to the landscape by the power of Buddhist teachings and through the actions of Buddhist practitioners themselves. This is why great walls of *mani*[6] stones and prayer wheels cover the landscape, meant to give it symbolic form and meaning in the context of human spiritual battles embodied as relationships between person and nature. For these reasons, many Buddhists and Bonpos in Mustang also leave Shaligrams inside or on top of roadside *stupas* that mark crossroads and places of particular physical or spiritual danger. Their clockwise spirals (or counterclockwise for Bon) encapsulate the divine movement of the universe and guard the traveler against misfortune by anchoring the vengeful spirits of the land within the peaceful and orderly motion of the cosmos.

Living History

The name "Mustang" is the Nepali cognate of the Tibetan *monthang*, a term meaning "plain of aspiration". Founded as the Kingdom of Lo in the late fourteenth century, it emerged as an independent kingdom in AD 1440 (Dhungel 2002: 4). Yet from its founding until the Gorkhali conquests of Jumla in 1789, and in reality, until the Chinese occupation of Tibet in the 1950s, Lo retained strong cultural and political ties to the ancient kingdoms of western Tibet, namely Zhang Zhung, Guge, and the Gung Thang region of present-day Ngari Prefecture (Tibetan: *mnga' ris skor gsum*; Craig 2001).

carried out primarily in Nepali and Lower Mustangi Tibetan. Transcribed, with help from a local translator, from a combination of recorded dialogue and fieldnotes written immediately after the conversation.

6 Mani stones are stone plates, rocks and/or pebbles, inscribed with the six-syllable mantra of Avalokiteshvara *Om mani padme hum*, (hence the name "Mani stone"), as a form of prayer in Tibetan Buddhism. They are intentionally placed along the roadsides and rivers or placed together to form mounds or cairns or sometimes long walls, as an offering to spirits of the place. Rizvi, Janet. 1998. *Ladakh, Crossroads of High Asia*. Oxford University Press. 1st edition 1963. 2nd revised edition 1996. Oxford India Paperbacks 1998. 3rd impression 2001. p. 205.

Even today, people often refer to Upper Mustang as "Lo" and the peoples who live there as the Loba (or conversely, and sometimes derogatorily, Bhotia), who are distinguished from the peoples of Lower Mustang not just by a militarized border but by language, lineage, and ritual observances. By the late fifteenth century, the Kingdom of Lo was already actively engaged in trade and cultural exchange with the Kathmandu Valley, Lhasa, India, and Persia. In 1769, King Pritvi Narayan Shah, the first king of unified Nepal, swept through the Himalayas with his Gorkha army and conquered the Kathmandu Valley. At the same time, Lo and many of its connecting principalities (which would eventually become Mustang) remained closely connected with Tibet but eventually sided with Nepal during the Nepal-Tibet wars of the eighteenth and nineteenth century. Later, Lo was politically incorporated into Nepal, but the region's cultural and linguistic mores still reflect its close associations with Tibet.

Lower Mustang begins with the region of Thak Khola, an area roughly encompassing the upper Kali-Gandaki valley, which lies between the peak of Dhaulagiri (8,167 meters) and Annapurna I (8,078 meters). Most scholars and cartographers place the southern border of Thak Khola at the village of Ghasa, the first village that marks the narrow canyon that accesses the Kali Gandaki valley from the south. The northern border of Thak Khola is variously placed in the village of Tukche, in Jomsom, or occasionally as far north as Kagbeni, or just about anywhere south of the border with Upper Mustang (Fisher 2001: 24-25). Other scholars suggest that Thak Khola is better divided into two parts, a southern part encompassing the areas of the Panchgaon (Nepali: *panchgaun* or "five villages" which were originally Thini, Syang, Marpha, Chimang, and Chairo/Tsherog but now include quite a few more) and Thaksatsae (*Thāksātsae*, literally "the seven hundred Thak") and a northern part surrounding the area known as the Baragaon (Nepali: *baragaun* or "twelve villages" which includes many more villages than twelve but incorporates the areas around Jhong/Dzong, Jharkot, Kagbeni, Ranipauwa, and Chusang).

About an hour's walk just north of Jomsom, the Kali Gandaki is joined from the east by the Panda Khola, a small river that marks the boundary between Panchgaon and Baragaon and must be crossed when traversing from Kagbeni back to Jomsom. The Bon village of Lubra is also considered to be a part of Baragaon, though the land on which the settlement was founded in the thirteenth century originally belonged to Thini (see Ramble and Vinding 1987: 18). On the opposite side of the Kali Gandaki are two more villages, Dangardzong and Phelag. Continuing north, closer to the river, is the settlement of Pagling. The village is said to be the most recent in Baragaon, having been settled by one family from each of the existing communities. A short distance north of Pagling, then, on the left bank of the river, is the village of Kagbeni, whose name is derived from its position at the confluence (Nepali: *beni*) of the Kali Gandaki and the Dzong Chu rivers, the latter indicating a mountain stream coming down out of the Muktinath Valley which runs parallel to and north of the Panda Khola.[7]

At the head of this valley stands the temple of Muktinath, the pilgrimage site at the center of Shaligram veneration in Mustang, although the entire temple complex also includes shrines that are visited and revered by a number of Hindu, Buddhist, and Bon traditions. The peoples of Lower Mustang are generally categorized under the ethnonym "Thakali", though divisions within and among various groups and villages, such as between Thaksatsae and the Panchgaon, continue to challenge the meaning of the term and of the ethnic cohesion it might imply (see Fisher 2001). Regardless, the diverse communities living in Mustang retain a wide variety of strong cultural, linguistic, and political ties to Tibet as well as migratory and economic ties to other areas of Nepal and India. Indeed, tourist literatures designed to encourage tourism to Mustang still refer to it as the "lost" or "hidden" kingdom of Tibet, as though it were a place suspended in time where a "pure" and "untouched" Tibetan culture could still be experienced intact. But as the people and history of Mustang reveal, no time or place is ever so static.

This rising isolation from the historical trends of mobility has led to many of the continuous contentions between religious pilgrims attempting to travel to Mustang, residents striving to build a working economy in the region, and government officials who want to both culturally "preserve" Mustang and keep it politically separate. This is why access to the Kali Gandaki River, where the vast majority of Shaligrams are collected, is often fraught with difficult travel, expensive permits and transportation, and highly unpredictable local conditions. As I left Jomsom for the first time, heading northeast towards Kagbeni, I followed the contours of the Kali Gandaki riverbed. There I encountered a group of pilgrims who had been out on the river all morning in search of Shaligrams.

"Look here!" one of them shouted as I carefully made my way over to where several members were conducting an impromptu *puja* and *aarti* (lamp offering) on the edge of the river. The pilgrim gently touched the Shaligram to his forehead before extending it towards me in his open palm. "This is Sri Kurma

7 The villages on the south side of the valley are, in ascending order, Khyenga, Dzar, Purang, and Chongkhor. On the northern side are two others: Putra and Dzong. The communities of the Muktinath Valley are sometimes referred to collectively as Dzardzong Yuldrug.

(the tortoise incarnation of Vishnu), revealed just an hour ago! He walked from the mountain all the way down here to us. I can see by the *chakra* (spiral) on his back that Vishnu is merciful with me today. The river has been so strong with the rains and I am grateful for the stability!"[8] Within the social and political world of the Nepal-Tibet Himalayas, Mustang is indeed a complicated living landscape: a landscape that both begins the journey of Shaligrams into religious consciousness and physical movement and began my own journey into the world of Shaligrams. When I returned to Nepal again for the long term in June of 2016, I hadn't even begun to realize the distance yet to go.

An Ethnography of Mobility in Time and Place

As any researcher or trekker who has ever been to Mustang will tell you, getting there is quite an entire adventure unto itself. On reflection, the endless difficulties and challenges I faced traversing the distances from Calcutta (Kolkata) to Kathmandu to Mustang to Muktinath speak to the conundrum of an ethnography of mobility. Ethnography is, by design, iterative, combining responsivity with observation, analytics with experiences, and a determination to attune oneself to circumstances as they present themselves. And in this case, all of this had to be done while moving through any number of transient contexts. Looking back on these experiences, I later realized that many of my most informative interviews and interactions were carried out on pilgrimage buses, sitting at jeep stands, on horseback, or walking breathlessly up a mountain road. To recall Robert Louis Stevenson's aphorism: "to travel hopefully is a better thing than to arrive". But the continuous shift between national, geographical, and religious contexts can quickly become overwhelming.

My focus on the ethnographic object, the Shaligram, was only as revealing as the people, practices, and places could make them in any given moment, and I was determined to attend to movement just as much as I would attend to place. The notion of progress along a pilgrimage route presents something of an intellectual quagmire. It embodies a belief in the "destination" which, while important, ignores some of the more pressing concerns of actually

8 The pilgrim here was referring to the story of Kurma, Vishnu's second incarnation, who bore Mount Mandara on his back while the gods and asuras (demons) churned the Ocean of Milk to gain the elixir of immortality. As is common in Shaligram practice, interpreting the religious story attached to the stone often reveals some relationship to the devotee's current circumstances, in this case, the difficulties of trying to find sacred stones in a river swelled by glacial melt and heavy rains. Conversation in Hindi. Transcribed from fieldnotes taken shortly after the encounter.

getting there and of the processes of meaning-making that take place not in arriving but in traversing. As a point of departure, then, from more typical literatures on pilgrimage, this work takes the encounters and co-mobilities of persons and objects together as the focal point for undertaking the journey in the first place and not specifically in a need to arrive at the temple of Muktinath at the conclusion of the pilgrimage. For Shaligram devotees, the object/person movement through the sacred landscape is truly what is meant by "undergoing pilgrimage" and not specifically implying that an individual is intent on "going somewhere". This is because, though the places of Shaligram pilgrimage are significant, they are not the primary goal of this particular community of devotees. Having been to the Kali Gandaki or to Muktinath was not the end of the journey but the beginning of one.

There was simply so much potential data in any one place, linked to even wider networks of information above and beyond the presence and use of a Shaligram wherever I might encounter one. As Sondra Hausner once described her work with Himalayan *sadhus*, "all circumstances, like all experiences, are endlessly interpretable" (2012: 7). I had to remember that at each moment I was sampling what amounted to a few brief snapshots in the larger lifetime of a person, itself a fragment of the time endured by Shaligrams, themselves a brief instant in the whole of geological and mythological time. I also needed to understand how and why it was that Shaligrams were meaningful and profound in so many variable contexts across so many different variations of religion, community, nationality, and caste. So how, then, was I going to take these conversations, detailed observations, and participatory experiences and make them interpretable?

In the case of Shaligrams, I knew I would need to consider the relationships between individual voices and the communal systems they reference, because what people do and what cultural and religious ideals they strive after are also not always consonant. Narratives had a tendency to come in fragments. As I came to know more pilgrims and devotees, either on the road or in their homes, I found that narrative elements discerned in the field had ways of coming back. While I was still in Mustang, many of my research participants (who had long returned home from their pilgrimages) began to send me videos of their home practices, photos and recordings of local festivals and temple celebrations, and lengthy descriptions of *puja* rituals and Shaligram veneration events. Some of that information has been included in these pages. In the end, there is no final truth in anthropology. Every new anthropologist brings something of his or her own to the field: a new method or technique, a new perspective, a new observation, even a few new mistakes. The past changes and shifts while the researcher remains on a perpetual journey to know a world that can never fully be known. We are caught up in the river of culture carrying us and everyone else along.

Because the buying and selling of Shaligrams is typically forbidden in Hindu scripture (specifically the exchange of a sacred stone for money), as a matter of contrast, I also spent several months in the company of Shaligram merchants in the village of Ranipauwa, in Jomsom, in Pokhara, and in Kathmandu. Sometimes Hindu, but more often than not Buddhist, many of these individuals have been engaged in Shaligram ritual practices themselves for decades, and some even started out as pilgrims to Mustang, bringing rare stones out of the mountains in the years prior to 1992. The significance of being either Hindu or Buddhist as a Shaligram merchant was occasionally tied to the Puranic restriction on placing monetary value on sacred stones (which many Hindu sellers tended to avoid but which is not a central prohibition for Buddhists) but also often had to do with individuals' own understandings of Shaligram mobility and on the current political issues facing pilgrimage to Mustang. For Hindu sellers, Shaligram markets could continue to extend Shaligram mobility around political restrictions on pilgrimage travel and for Buddhists, oftentimes Mustang was the home from which they had now been displaced. Shaligrams, then, became a part of the larger movement of commodified Hindu and Tibetan Buddhist culture. As merchants knowledgeable in their products, many of them also have extensive training in the interpretation of Shaligrams, so they can offer pilgrims specific stones tailored to their needs and requests.

Navigating the contradiction between never placing a monetary price on a sacred stone and the now global demand for buying Shaligrams (resulting from the inability of many devotees to undertake Shaligram pilgrimage), many of these sellers occupy an ambiguous space between practitioner and outsider. Some sellers unapologetically buy and sell Shaligrams as commodities, putting them up for sale to pilgrims and tourists alike in the bazaars of Jomsom, Pokhara, and Kathmandu. Others are more circumspect and request "donations" to cover their time and travel costs so that they may continue supplying new Shaligrams each season; the stone, however, is free. In the context of global markets, some Shaligram devotees see merchants as part of the inevitable degradation of Kali Yuga[9] (the decline of the current age

9 Kali Yuga ("age of Kali", or "age of vice") is the last of the four stages (or ages or yugas) the world goes through as part of a 'cycle of yugas' (i.e. Mahayuga) described in the Sanskrit scriptures. The other ages are called Satya Yuga, Treta Yuga, and Dvapara Yuga.
Kali Yuga is associated with the demon Kali (not to be confused with the goddess Kālī). The "Kali" of Kali Yuga means "strife" or "discord".

and a typical euphemism for modernity) and continue to speak out against their collections of stones priced in the hundreds to thousands of dollars. These devotees then view the potential shift between kinship networks of exchange to commodified networks of exchange as a considerable danger for Shaligram mobility as a whole (see Chapter 8). For other devotees, many of whom cannot afford to or cannot physically make the pilgrimage to Mustang, these sellers, including several who now advertise Shaligrams for sale online, are a godsend, representing another kind of mobility inherent to Shaligram stones. One way or the other, they say, a Shaligram will go wherever it needs to, by whatever means necessary.

The conversations that came most readily to and among Shaligram devotees were not actually about movement, however, but about space. We were often moving together, along pilgrimage paths or in and around religious sites, and it quickly became apparent to me that such movements were always contextualized within structures and landscapes linked through elaborate social networks of spiritual embodiment. Each place we might stop was a potential moment of *darshan*, a term derived from the Sanskrit word meaning "to see", a kind of ritual viewing of and interacting with divine forces and persons. Within a *darshan,* material and ritual practices are meant to make the deity available to the senses and to bodily experiences (Eck 1998). Because images and material icons have agency in the *darshan*, it is vital to understanding the role of Shaligrams as they move outwards from their places of origin and into the lives of devotees.

Most Hindu deity altars, including temple and home altars, are often arranged with great care and include a wide variety of deity pictures or icons (*murti*), deity clothing and accessories, miniature animals or people, photos of ancestors, and, of course, sacred stones. In many cases, these arrangements reflect specific mythological events or are meant to distill larger mythological narratives into a single perpetual moment in ritual time. For example, one particular home deity altar belonging to a Gaudiya Vaishnava woman I met while working in West Bengal contained an entire miniature diorama of the story of Gajendra, a tale related in the Bhagavata Purana of the rescue of a devoted elephant king by Vishnu from the clutches of the crocodile Makara. While the roles of Gajendra and Makara were filled, in this case, by two intricately carved marble statuettes, the role of Vishnu was supplied by a Sudarshan Shaligram.[10]

10 As Gajendra was being attacked by the crocodile, he began to pray and call out to Vishnu for salvation. Upon hearing his devotee's call and prayer, Vishnu rushes to the scene and finds Gajendra near death. As Gajendra sees Vishnu coming, he lifts a lotus with his trunk as an

Temple complexes are also carefully arranged, both architecturally, using a religious system called Vastu Shastra, and in terms of the placements of their deities. While Hindu temple architectures and arrangements have been described in greater detail elsewhere, some of the principal themes to note are that of the taming and moderating of powerful natural forces and the recreation of sacred landscapes that lie elsewhere. The concept of the *dham*, literally meaning the "seat" or "abode" of a deity, is ubiquitous in Hindu and Buddhist worship. *Dham* refers to the sacred landscape that literally houses the deity in a place of pilgrimage as well as to the recreation of those landscapes within other temples such that those temples may be said to actually "be" those places of pilgrimage as opposed to simply symbolizing them. Many temple devotional events also require various circumambulations around deity altars or from one altar to the next (a short-form pilgrimage). It is also not unusual for smaller representations of larger temple deities (called *utsav murti*, or sometimes "extensions") to be carried along during these movements or for these smaller deities to be brought out into the village to "walk around" and "visit".

The case of Shaligrams, however, might include any one of these roles or even all of them simultaneously. A Shaligram may be both home deity and pilgrimage companion, playing a role in a mythological diorama or receiving offerings of worship during *puja*, acting as a temple deity as well as the one that travels throughout the village region. Shaligrams are never considered *utsav* or "extensions", however, as they are completely self-manifest and self-contained forms of the divine. It is therefore not unusual for temple deity arrangements to reflect mythological narratives or the life journeys of saints and gurus. Larger landscapes are viewed similarly, with the placement of sacred sites along roadways, near rivers, and on mountaintops that continuously recapitulate mythic events and direct the journey of mankind within the karmic wheel of life. This is why questions such as "Where are you coming from?", "Where do you live?", "How did you get here?", and "Where are you going?" took on a rather different valence for Shaligram devotees. Travel routes and methods (did you walk, or did you hire a car?), places of religious gathering (a roadside shrine or the temple of Muktinath itself), and pilgrimage circuits (Did you go to Mount Kailash first? Did you start at Pashupatinath?) clearly help to define the temporal and spatial relationships of communities of devotees, pilgrims, and Shaligrams alike. Despite not sharing their lives in one particular place, the community of

offering. Seeing this, Vishnu is pleased by Gajendra's devotion and decapitates the crocodile with his principal and iconic weapon, the Sudharshana Chakra.

Shaligram worshippers continues to produce and reproduce itself as one of perpetual movement in time and space.

Pilgrimage in and of itself is significant, especially for those who wish to obtain Shaligrams according to the “proper” methods of veneration, but so are the final spaces of worship in destination temples and in-home shrines. Again and again, I would hear stories from devotees that usually began with their motivations for pilgrimage (their fathers and grandfathers had venerated Shaligrams; they wanted Shaligrams for their own children; they needed a specific Shaligram to attend to a struggling child, a sick relative, or an elderly parent expected to pass soon) but did not end with the story of pilgrimage itself. Rather, pilgrimage framed the middle part of the narrative, while coming home to install the Shaligrams into their communities and homes represented the true culmination of the story that brought their tale to the present day. What was vital about Shaligrams wasn’t just their mythological formations or their geological antiquity, it was the way in which divine persons, embodied as stone, continued to live on from ancient times and, consequently, how those divine persons linked together ancestors, descendants, and community lineages in the present. In short, the Shaligram lived a life.

Keeping Up with Shaligrams

Shaligram stones are “born” from the river, taken into homes where they are fed and cared for, exchanged in real and symbolic marriages, and then passed down to the next generation or given a symbolic “death”. In “death”, a Shaligram could be either retired to a temple, returned to the river, or placed into the hands of the dead just prior to cremation. The Shaligram would then be ready for its karmic “rebirth”, appearing to a new devotee arriving at the river for pilgrimage or to a destination temple where it might be given as a gift. For devotees, it was important to move as the Shaligram does, from place to place and from person to person in a never-ending cycle of relationships marked by birth, marriage, children, and death. Later on, I began to wonder whether this collective inclination to describe Shaligrams in this way – as symbolic manifestations of the movement of life itself – might ground a methodology that viewed the links between far-reaching people, places, and objects as a way to understand how those links created and maintained communal identities.

When I first posed the possibility of linking the physical movement of Shaligrams with the temporal movements of life to Shaligram devotees, they explained to me that this was, in fact, how the entirety of the cosmos existed: a continuous and eternal cycle of material creation caught up in the persistent rhythms of time. In other words, it was the nature of all things material – be it stone, human, or deity – to be born, to live, and to die in this manner, regardless of how much or how little time it took to do so. Anthropologically speaking, it was also easy to understand these concurrent understandings of movement as one of the basic elements by which practices – and cultures in general – are able to reproduce themselves.

Unlike most multi-sited ethnographies, this work does not focus on any one particular community moving from place to place. Rather, my methodology here required something different: to follow the movement of a specific object as it transitioned between national and religious contexts, between different cultural contexts, and through various political and economic conflicts. One of the premises of this research is that, despite extensive geographical distances and cultural differences, Indian and Nepali Shaligram devotees are participating in and reproducing larger communal structures in conversation with a broader spiritual framework that deeply influences and affects political life in South Asia. And it is these communal structures, dependent on ritual mobility and access to sacred landscapes, that are coming into conflict with politics of division and isolation as well as the rhetoric of "cultural preservation" and notions of "cultural purity" in academic and development discourses today. In some sense, then, this is an ethnography of the lives of stone persons and their communities.

India is, unsurprisingly, viewed as the traditional locus of Shaligram practice, but I did not want this research to be limited to the subcontinent. This was not especially difficult given that both Hindu and Buddhist views of space make specific distinctions between national boundaries and religious ones. As more than one of my research participants described, the border between India and Nepal, the border between Nepal and Tibet, as well as the borders within India and Nepal were really matters of politics and not matters of spirituality. Or as another Hindu devotee described it, "people are much the same as plants. Plants grow where they grow. They don't pay attention to government borders. Neither do people really."

Most of the people I encountered in my work (religiously affiliated or not) contended that there was really no spiritual distinction between Nepal and India and that this was obvious due to the incontrovertible links between their places of pilgrimage, the similarities between their gods (conflating multiple deities together is an accepted practice in most places), and the necessity for people everywhere to access them. These links then manifest in the symbolic connections between sacred spaces. Just as Kathmandu's Bagmati river as well as Mustang's Kali Gandaki are symbolically associated with the Indian Ganges River, sites like Pashupatinath and Mount Kailash are connected to entire circuits of prominent Shiva temples throughout South Asia (Hausner 2012: 13; Eck 2012). Gods share similar qualities in that the icon of Avalokiteshvara at Muktinath is also Vishnu, Brinda in the story of the creation of Shaligrams is also Tulasi, and the Hindu manifestation of the Kumari (*Shakti*) goddess in Nepal always appears as a Buddhist girl.

I chose my region of focus – Northern India to Mustang, Nepal – as a way to encompass as much of the movement of Shaligrams as I reasonably could at a given time. As a South Asianist, this also gave me the opportunity to focus less on nations as separate categories of study and more on the interactions between nations and the ways in which ideas of "nationhood" are debated, impugned, and challenged by lived experience. This would then include devotees from multiple Hindu, Buddhist, and indigenous religions hailing from a variety of countries of origin, Shaligram pilgrims from all over the world (from South India to Australia), and individuals engaged in Shaligram trade. This was also partly facilitated by the fact that, for Indians and Nepalis, the border is an open one with no required visas for travel between the two countries. For Mustang, however, this posed a much different problem because, while the border between Upper and Lower Mustang is open for Nepalis, it is highly restricted for everyone else.

My decision to work in both India and Nepal was also consistent with the routes typically taken by the Shaligram devotees that I worked with, almost all of whom were well-versed in crossing national boundaries for the sake of religious realization. Speaking Hindi was the most vital part of communicating with the vast majority of my research participants, but the moderate amount of Nepali I learned while living in Kathmandu proved to be equally essential for this research, especially while in Mustang District and elsewhere in Nepal. Since I never quite mastered the Tibetan dialects common to the high Himalayan areas I frequented, I was also grateful for the fact that many of those I worked with, themselves engaged in the Shaligram pilgrimage economy, also spoke a combination of Nepali and Hindi. Many of my Indian informants were also members of the Hindu diaspora in the USA, in the UK, and in Australia and so spoke a combination of Hindi and English. Throughout this research, the vast majority of interviews were therefore conducted with some combination of Hindi, Nepali, and English.

The Practice of Shaligram Ethnography

The conversations I had with my research participants were predominantly very informal. On the rare occasions when I scheduled interviews, they were mostly conducted with ritual specialists or *gurus* whose time required appointments or other kinds of formal interactions before I was allowed to speak with them. Fieldwork mainly involved walking or sitting with Shaligram devotees, oftentimes during *puja* or more usually with pilgrims along a pilgrimage road or settling into an evening meal, watching their interactions, listening to their stories, and discussing their life experiences with them. Despite the fact that we were both usually strangers to the land we traveled, I was still treated as a guest or later on as a friend and confidant. While working among village residents, I continued to assist in as many household chores as I could (washing dishes, helping to prepare meals, or gathering ritual implements), but I was often admonished to cease such work and to accept tea so that we might sit and talk. This extended to ritual events as well.

As a part of a larger crowd, I usually stayed out of the way and accepted *kumkum* and sandalwood *tilaks* (forehead markings) or *prasad* (food or items which come from a deity or holy person) if it was offered. As a researcher, and therefore as someone intent to learn as much as possible, I also did not present myself as an expert in Shaligram practices. It has never been my intent to supplant the role of *gurus* or ritual specialists and I endeavored to keep my position as one of a student, no matter how long I spent studying texts, interpreting Shaligrams, or practicing rituals. Even when I knew the *mantras* required for the performance of specific rituals or already understood the characteristics of a particular Shaligram as identifying a specific deity, I did not offer this information unless asked.

Very few of my interactions with Shaligram devotees were tape-recorded. Rather, I preferred to take notes by hand in the moment of conversation and summarize and elaborate them in further detail later on. I also consistently maintained two separate notebooks: one that contained the written account of my observations and the second that was reserved for drawings of Shaligrams and their accompanying details and mythological explanations. It was this second notebook that ultimately proved to be my point of entrance into many deeper and more long-term relationships.

Drawing Shaligrams became an important aspect of this work for two reasons. Firstly, photography was often forbidden within the confines of sacred spaces, such as temples or within the inner sanctum of a shrine, and secondly, drawing specific Shaligrams became an invaluable pedagogical technique for learning how to read and interpret stones according to their particular characteristics. As happens with many ethnographers in the field (Dejarlais 1992; Hausner 2012), word quickly spread among pilgrims and devotees that there was a researcher from America writing a book about Shaligrams and who had creating detailed drawings of their many different types. On many occasions, devotees asked to see this notebook, a request I was more than happy to oblige. Ultimately, this extensive packet of drawings and sketches proved to be exceptionally useful in starting conversations about Shaligram practices and provoking memories from gathered pilgrims or family members. I was also genuinely interested in their feedback as well as in seeing how my representations corresponded with what people wanted me to understand.

These conversations were also useful for in-field fact-checking and helped me to see where different people and different groups diverged from the information I had recorded previously. In the end, my research participants' investment in this work focused primarily on this portion of the work, and quite a few of them requested that, should any final publication result from my time in Nepal, that printed copies of these drawings or photographs of the original Shaligrams, along with their corresponding descriptions, be made available to them.

Combining drawing and writing, photography and textual references, the process of drawing out analytical connections from my informants' multiple threads of narrative and experience has continued to evolve over time. However, I have tried to preserve the numerous voices and perspectives of Shaligram practitioners themselves as the core of my ethnographic and theoretical discussions. As one Shaiva *sadhu* explained, sitting outside the gates of Muktinath as he had off and on for some twenty years or so: "The beauty of Shaligram is that it can be many things to many people. This doesn't mean that you should not learn to read them properly but that whatever Shaligram is to you may not be what Shaligram is to me. And that is ok, because Shaligram always is what it must be." While at the time I took him to mean that the nuances of Shaligram interpretation may sometimes be left to individual perceptions or insight, I later began to understand that what he was really referring to was a far-reaching sense of the formlessness of the divine: where the superficial nature of the material object and of the narrative was meant only to lead one to deeper understanding, not to be that understanding in and of itself.

Taking the concept of continuously layered meanings as a methodological starting point, I found it helpful to approach both the conversations I recorded and the texts I included as referential moorings in an otherwise continuously shifting narrative. This way, I would be careful not to incorrectly characterize social change as something that moves between static forms or portray mobility as instability. Instability is, after all, a privileging of the stable, and the stable largely depends on one's point of view. As I gradually came to understand the broader contexts within which my research participants expressed their understandings or experiences, I tried to analyze their explanations in ways that most closely fit with what they were trying to tell me at the time, while also working under the realization that these conversations were linked to even larger conversational frameworks and issues being shared over great distances and across numerous borders.

Tangled Up in Texts

The highly mobile contexts of Shaligram use and practice presents something of a challenge for locating discussions and analyses of these practices within potentially productive theoretical frameworks. This is because, as should be clear by now, Shaligrams at any given point may conceivably touch on a wide variety of issues and topics that, for the sake of space and argument, are not fully realized here. This work as a whole revolves around three thematic issues: mobility, landscape, and object-personhood. These three frames of reference then inform a project grounded in the anthropology of religion, where religious practices are analyzed in relation to social institutions and compared cross-culturally. By focusing first on the nature of the sacred and of the sacred object, this work speaks to the paradox of Shaligram personhood and their incorporation as divine entities and as erstwhile family members into Hindu communities.

Every Shaligram devotee I met was able to explain the necessity of including Shaligrams in daily life and incorporating them into any worship using "man-made" icons due to the fact that Shaligrams were self-manifest and therefore carried their divinity with them in a purer form than humans could possibly create. This divinity, however, was not always readily visible except to the trained observer. What was still necessary for proper veneration included extensive social networks of ritual specialists and learned gurus who could "read" a Shaligram, as one might a text and discern its identity, needs, and intents. An "unidentified" Shaligram was always Vishnu (regardless of its appearance), but identification of the exact divine nature of a Shaligram nearly always preceded its ritual incorporation into family or temple life.

Though described as bodies themselves, Shaligrams are nevertheless said to transcend human boundaries of caste, class, and gender. They are said to link places, communities, and generations together without regard for the superficial and temporary characteristics of human bodies, which by their nature must be born and then die. A Shaligram has no need to endure this karmic cycle of birth, death, and rebirth, of course, even though they do; and this is further proof that God Himself manifested as stone so that He would join with His devotees just as they were and experience life for Himself in the care of generations of devout families. As a result, the continuing troubles of accessing the Mustang pilgrimage (be they political or economic) have become not only an affront to religious sovereignty but a transgression against the divine workings of time itself. And in the end, defining and maintaining the mobility of Shaligrams, and by that token the people involved in their veneration, is paramount to ensuring the continued longevity of the practice and of the community. But this does not mean that Shaligram practices cannot be viewed in light of other, more localized concerns of religious or ritual practice, religious hybridity, or national unification and development. These concerns, however, though informative of this work are secondary to the main arguments.

One of the issues concerning this ethnography lies in the literature. While there currently exists a significant body of research regarding pilgrimage, shamanism, and ritual practices in Nepali society (see Hitchcock 1976; Gellner 1994; Guneratne 1999; and Maskarinec 1992), there are only two major published works available that are specifically dedicated to the study of Shaligram stones: the *Śālagrāma-Kosha* by S.K. Ramachandra Rao (1996) and the *Śāligrāma Purāṇa* by Ram Charan Sharma (2000). Both of these works, released in several volumes in the late 1990s and early 2000s, contain detailed descriptions of Shaligram myths and stories as well as numerous drawings depicting various types of Shaligram stones but contain little in the way of analytical explanation and no commentary at all in terms of Shaligram traditions located in specific times or places. More generally, both of these works are primarily concerned with the religious and mythological significance of Shaligrams as supported by references in Hindu scriptural literature but are not ethnographic in terms of practices, pilgrimage, or ritual use.

Other works on Shaligrams tend to fall into one of two categories: colonial-era geological surveys with very brief commentaries on "Shaligram cults" and pilgrimage literatures. In an effort to address ongoing understandings and interpretations of Shaligrams, these types of work will be addressed where appropriate, including the commentaries of Gustav Oppert (French: 1901) and Joseph Kohl-Bonn (German: 1936), modern paleontological work on ammonites in the Himalayas (Page 2008; Sakai 1989; Enay and Cariou 1999), and local pilgrimage literatures, such as *Muktichhetra Mahatmyamam* by Madhu Sudhan Ramanujadas (2003), and other compilations of Puranic references. The considerable dearth of ethnographic work, however, means that addressing questions of the ritual and economic roles of Shaligrams in the building of cross-cultural religious identity, in the negotiation of economic development strategies based on pilgrimage and the exchange of the stones, and in manifestations of South Asian political inequality is something of a new frontier. Given that this research must also address the movement of sacred objects between and among different nationalities and between and among different religious affiliations and identities, mobility is taken not as a transient process necessary for getting from place to place but as an expression of identity and meaning in its own right.

Up until this point, I have also avoided referring to Shaligram practices as "syncretic". Religious syncretism has long been a contentious analytical method in the social sciences, and I have chosen not to use it to describe Shaligrams for two reasons. One, because the boundaries between religious traditions in Nepal (and in India as well) overlap to such a degree that they become virtually non-existent and two, because the methods by which any particular practice or object can be said to be "Hindu", "Buddhist", or "Bon" is often a matter of perspective and belies more about the person making the claim than it does about any actual kind of objective "belonging" to a definable tradition.

I have drawn principally on the works of Charles Stewart, Rosalind Shaw, and Melville Herskovits in understanding religious syncretism as a method of exploring difference (Stewart and Shaw 1994; Herskovits 1958) but also on the works of Stephan Palmié, who argues convincingly for the reconsideration of the very notion of syncretism/hybridity in favor of viewing religious blending and exchange as an integral part of the nature of religious practice itself (Palmié 2013). This is especially useful here in that this analysis speaks against the literatures of "pure culture" common to travel brochures, international development policies, historical revisionism, and desktop photography books that portray Nepal and India as lands of "untouched history", "pure peoples", and "ancient cultures unchanged by time" and where mobility in terms of pilgrimage, tourism, and migration is viewed as polluting – a "leakage" into the pristine nature of resident cultures. This is why my use of religious syncretism is not meant to indicate any actual kind of ongoing hybridization of otherwise "pure" traditions of origin. Rather, any use of religious syncretism here addresses the fact that labeling religious overlap is often a part of many people's perspectives (both ethnographic and academic), rooted in specific contexts, that categorizes religion according to relevant contexts at the time.

Questions of alternative kinship networks and object-personhood are also central to this research. In that vein, I take Lucinda Ramberg's work on categories of divine kin-making among the *devidasis* of India (2014) as an important starting point in theorizing kinship beyond systematic lenses (Lèvi-Strauss 1969) or culturalist viewpoints (Schneider 1984), all the while taking into account the nature of persons in general and their understandings of their role in the world (Strathern 2005). However, due to the object-nature of Shaligrams, this work must also dispense with the privileging of relatedness through the conjugal pair, as Ramberg does, or the biological tree of relationships mapped out between parents and their children (Hayden 1995; Rubin 1975; and Weston 1997). This is because the designation of "fictive" in terms of Shaligram kinship with communities and families does not adequately capture the depth of relationship and relatedness that is so often expressed between practitioners and their deity-stones and would unfairly focus attention on the already dominant family forms and genealogical categories that place biological and genetic ties at the forefront of determining who and what counts as "family". Along with a number of new kinship studies in the anthropology of relatedness, Shaligrams as kin then opens up new possible ways of enacting and valuing relationships more broadly and continues to trouble our current understanding of what we truly mean when we say blood or affine (Carsten 2000; Hayden 1995; Weston 1997; and Franklin and McKinnon 2001).

These works and others frame the need for an in-depth understanding of how the use of Shaligrams relates to political, social, and economic transformation in Nepal and what effect Shaligram pilgrimage has on Muktinath and the surrounding region. In the end, by combining theoretical frameworks in religious anthropology, the social and religious history of images and landscapes, and previous work on kinship, personhood, and mobility, my inquiries here work to conceptualize the production and exchange of sacred stones as a method of identity and community building across boundaries of caste, nationality, gender, and ethnicity particularly in light of increasing Westernization, fears of "foreign" claims to "domestic" religious sites, and other recent political and social changes in modern South Asia.

The View from Ten-Thousand Feet

It takes more than an ammonite fossil to make a Shaligram a Shaligram. As living deities, they are born of the sacred landscape of Mustang, Nepal – itself a site of multiple, intersecting cosmologies and political tensions.

As a popular destination for pilgrimage, Mustang is home to a number of sacred sites belonging to Hinduism, Buddhism, and Bon, all of which have blended, hybridized, and become ritually syncretic with one another throughout the region for centuries. The land of Mustang is also routinely characterized through bodily metaphors, wherein the Kali Gandaki River Valley is simultaneously the location where the Hindu deity Vishnu manifests himself as a sacred stone within the womb of the goddess Tulsi (Or is cursed to do so, depending on which version of the story you encounter), as well as the place where the body of a great Buddhist/Bon demoness (*sinmo*) is continuously placated and held in check by ritual practices and sacred architecture. The cultural make-up of Mustang is also equally multivalent, with numerous ethnic and linguistic divisions co-existing within several village conclaves that stretch all along the river; from the relatively accessibly reaches of Lower Mustang in the south to the heavily restricted and militarized border of Upper Mustang in the north.

A Shaligram is also characterized by the movements of people, especially the pilgrims and ritual practitioners who travel to Mustang to seek them out and to take them home. To approach an ethnography of Shaligrams, one must then take mobility, time, place, and access into account. Encountering any particular Shaligram or any particular Shaligram practitioner might be revealing in that moment or that context but would likely miss many of the broader socio-cultural systems at work. Systems that could often only be glimpsed in the traveling itself. As such, the narratives of Shaligram practice tend to come in bits and pieces and are heavily dependent on time, place, and individual but together they begin to form a pattern that networks geological and demographic mobility with ritual personhood, history, and community identity. Assembling these narratives then, into a cohesive conceptual whole requires a methodology that takes into account everything from pilgrimage treks to in-home ritual worship gatherings, from community festivals to temple altar setups.

Ultimately, it is the consolidation of movement and ritual that then enlivens Shaligrams and begins the process by which each sacred stone "lives" a life as a member of a household and a community. Shaligrams are "born" of the river and welcomed, with much celebration, as new members of the family once they appear during pilgrimage. When they are brought home, they are bathed, fed, given gifts and offerings, and brought along to significant events (such as weddings, births, or other milestones) in the same manner as one might any other member of the family. And in the end, they must also "die;" a moment wherein they will either be cremated on a funeral pyre along with a deceased person or passed down to another generation where they will be reborn into the new family that inherits them. In this way, Shaligrams live the same kinds of karmic lives that their human counterparts do.

As a result, this ethnography is not only multi-sited (Nepal, India, the USA, the UK) but is, in a sense, multi-species. It is my intent here to demonstrate how the lives of divine persons (Shaligrams) are carried out through the same social processes and spiritual agency as human persons, despite their obvious physical differences. The result of this is an ethnography that follows a set of Shaligram stones, and their families, through time and space; beginning with the pilgrimage routes of Mustang and then following them outwards to their destinations in Nepal, India, and among the South Asian Diaspora in the US and the UK. I also note the importance of this methodology especially because Shaligram devotees do not themselves form a distinct community and tend to incorporate any number of different nationalities, ethnic identities, political affiliations, and even religions all under the unifying framework of Shaligram pilgrimage and practice (*seva*). What binds this ethnography together then, is a shared spiritual world and set of material ritual practices contextualized by a unique entity that is itself both object and divine person.

We therefore begin in Mustang, where the Shaligrams themselves begin. The history of Mustang, and of Shaligram pilgrimage, is, however, hardly straight-forward and the complex interweaving of invasions and conquests, migrations, trade, and political conflict have resulted in a region whose identity remains contested and unstable. This history then contextualizes modern Shaligram pilgrimage practices and reveals how various claims to religious belonging and national origin are used as starting points for political resistance and for concerns about the growing difficulty in reaching restricted pilgrimage sites. Such restrictions, many people note, that may now threaten the continuity of the Shaligram tradition itself.

Bibliography

Carsten, Janet. 2000. *Cultures of Relatedness: New Approaches to the Study of Kinship.* Cambridge: Cambridge University Press.

Chetri, M., L.K. Maskey, N.R. Chapagain, and B.D. Sharma, eds. 2004. *Mustang – The Land of Fascination.* King Mahendra Trust for Nature Conservation, Nepal.

Craig, Sienna. 2001. "A Tale of Two Temples: Culture, Capital, and Community in Mustang, Nepal." Paper Presented at the New York Conference on Asian Studies in Ithaca, NY.

—. 2008. *Horses Like Lightening: A Story of Passage Through the Himalayas*. Somerville: Wisdom Publications.

Dejarlais, Robert R. 1992a. *Body and Emotion: The Aesthetics of Illness and Healing in the Nepal Himalayas*. Philadelphia: University of Pennsylvania Press.

—. 1992b. "Yolmo Aesthetics of Body, Health and 'Soul Loss'," *Social Science and Medicine* 34 (10): 1105-1117.

Dhungel, Ramesh K. 2002. *The Kingdom of Lo (Mustang): A Historical Study*. Kathmandu: Tashi Gephel Foundation.

Eck, Diana. 1986. "Darshan of the Image," *India International Centre Quarterly* vol. 13, no. 1, IMAGES (March): 43-53.

—. 1998. *Darśan: Seeing the Divine Image in India*. New York: Columbia University Press.

—. 2012. *India: A Sacred Geography*. New York: Three Rivers Press.

Enay, Raymond, and Elie Cariou. 1997. "Ammonite faunas and palaeobiogeography of the Himalayan belt during the Jurassic: Initiation of a Late Jurassic austral ammonite fauna," *Palaeogeography, Palaeoclimatology, Palaeoecology* 134: 1-38.

—. 1999. "Jurassic ammonite faunas from Nepal and their bearing on the palaeobiogeography of the Himalayan belt," *Journal of Asian Earth Sciences* 17: 829-848.

Fisher, William F. 1987. *The Re-Creation of Tradition: Ethnicity, Migration, and Social Change Among the Thakali of Central Nepal*. PhD diss., Columbia University.

—. 2001. *Fluid Boundaries: Forming and Transforming Identity in Nepal*. New York: Columbia University Press.

Franklin, Sarah, and Susan McKinnon. 2001. *Relative Values: Reconfiguring Kinship Studies*. Durham: Duke University Press.

Gellner, David N. 1994. "Priests, Healers, Mediums and Witches: The Context of Possession in the Kathmandu Valley, Nepal," *Man* 29 (1): 27-48.

Guneratne, Arjun. 2002. *Many Tongues, One People: The Making of Tharu Identity in Nepal*. Ithaca, NY: Cornell University Press.

Hausner, Sondra L. 2007. *Wandering with Sadhus: Ascetics in the Hindu Himalayas*. Bloomington: Indiana University Press.

Hayden, Corinne P. 1995. "Gender, Genetics, and Generation: Reformulating Biology in Lesbian Kinship," *Cultural Anthropology* 10 (1): 41-63.

Herskovits, Melville. 1958 [1941]. *The Myth of the Negro Past*. 2nd ed. Boston: Beacon Press.

Hitchcock, John T., and Rex L. Jones. 1976. *Spirit possession in the Nepal Himalayas*. New Delhi: Vikas Publishing House.

Kohl-Bonn, Joseph. 1936. "To the Indian Stone Cult," *Journal of the German Oriental Society* vol. 90 (2): 432-440. ("Zum indischen Steinkult," *Zeitschrift der Deutschen Morgenländischen Gesellschaft*).

Levi-Strauss, Claude. 1955. "The Structural Study of Myth," *The Journal of American Folklore* vol. 68, no. 270, Myth: A Symposium (Oct-Dec): 428-444. Published by American Folklore Society.

—. 1969. *The Elementary Structures of Kinship.* Translated by James Harle Bell and John Richard von Sturmer. Boston: Beacon.

Maskarinec, Gregory G. 1992. "A Shamanic Etiology of Affliction from Western Nepal," *Social Science and Medicine* 35 (5): 723-734.

Mauss, Marcel. 1954. [1925] *The Gift; forms and functions of exchange in archaic societies.* Chicago: Hau Books.

Messerschmidt, Donald A. 1984. "The Thakali of Nepal: Historical Continuity and Socio-Cultural Change," *Ethnohistory* 29: 265-80.

—. 1989. "The Hindu Pilgrimage to Muktinath, Nepal. Part 1 and Part 2: Natural and Supernatural Attributes of the Sacred Field," *Mountain Research and Development* vol. 9, no. 2 (May): 89-104.

Oppert, Gustav. 1901. "On Shaligrams: Sacred Stones of Indian Aborigines Become Emblems of the God Vishnu," *Review of the History of Religions* vol. 43: 325-332. (Sur Les Sālagrāmas: Pierres Sacrées Des Aborigènes De L'inde Devenuesemblèmes Du Dieu Vishnou. *Revue de l'histoire des religions*).

Page, Kevin N. 2008. "The Evolution and Geography of Jurassic Ammonoids," *Proceedings of the Geologists' Association* vol. 119 (1): 35-57.

Palmié, Stephan. 2013. "Mixed Blessings and Sorrowful Mysteries: Second Thoughts about "Hybridity," *Current Anthropology* vol. 54, no. 4 (August): 463-482.

Ramanujadas, Madhu Sudhan. 2003. *Muktichhetra Mahatmyamam.* Kathmandu: Dhaulagiri Osset Press.

Ramberg, Lucinda. 2014. *Given to the Goddess: South Indian Devadasis and the Sexuality of Religion.* Durham: Duke University Press.

Ramble, Charles, and Michael Vinding. 1987. "The Bem-chag Village Record and the Early History of Mustang District," *Kailash* 13, nos. 1-2: 5-47.

Rao, S.K. Ramachandra. 1996. *Shaligram Kosh* [*Śālagrāma – Kosha*]. Sri Satguru Publications. Delhi: Indian Books Center.

Rubin, Gayle. 1975. "The Traffic in Women: Notes on the 'Political Economy' of Sex." In *Toward an Anthropology of Women,* edited by Rayna R. Reiter, pp. 157-210. New York: Monthly Review.

Sakai, Harutaka. 1989. "Rifting of the Gondwanaland and Uplifting of the Himalayas Recorded in Mesozoic and Tertiary Fluvial Sediments in the Nepal Himalayas." In *Sedimentary Facies in the Active Plate Margin,* edited by A. Taira and F. Masuda, pp. 723-732. Tokyo: Terra Scientific Publishing Company.

Schneider, David Murray. 1984. *A Critique of the Study of Kinship.* Ann Arbor: University of Michigan Press.

Sharma, Ram Charan. 2000. *Shaligram Purana*. S.R.C. Museum of Indology & Universal Institute of Orientology Trust, 199. Digitized 2009. Ann Arbor: University of Michigan Press.

Shaw, Rosalind, and Charles Stewart, eds. 1994. *Syncretism/Anti-Syncretism: The Politics of Religious Synthesis* (European Association of Social Anthropologists). London: Routledge.

Strathern, Marilyn. 2005. *Kinship, Law, and the Unexpected: Relatives Are Always a Surprise*. Cambridge: Cambridge University Press.

Weston, Kath. 1997. *Families We Choose: Lesbians, Gays, Kinship*. Rev. ed. New York: Columbia University Press.

3 Picked-Up Pieces

Constructing a History of Mustang

Abstract
The history of Mustang, Nepal is complicated and can vary significantly depending on the textual sources one uses. For local Mustangis and pilgrims, however, issues of place, space, and time are a vital part of what it means to be Hindu or Buddhist as well as Nepali, Indian, or Tibetan, even though these categories remain continuously blurred and fluid. Beginning with the paleontological history of Mustang's extensive fossil formations and ending with an overview of the political history of the region, this chapter focuses on the ways in which historical narratives have affected access to the Kali Gandaki River Valley, and to Shaligrams specifically, since the earliest days.

Keywords: Muktinath, fossil, Himalayas, paleontology, history

> "may came home with a smooth round stone
> as small as a world and as large as alone."
> – *e.e. cummings*

Dawn had barely broken over the horizon as Bikas Shrestha and eight other Hindu pilgrims made their way along the narrow mountain road between Kagbeni and the village of Ranipauwa. Having spent nearly three days searching for Shaligrams along the banks of the Kali Gandaki River below, they were eager to reach Muktinath temple by no later than mid-day and begin their ritual bathing in the 108 waterspouts of the Vishnu *mandir*. Bikas was especially excited as he clutched the embroidered bag that now held his four most recent Shaligrams, each of which he hoped to lay at the feet of Sri Muktinath (Vishnu/Avalokiteshvara), the principal deity of Muktinath temple, during the afternoon *darshan*. "It is said," he began breathlessly,

Walters, Holly, *Shaligram Pilgrimage in the Nepal Himalayas*. Taylor & Francis Group, 2020
DOI: 10.5117/9789463721721_CH03

still struggling against the thin high-altitude air, "that wherever there are twelve or more Shaligrams, that place is the same as the *dham*. It is no different than Muktinath. These four make fourteen for me now, so I think that this will be my last pilgrimage. I have enough for my children to take when they are older and two that I will give to my *guru*. Wherever I go now with them, it is pilgrimage to Muktinath."[1]

Several hours later, as we approached the temple complex just beyond the village, we passed a small group of Buddhist nuns of the Nyingma order on their way to the Vishnu *mandir*. "Namaste, *aani*!" Bikas called out, hurrying the others along more quickly as the nuns began to open the temple doors in preparation for *darshan*. Nearly twenty more Hindu pilgrims waiting in the courtyard gathered closely around. "Isn't it strange," I asked, "that a Hindu site of pilgrimage should be attended to Buddhist nuns?" "Not at all," Bikas replied, his eyes carefully trained on the inner sanctum of the temple ahead of us. "Buddhists are Hindus, you see. They are really no different. Lord Buddha was an incarnation of Vishnu and appears in Shaligram from time to time. So, it is not surprising that Buddhists also come to Muktinath."

The landscape of Mustang, Nepal has two kinds of histories. First, there is the lengthy political history of migration, exchange, and identity reconciliation between the region as a "lost kingdom of Tibet" and then as part of the Nepali state in the years of territorial consolidation following the Gurkha conquest of 1768. In particular, this historical narrative encompasses the control of the salt trade by the Thakalis (the most populous Mustangi ethnocaste) and the eventual political isolation of Mustang in the years following the Chinese invasion of Tibet in 1950. Second, there is the mythological history of Mustang, which includes not only the scriptural references to the land of Shaligram (Śālagrāma)[2] but also the narratives that people, both native-born and pilgrims, establish about themselves and their own positions in the world. This shaping of historical narrative through a mytho-poetic lens is part of the strategies of adaptation within an ever-shifting socio-political environment not just in Mustang and in Nepal but throughout South Asia. Moreover, since the 1950s, Mustangi narratives about themselves have been deeply affected by the narratives that scholars tell about them.

1 This and the following conversation were conducted in a combination of English and Hindi. Dialogue was transcribed from fieldnotes recorded during and after the conversation.

2 When referring to the land of Shaligram or the *dham* of Shaligram, I have intentionally used the diacritic transliteration throughout this work. This is both to differentiate the spiritual landscape from the stones themselves and because practitioners quite often pronounce the two words slightly differently, such as opting for the short "a" in the second syllable rather than the long "i".

Histories of Mustang are usually constructed from sources outside of the region, and because many villages and groups are themselves historically or currently migratory and have not kept extensive written records over long periods of time, their understanding of Mustang as a kingdom and then as a national polity are most often drawn from books and articles read in school or made available to them by local booksellers (most of these texts are also printed in English). There are also a number of pressing issues behind the competing mythological narratives of the multiple religious traditions that can be found from Jomsom to the Baragaon to Lo Monthang. Understanding how such histories are negotiated requires us to examine the interrelatedness of the social, economic, ritual, and political circumstances of the pilgrimage landscapes of Mustang that confronts the various agents involved in the telling and retelling of specific narratives as well as the influences of pilgrims and other travelers on the telling of these stories elsewhere. But as William Fisher notes, in regard to Mustang, "not all narratives are equal". This is not to say that each version does not have its own particular kinds of merit but that various "inequalities arise in different ways: they appeal to different audiences; they adhere in varying degrees to the so-called facts; some are more persuasive than others, and some are more widely recognized as 'coherent'" (2001: 45).

There are two particular narratives prevalent among the pilgrims who frequent Mustang in search of Shaligrams: the narrative of mobility and landscape and the narrative of syncretism and hybridity. While neither pilgrims nor local peoples themselves typically use terms like "syncretic" to describe their own religious practices, many are acutely aware that scholars, politicians, and tourists to the region are especially keen on labeling art, architecture, and practices as being Buddhist (Tibetan) or Hindu (Nepali or Indian) or Bon (indigenous) in ways that local peoples or pilgrims don't necessarily see. In fact, calling aspects of religious practice, temples, or the landscape by these terms is viewed more as a political statement than a religious or historical one, aimed at outside audiences in the development and environmental conservation world (NGO or government) or scholars tasked with the agenda of cultural preservation, which thereby weighs some of these narratives with the promise of a possible economic windfall in the form of project funds.

Landscape plays a primary role in this analysis as both a location where the connections between place, space, identity, nationalism, history, and memory are embodied and enacted and as a cultural process in and of itself. Drawing on perspectives that include discussions of landscapes in the context of preservation and tourism, as the focus of development narratives, as mythological constructs, and as spaces wherein socio-political and ethnicized communities are imagined and created (Stewart and Strathern 2015; Hirsch and O'Hanlon 1995), this chapter addresses the changing paradigms of mobility and nationalism along the pilgrimage routes of the Kali Gandaki as a vital context, influence, and origin point for modern Shaligram practice. More generally, this chapter presents an anthropological overview of the political and cultural issues currently facing Mustang, describing its history of conflict, migration, and religious blending as a way to address the conflicts between sacred and political landscapes from which Shaligrams are produced, collected, and exchanged. My point here is that the political and ethnic fluidity and hybridity of Mustang in general has produced a long trajectory of boundary negotiations and political isolation from which deep connections of religious material practice and mobility as power have arisen.

In both narrative cases, issues of place, space, and time are paramount to the identities of residents and pilgrims alike. But who tells these stories and for what purpose? How and why do some people adhere to distinct terms like Buddhist and Hindu or find them useful even when the blending of ritual practices and religious identities may be so fluid as to be virtually indistinguishable? Why do so many scholars react skeptically to the Hindu "origins" of specific places and practices or find the idea of largely Tibetan Buddhist influences of cultural and social change in Mustang so appealing? How do these narratives fit with verifiable histories of the region and to whom is this evidence more or less convincing? To answer these questions this chapter considers the what, why, how, where, and by whom certain narratives of sacred landscape, the sovereignty of mobility, and political participation become privileged over others. The relationship of mythological events within these narratives to historical moments and specific historical conditions within the narratives are considered here against other possible narratives and interpretations offered by Mustangis and pilgrims alike. As an issue of syncretism, however, I take to heart Stephan Palmié's challenge to rethink "syncretism" and "hybridity" (2013) not as a blending of actual categories of objects or practices but as conceptual labels that various people might employ in their own efforts to locate otherwise multivalent religious practices within certain cultural or political frameworks.

The land of Mustang has long been portrayed as a land of "pure Tibetan culture". From travel literature and coffee table books to history books and ethnographic accounts, the privileging of Buddhist origins has been used to paint Mustang as a kind of land outside of time: frozen, hidden, and untouched. The narrative of syncretism then serves this image especially well in that Hindu, Bon, and Buddhist "hybridity" easily implies the prior assumptions of something there that is "pure". And also vice versa, where "purity" cannot exist except through the presence of "hybrids" which in many cases can then be labeled as "foreign corruption" or "globalized pollution". This does not mean to say that hybridity narratives always necessarily carry the assumption that there were ever such things as "pure", clearly bounded, internally homogeneous, and monolithic "cultures" that only now are encountering one another in their most essential states (especially not in a region such as Mustang) but that the use of "syncretism" and "hybridity" as concepts are often employed in circumstances where divisions between identities are socially or politically relevant to the cultural contexts of the moment. This use of syncretism echoes Reinhardt Koselleck's use of "asymmetrical counter-concepts" such as "tradition" and "modernity" (1985)[3] or, in the case of Shaligram pilgrimage in Mustang, "purity" and "mixture". In the end, understanding the epistemological implications in labeling Mustang, its peoples or cultures, its history, or its Shaligram pilgrimages as "belonging" to one specific nation-culture or another is both troublesome and necessary.

The linking of Shaligrams as sacred object-persons and as focal points of mobility (for both pilgrims and the stones) across sacred landscapes, as opposed to ideal texts and traditions, is revelatory because it highlights moments where "pure" Tibetan culture (read: Buddhist) comes into conflict with practices labeled as Hindu. The history of Mustang (and of Muktinath) is often framed by folk-classificatory systems of religious belonging. Though it might seem odd to refer to "Tibetan" or "Hindu" as a kind of folk label, pilgrims and even Mustangis themselves often describe these markers using a blend of scholarly, political, and essentialist travel definitions of what these cultures are and what they represent. Bruno Latour (1993) refers to these kinds of categories as the lush jungle of "intermediaries" or "inhabitants of the middle kingdom". But when history books, travel brochures, and art magazines continuously visualize peoples, landscapes, and religious practices as belonging to certain categories or types and call their points of interaction "hybrid", pilgrims and Mustangis themselves then come to create new and somewhat amalgamated forms of categorization that incorporate both printed narratives and oral ones. My use of "hybridity" and "syncretism" is therefore not situated within the pedagogical reconstruction of hybridity and purity that so often characterizes past writings on Mustang. Rather, I aim to demonstrate how Shaligram pilgrimage (as well as the general histories of Mustang) defy static preservationist interpretations of peoples and cultures.

3 Koselleck, Reinhardt. 1985. *Futures past: On the semantics of historical time.* Cambridge, MA: The MIT Press.

A Brief Fossil History of the Himalayas

Abheesh Maharana, a Hindu pilgrim from Kerala, scowled up at the sign displayed prominently on the permit check-in building in Jomsom, Nepal. In large, English block letters it read, *Ammonites and the Tethys Sea: A History of Mustang*. After reading for several minutes, he turned to me. "I am an educated man," he said, "I went to college. I know it's a fossil. We learned all this in school when I was a boy, you know. How the Himalayas were formed, how fossils are made, how old the world is. This is all very common. It's the history they teach you about your home. But why would they put this here? This is Kali Gandaki, not the school. Is it for tourists or do they think pilgrims should read this? Why wouldn't they have a sign about Shaligram here then? It doesn't say a thing about them. This makes no sense to me."[4]

The history of Mustang is complicated, contentious, and a source of relative animosity between scholars, local peoples, and pilgrims but not more so than the geological history of the Himalayas. By and large, Shaligram practitioners are well aware of the scholarly discourses surrounding mountain paleontology, plate tectonics, and the previous existence of Shaligrams as ammonites. For many of them, the scientific discourses, however, are not necessarily to be rejected out of hand but instead comprise one part of the story of Shaligram origins. These discourses are also often paired with textual authority and the stories of Shaligrams in Hindu Scriptures to further prove the spiritual authenticity of their practices. For example, the geological date of 175 million years is taken as evidence of the great antiquity of Shaligrams in a time when the world as we know it was just beginning to form and when the stories relayed in the sacred texts were concurrently taking place. The uplift of the Himalayas from the Tethys Sea is viewed as analogous to the story of Samudra Manthan, the rise of the primordial mountain out of the Ocean of Milk, and the weathering and wearing of Shaligrams out of the high-altitude fossil beds as a continuation of the story of Vishnu forever becoming stone in order to appease the goddess Tulsi/the Kali Gandaki. But more importantly, the geological history of the Himalayas constitutes the beginning of and a foundation for a series of narratives and events that will begin the life cycle of the Shaligram stone: a life cycle that, at this point, is entirely in the hands of the Shaligram itself.

The multiple ontologies of fossil, person, and deity also mediate contentions between contemporary debates on the roles of science and religion more broadly. Take, for example, the plethora of books and articles written

4 Hindi. Transcribed and translated from fieldnotes recorded shortly after.

in the last thirty years or so on the relationship of the fossil record and paleo-hominid research to the Abrahamic faiths: from Christian Young Earth Creationism (Ian Barbour, Arthur Peacocke, and John Polkinghorne to name a few), to Jewish cosmology (Nathan Aviezer),[5] and to Islam and evolution (T.O. Shanavas[6] and Reza Aslan[7]). Despite widely disparate religious traditions, the central contention of most of these works remains the same: does Darwin's theory of evolution (and by that token, the fossil record) support or discredit the religious conceptualization of the creation of the world by God? Similar to the fossil folklores of South Asia, religious stories are equally invoked. Do large fossil amalgamations (such as Africa's Karroo Formation or Canadian Edgar Nernberg's fossil fish) supply evidence for Noah's flood?[8] Does the "Cambrian Explosion" of life roughly 600 million years ago indicate the work of Allah, who created all life in its final forms all at once?[9] Or were fossils simply placed on Earth by God, in defiance of religious texts, to test the loyalty of the faithful?[10] Similar debates take place in Hinduism, Buddhism, and Jainism as well, and at their center is the fossil ammonite called Shaligram.[11]

5 Nathan Aviezer. *Fossils and Faith: Understanding Torah and Science*. 2001. KTAV Publishing House, Inc., Hoboken, New Jersey.

6 Shanavas, T.O. *Evolution and/or Creation: An Islamic Perspective*. 2005. Xlibris Publishers.

7 Aslan, Reza. *No god but God: The Origins, Evolution, and Future of Islam*. 2005. Random House; Later Print Edition.

8 Mark Isaak, "Problems with a Global Flood," at: http://members.shaw.ca/ and John Woodmorappe, "The Karoo vertebrate non-problem: 800 billion fossils or not," Answers in Genesis, CEN Technical Journal, 14(2), 2000. Online at: http://www.answersingenesis.org/ Accessed 19 September 2016.

9 "The Origin of Life – An Islamic Perspective." Accessed 19 September 2016. https://www.missionislam.com/knowledge/orignlife.html.

"The Fossil Record Refutes Evolution." 2006. Accessed 19 September 2016. http://www.islamweb.net/en/article/111486/the-fossil-record-refutes-evolution.

"Islam and Evolution: A Letter to Suleman Ali. Nuh Ha Mim Keller." 1996. Accessed 19 September 2016. http://www.masud.co.uk/ISLAM/nuh/evolve.htm.

10 "Jurassic Judaism," from Ohr Somayach's *Torah and Nature*. Accessed 19 September 2016. http://ohr.edu/ask_db/ask_main.php/238/Q1/.

Slifkin, Nathan. *The Challenge of Creation: Judaism's Encounter with Science, Cosmology, and Evolution*. ZooTorah/Lambda Press, Brooklyn, 2010, section two, "Cosmology," pp. 157-190 for a discussion of these beliefs.

"Creation, Evolution, and Intelligent Design." Rabbinical Council of America. 2005. Accessed 19 September 2016. http://www.rabbis.org/news/article.cfm?id=100635.

11 Ammonites are not the only fossils with both a scientific and religious history. One of the more famous examples are the trilobites (a branch of extinct marine arthropod so named for its "three lobed" appearance). In fact, one such trilobite fossil, which resided in the *grotte du Trilobite* (whose name aptly comes from a trilobite fossil found in it by Dr. Ficatier from Auxerre)

Abheesh sighed and paused to rewrap his pilgrimage garments and adjust his coat. "There should be another sign up next to that one, don't you think? It should say what Shaligrams are and why you should be respectful when you come here. It's OK to call it fossil but they should also have pilgrim books [meaning pilgrimage guides]. I should leave some at the checkpost for people to read. This is only half the story."

One of the most popular sayings in geology describes layers of fossil and rock as pages of a book, and the paleontologist and geologist as apt readers, turning the pages of Earth's history with each strike of the hammer. But this isn't just a book we're reading; it's also one we're writing. Prior to 1950, few geological observations were made in Nepal, owing largely to political and geographic isolation. Following 1950, when foreign visitors were once again provisionally allowed back into the country, Nepal soon became a significant focus for Himalayan geology. However, this particular time in the history of geology, referred to as "descriptive geology", considered mapping as the primary objective of research and publication rather than in-depth stratigraphic analysis or correlative dating.

While several notable monographs and the first geological maps of the Nepal Himalayas were produced during this time, most surveys tended to focus on the rich fossil deposits of the "Tibetan" sedimentary zone in the north rather than the comparatively fossil-poor meta-sediments of the Lesser Himalayas to the south. Building on the first references to Himalayan geology in the voluminous *Himalayan Journals* of Joseph Dalton Hooker (1854), who styled himself "a naturalist" and intellectual descendent of Charles Darwin, many of these early researchers took great pains to describe rolling foothills, difficult footpaths, and awe-inspiring rock formations (see Medlicott 1875; Auden 1935; Bordet et al. 1968; Hagen 1969; Frank and Fuchs 1970; and Colchen et al. 1986).[12] This focus on mapping and surveying reflected many of the widely differing interpretations and conflicting views

within the Caves of Arcy-sur-Cure in Burgundy, France may have acted as a religious icon for Paleolithic European cave dwellers some 28,000 years ago.

12 Medillicot, H.B., 1875, Note on the geology of Nepal. Rec. Geol. Surv. India, 8/4, 93-101.
Auden, J.B., 1935, Traverses in the Himalaya. Rec. Geol. Survey India, v. 69, pp. 123-167.
Bordet, P., 1961. Recherches géologiques dans l'Himalaya du Nepal, region du Makalu. Paris (CNRS).
Bordet, P., Colchen, M., Krummenacher, D., Le Fort, P., Mouterde, R., and Remy, J.M., 1968, Esquisse géologique de la Thakkhola (Nepal central). Paris (C.N.R.S.).
Hagen, T., 1968, Report on the geological survey of Nepal. Geology of the Thakkhola. Denkschr.
Frank, W., and Fuchs, G.R., 1970, Geological investigations in West Nepal and their significance for the geology of the Himalayas. Geol. Rdsch., v. 59, pp. 552-580.
Colchen, M., Le Fort, P., and Pêcher, A., 1986, Recherches géologiques dans l'Himalaya du Nepal: Annapurna – Manaslu – Ganesh Himal. Paris (C.N.R.S.)

of the investigators privileged to work in the region at the time, most of whom favored studying nappe structures (the visible edges and surfaces of thrust faults) over block tectonics.

With the advent of plate tectonics in the late 1960s, the Himalayas quickly became distinguished as the "collided range". This led to a new advent of microstructural, mineralogical, and geochemical studies in the search of the stress and heat effects that characterize tectonic subduction and collision. This is how metamorphism and magmatism became dominant aspects of later geological study in the Himalayas and in Nepal particularly. With it went a shift of emphasis from the field to laboratory work and therefore from observation to interpretation, from mapping to modelling, and from practice to theory (Stöcklin 2008). The last thirty years of research, however, have shifted from both earlier approaches in that they have been characterized more by a growing attention to the human effects of geologic instability in the region and in the strengthening and diversification of geological institutions in Nepal, especially with the creation of a National Seismological Centre in Kathmandu.

In 1939, Heim and Gansser[13] designated the Tibetan zone north of the Central Crystalline, a term for the band of metamorphic uplift across the Higher (Greater) Himalayas on the southern edge of the Tibetan Plateau, as the "Tethys Himalaya". In doing so, they described it as "made essentially of marine deposits [...] squeezed and pushed up out of an old sea". With this, both they and subsequent scholars began to refer to the ancient landscape of the Himalayas exclusively as the "Tethys Sea", a name that had been originally conceived in the late nineteenth century by Austrian geologist Eduard Suess as a long-stretched, narrow seaway that once separated the old Laurasian and Gondwanaland continents out of whose ultimate break-up and collision the Alpine-Himalayan chains were born. Suess' "Tethys" then became a classical example of a geosyncline (a large-scale depression in the earth's crust containing very thick deposits). Additionally, the history of the Alpine-Himalayan chains was standardized into a series of subsidence, compression, and inversion events that would be used to describe the Tethys for years to come. It was this widely accepted view that was then profoundly revolutionized after the mid-1960s when the theory of plate tectonics arrived on the scene.

As a geosyncline, the Tethys was believed to have been a narrow seaway, but in the view of plate tectonics it became a gigantic ocean some

13 Heim, A. and Gansser, A., 1939, Central Himalaya. Geological observations of the Swiss expedition 1936. Mém. Soc. Helv. Sci. Nat., v. 73/1, 245 p.

6,000 kilometers wide between what is now India and Tibet. Where previously the Tethys was imagined as the result of crust down-buckling during a massive (if slow-moving) impact, now it was thought to have originated from the upwelling and spreading of underlying mantle materials. Previously, deep (eugeosynclinal) and shallow (miogeosynclinal) marine deposits overlying unspecified layers of crust were distinguished in the Tethys. Now, the Tethys became a specific "oceanic" feature, complete with oceanic sediments and an oceanic crust. This shift in theoretical understanding is important because, in the geosynclinal concept, the Alpine-Himalayan orogenic belt had been unified since the birth of the Tethys. Now in the plate-tectonic view, this section of significant uplift was recognized as the strange composite of two continental margins, a fluid boundary of rock melted down by indescribable forces.

Once thousands of kilometers apart, it was believed that the two sections had nothing to do with each other in their structural development prior to collision, but now their borders and boundaries were intimately intertwined. It is this view of the geological Himalayas that is prominently displayed on signs and on trekking brochures throughout Mustang and that forms the basis for many calls to step up geological research in the river valley. For Shaligram pilgrims, there is a pervasive sense that the geological narrative is also being leveraged by regional governments to appeal specifically to Western scholars and tourists and that, if left unchecked, will eventually result in further restrictions on the movement of Shaligrams in and out of the Kali Gandaki. More specifically, they fear that if Shaligrams are further classed as protected state resources or museum-worthy historical items, the practice of taking stones in pilgrimage will end.

Theories regarding the enormous width of the Tethys Ocean and its disposal by subduction did not result, however, from any new geological discoveries in Tethyan rocks but largely from new forms of geophysical data (such as palaeomagnetics), which was primarily obtained outside the Tethyan realm (Stöcklin 2008). Unfortunately, while it is true that these and other premises of plate tectonics have not remained uncontested, broader discussions of these issues lie outside the scope of this work (see, for example, Lavecchia and Scalera 2003). What concerns me here is the consequences that plate tectonics had for geological investigations and subsequent paleontological narratives in Nepal. In the end, the Himalayas became a sort of scientific test case for the concept of the "collided range" (Le Fort 1975), and this theory required that all orogenic events – deformation, metamorphism, magmatism – were the consequence of subduction and/or collision. Both geological and paleontological inquiries in the Himalayas have been subsequently fraught with disagreements regarding precisely how specific layers should be dated and correlated with similar layers elsewhere. And, more importantly, it continues to affect the ways in which ammonites are understood as markers of geologic time and biological space in the Himalayas today – correlations, for example, that have specific implications for what kinds of ammonites elsewhere could be classified as Shaligrams and what could not. Lastly, while Heim and Gansser's 1939 study is one of a few early twentieth-century surveys credited with introducing the ammonites of the Himalayan Paleozoic-Mesozoic sedimentary section to geological science, it was by no means the first time Shaligram ammonites had been introduced to Western discourses.

Shaligrams are mentioned in South Asian inscriptions and texts roughly as far back as the second century BCE but according to Hagn (1977, 1988), Shaligrams had already been introduced into Europe as far back at the late 1600s or early 1700s. In one such example, he describes the French Jesuit missionary and Sanskrit scholar, Father Jean Calmette (1693-1740) who once wrote a letter to his superior about his encounters with the *caillou vermoulu* (worm stones) or, conversely, the *caillou perce* (perforated pebbles) in south India (possible somewhere in Tamil Nadu or Andhra Pradesh). Additionally, in 1767, it was then the Spanish naturalist Dávila who recognized that the key components in the descriptions of such stones were fossil ammonites. In 1901, Gustav Oppert also described Shaligrams in a lecture presented to the Congrès International d'Historie des Religions on 7 September 1900. In his lecture, however, he argues that the veneration of Shaligrams as icons of Vishnu was only just beginning to take hold within an older practice that worshipped Shaligrams as manifestations of feminine divine energy (according to him, as a result of conquering "Aryan" groups imposing foreign beliefs on indigenous cultures). He states that "the aborigines of India believe that the *śālagrāma* represents their supreme deity, the female energy, Prakriti, which is introduced by Kapila in his philosophical system called the Sâṅkhya.... *śālagrāma* are dedicated to the principle of Sakti, where it represents the Kundalini Bhavani and other goddesses. It is even said that the great goddess Mahadevi remains in the *śālagrāma*."[14]

14 Gustav Oppert also argued that while Saligram represents the feminine aspect, the Shiva lingam is symbolic of masculine energy. Swami Vivekananda rebutted this explanation and traced the reference to the Atharva Veda: "The worship of the Siva Lingam originated from the famous hymn in the 'Atharva Veda Samhita', sung in praise of the Yupastambha, the sacrificial post which gave place in time to the Siva Lingam and was deified to the high Devahood of Sri Sankara." See pages 325-326.

Any discussion of Shaligrams in the West would also be remiss not to mention Joseph Kohl-Bonn's 1936 article *Zum Indischen Steinkult* (On Indian Stone Worship) where he describes Shaligram veneration not only among Hindu Vaishnavas but Jains as well. While this text describes Shaligram worship as, at least in this case, entirely referential to Hindu scriptures (especially the Puranic texts), he does note that the practice of painting Shaligram stones with faces and other representative *murti* was already common in his regions of travel. He also notes that Jain worship differs in both veneration ritual and in the Shaligram characteristics most valued (such as number of whorls in the spiral and the color of the stone), though he remains fascinated that the practice of Shaligram veneration itself was equally present in many of the religious traditions he had encountered. I've chosen to include brief discussions of these works here mainly for two reasons. First, many of these short articles form the foundations for much of what is still understood in Western cultural and scientific discourses about Shaligrams, and secondly, all of these authors (like most geologists and paleontologists today) viewed Shaligrams primarily, and unsurprisingly, as ammonites first and religious objects and deities second.

Widely abundant and distributed throughout the globe, ammonites are probably the most famous marine fossil of the Jurassic epoch. Today, ammonite paleontology plays a fundamental role in Jurassic stratigraphy and correlation. With a high frequency of occurrence and wide distribution across multiple continents, ammonites also provide valuable insights into Jurassic marine biogeography as well as into evolutionary and other larger palaeobiological processes through vast stretches of geologic time. Starting from the mid-Devonian, ammonoids were extremely abundant, particularly during the Mesozoic era. Many genera evolved rapidly and died out rather quickly, becoming extinct in just a few million years. Subsequently, due to the speed of their evolution and their widespread distribution throughout much of the globe, ammonoids make excellent index fossils (fossils used to define geologic time periods), and it is often possible to link rock layers in which ammonites are found to layers of the same time in other places throughout the world; a point that many Shaligram practitioners are well aware of.

Most practitioners, however, view the global relationships of fossil ammonite populations in two parts. First, that the form of a Shaligram – i.e., the stone body itself – is undoubtedly and inexorably linked to the formation of the world and can therefore reveal details about Deep Geological Time and the creation of the universe but the manifestation of the Shaligram, its deity and personhood, is transcendent of purely physical processes. Therefore, while "ammonites" might be useful in pinpointing global movement in time, "Shaligrams" are divine presences unique to Mustang. And second, ammonites found elsewhere in the world are not necessarily sacred and certainly are not Shaligram (though there is a level of debate on this point). This is not only because they have not undergone the same specific processes of erosion and river wear as Shaligrams have but because they have not done so specifically in the Kali Gandaki, in the *dham* of Śālagrāma.

The *Treatise on Invertebrate Paleontology* (Part L, 1957)[15] divides Ammonoidea, here regarded as one overarching taxonomic order, into eight suborders: *Anarcestina, Clymeniina, Goniatitina,* and *Prolecanitina* from the Paleozoic; *Ceratitina* from the Triassic; and *Ammonitina, Lytoceratina,* and *Phylloceratina* from the Jurassic and Cretaceous. In many subsequent taxonomies, these are sometimes rearranged as orders themselves within a subclass labeled Ammonoidea. In the Jurassic, roughly seven suborders of ammonites are recognized: *Phylloceratina, Psiloceratina, Ammonitina, Lytoceratina, Haploceratina, Perisphinctina,* and *Ancyloceratina* (see p. 2008). In Nepal, several successive faunal assemblages in the Jurassic (Tithonian-Berriasian[16] age) strata of what paleontologists refer to as the Tethyan Himalayas (in the Thak Khola region) reveal a series of subsequent ammonite genii in chronological succession: *Spiticeras, Blanfordiceras, Corongocer, V. denseplicatus, Hildeglochicera, Virgatosphinctes, Aulacosphintctoide,* and *Kossmatia* (Enay and Cariou 1997).

Biogeographical synthesis between the ammonite beds of Himalayan Nepal and many other notable ammonite fossil layers in the Indo-Malagasian, Indonesian, East Pacific, and Mediterranean has, however, been fraught with numerous problems, including uncertainties in published taxonomies and the challenges of relating age and species correlations between distant sites. Even now, extensive paleontological fieldwork in Mustang is difficult, and few studies have been published on the precise relationship of Kali Gandaki ammonites to ammonite-rich strata in other parts of the world. This ambiguity, then, in the precise linkages of specific ammonite taxa to the interpretive traditions of Shaligrams (i.e., the reading of specific deities

15 The Treatise on Invertebrate Paleontology (or TIP) published by the Geological Society of America and the University of Kansas Press, is a definitive multi-authored work of some 50 volumes, written by more than 300 paleontologists and covering every phylum, class, order, family, and genus of fossil and extant (still living) invertebrate animals.

16 In the geological timescale, the Tithonian is the latest age of the Late Jurassic epoch or the uppermost stage of the Upper Jurassic series. It spans the time between 152.1 ± 4 Ma and 145.0 ± 4 Ma (million years ago). It is preceded by the Kimmeridgian and followed by the Berriasian stage (part of the Cretaceous).

in Shaligram characteristics) forms one of several narrative threads that Shaligram practitioners use to link the discourses of Western science to the discourses of Vedic religion. After all, what is *Blandforiceras* really if not a reading of meaningful characteristics.

Himalayan Jurassic ammonites were first described by Dr. A. Oppel in an 1863 survey of the Spiti Shales and Gnari-Khorsum fossil layers in northwestern India. They were later made famous in Uhlig's series of monographs (1903-1910), which also finally identified Himalayan ammonites as morphologically distinct from their Mediterranean Tethys relatives (Enay and Cariou 1999: 829). There is quite a degree of disagreement, however, as to the exact relationship of ammonites in the Nepal Himalayas to ammonites found in India; more specifically, several scholars continue to question whether the ammonites of the Shaligram Shale Formations (as it is sometimes called) are actually related to the Spiti Shales or constitute an amalgamation of ammonite-bearing Late-Jurassic to Early Cretaceous formations either because the original biological spheres overlapped at various points or because the layered sections broke away from elsewhere in the world during several massive tectonic upheavals (Gibling et al. 1994).[17] However, despite recent studies that demonstrate significant gaps in the modern biostratigraphy of the region (see Krishna et al. 1982; Pathak 1993; and Pathak and Krishna 1995),[18] the Western Himalayas (specifically the Lahul-Spiti and Garhwal-Kumaon areas) remain the standard reference for Himalayan biostratigraphy and faunal divisions largely due to Uhlig's extensive monograph on the Spiti Shales.

17 Previous works on Jurassic rock stratigraphy in Nepal include Bordet et al. (1964, 1967, 1971), Gradstein et al. (1989, 1991, 1992), Gradstein and von Rad (1991), Gibling et al. (1994), which tried to set out a faunal succession. Other works by Ryf (1962), Helmstaedt (1969), Kamada et al. (1982), Matsumoto and Sakai (1983) are paleontological studies including descriptions of new species, but with either inaccurate or absent stratigraphic support.

18 Krishna, J., 1983a. Callovian-Albian ammonoid stratigraphy and palaeobiogeography in the Indian sub-continent with special reference to the Tethys Himalaya. Himalayan Geology 11, 43±72.

Krishna, J., 1983b. Reappraisal of the marine and/or "mixed'" Lower Cretaceous sedimentary sequences of India, palaeogeography and time boundaries. In: Cretaceous of India. Indian Association Palynostratigraphy, Lucknow, pp. 94-119.

Krishna, J., Pathak, D.B., 1993. Late Lower Kimmeridgian-Lower Tithonian Virgatosphinctins of India, evolutionary succession and biogeographical implications. Geobios M.S. 15, 227±238.

Krishna, J., Pathak, D.B., 1995. Stratigraphic, biogeographic and environmental signatures in the ammonoid bearing Jurassic-Cretaceous of Himalaya on the south margin of the Tethys. Himalayan Geology, Wadia Institute Himalayan Geology, Dehra Dun 16, 189±205.

Pathak, D.B., 1993. The First record of the Ammonite genus Hybonoticeras from the Himalaya and its biostratigraphic significance. Newsletters Stratigraphy, Berlin±Stuttgart 28 (2/3), 121±129.

Elsewhere in Himalayan Nepal, many regions include multiple mayaitid (ammonite) beds. Typically found above 4,000 meters, these fossil beds contain rich and diverse faunas and are noted for the occasional presence of *Kossmatia* ammonite layers, which are otherwise lacking in other Spiti areas (Enay and Cariou 1999) – possible further evidence in support of the amalgamation theory. Unfortunately, tectonic complications, such as the detachments of layers and disharmonic folding, have been one of the main problems in setting up a complete sequence of rocks and faunas in the Spiti Shale Formations and therefore a comprehensive biostratigraphy of the Mustang region remains incomplete. In fact, the Spiti Shale Formation itself is folded. In addition to this, owing to continued tectonic movements and shifting plant cover, certain layers and outcrops of Himalayan ammonites remain scattered throughout the mountains, and few, if any, contain significant stratigraphic changes or marker beds that allow for easy correlations between different sections or areas of exposure. Constant renewal of the outcrops, however, results from active erosion by the Kali Gandaki river and its tributaries. Consequently, fresh shale sections and nodules *in situ* are widely exposed, and new fossils are constantly fracturing out and rolling down the slopes and into the rivers that will eventually produce Shaligrams. Most of these nodules contain fossils, mainly ammonites, but a few also contain belemnites (e.g., *Belemnopsis gerardi*) and bivalves (*Retroceramus*), oftentimes occurring together within the same nodule and resulting in the rich diversity of eroded patterns present in Shaligrams today.

Shaligram Ammonites

The Nepal Himalayas contain a wide variety of ammonite species that belong to a number of different time periods, from *M. bifurcatus* and *M. apertusmantataranus* in the Ferruginous Oolite Formation[19] to Kimmeridgian[20] *Paraboliceras* assemblages to late Tithonian-Berriasian *Blanfordiceras* and *Proniceras* assemblages. Shaligrams, however, are generally comprised of three particular species of Jurassic ammonites: *Blandifordiceras*, *Haplophylloceras*, and *Perisphinctids* (both *Aulacosphinctus* of the Upper

19 A thin (roughly 3 meter) black shale marker bed which contains microscopic iron (ferruginous) particles which set it apart from the nearby Spiti Shales.

20 In the geologic timescale, the Kimmeridgian is a stage in the Late or Upper Jurassic epoch. It spans the time between 157.3 ± 1.0 Ma and 152.1 ± 0.9 Ma (million years ago). The Kimmeridgian follows the Oxfordian and precedes the Tithonian.

Kimmeridgian/Lower Tithonian and *Aulacosphinctoides* of the Upper Tithonian). Other Shaligram formations include belemnites (such as the Ram Shaligram) and the bivalve *Retroceramus* (such as the Anirudda Shaligram) but for the most part, "classic" Shaligram manifestations are by and large comprised of various black shale ammonites at assorted levels of erosion and wear.[21] *Blandfordiceras* species (lower Tithonian age) are widely distributed ammonites especially known for their tight but evenly balanced spirals and raised, biplicate (Y-shaped) ridges. Geologist Herwart Helmstaedt (1969)[22] was one of the first researchers to investigate the ammonites of the Thak Khola region (immediately south of Mustang) and, according to him, some fifty percent of all ammonites collected in Mustang belong to the *Blandfordiceras* genus. He is also credited with discovering and naming the new species *Blandfordiceras muktinathense*, though the name does not often appear in common usage (Dhital 2015: 288).

Haplophylloceras, by contrast, tends to include fewer rings in the formation of its central spiral and sports a distinctive chevron-like ridge pattern along the outer phragmocone (the back edge of the shell). Finally, *Perisphinctid* ammonites are recognizable by their evolute shell morphology with typically biplicate, simple, or triplicate ribbing. Larger shells may have simple apertures and smooth body chambers while smaller species tend to have lappets and ribbed body chambers (Arkell et al. 1957). *Aulacosphinctoides*, a member of the *Perisphinctidae* family, are also well represented in Shaligrams. These ammonites are characterized by an evolute shell with whorls broadly rounded, ribs sigmoid that mostly bifurcate (and occasionally trifurcate), and clearly defined lappets.[23] *Aulacosphinctoides* also closely resembles its Indo-Malagasian relative *Torquantisphinctes* but differs in that it has more rounded or depressed whorls and more sigmoid and frequently triplicate ribbing.[24]

The paleontological history of ammonites in the Himalayas is a complex one, and despite recent advances in the stratigraphical use of microfossil

21 Similar looking ammonites which occasionally appear in Shaligram discussions (but are not considered Shaligram) are Dactylioceras semicelatum from Whitby, North Yorks England; Toxaceratiode sp. From the Walsh River, Queensland, Australia; Crucilobiceras densinodulum from Charmouth, Dorset UK; Dactylioceras athleticum from Schlaifhausen, Forscheim, near Nuremburg, Germany; and Acanthoceras sp from Agadir, Morocco.

22 H. Helmstaedt. 1969. Eine Ammoniten-Fauna aus den Spiti-Schiefern von Muktinath in Nepal. Zitteliana 1:63-88 [W. Kiessling/M. Krause].

23 These are flanges that protrude from the final chamber at the front of the creature in adult male specimens [the microconch], which some speculate may have been used for sexual display. These features are not present on the larger female ammonites [the macroconch].

24 Sepkoski, Jack (2002). "Sepkoski's Online Genus Database." Retrieved 14 September 2016 and Phil Eyden (2003). "Ammonites: A General Overview." Retrieved 14 September 2016.

groups and non-palaeontological laboratory techniques in geological dating, ammonites continue to retain their pre-eminent position as one of the most reliable and accurate correlation tools available for marine Jurassic sequencing (not unlike dendrochronology to the archaeologist and paleo-ecologist). They also have a number of other uses. Ammonites have been recognized for their value in palaeobiogeography studies and in the study of evolutionary mechanisms and patterns, such as speciation and extinction over vast expanses of geological time. As Kevin Page notes, however, these latter studies are often hindered by incomplete understandings of ammonite correlation and taxonomy from the species level upwards (2008: 54). This situation is then exacerbated by the limited funding available for such research given the preference amongst many funding organizations and media outlets for more mysterious or sensational fossil groups and more fashionable (if transient) scientific theories and hypotheses. This is why the Shaligram traditions of South Asia have something to offer the world of paleontology, adding new dimensions of interpretation and importance to the image of the ammonite, perhaps even to cultural conversations about the modern meanings of fossils as a whole. For Shaligram practitioners, the constant scientific debate and limited amount of concrete detail for describing Shaligram ammonites in Mustang is both taken in stride and interpreted as further evidence of the entanglements of different kinds of "storytelling" when it comes to Shaligram origins and ontologies. Or, as my old friend and mentor Prasad Vipul Yash once expressed it, "They don't know and we don't know. Not all of it, anyway. They call it one thing, we call it another, but it's all the same thing. It just depends on what it is you want to know about the world."

Mustang, Historically Speaking

The history of Mustang is a reconstruction. For many of the current peoples of Mustang, written regional histories were non-existent up until the twentieth century, and much of what we know of the area of Mustang today is currently contained in oral histories, in the *rhabs*,[25] the four semi-mythological clan histories of the Thakali peoples of Lower Mustang, and the *bemchags*, or local village chronicles of the Baragaon and Panchgaon. With little scholarly access to these documents until quite recently, however, much of the current

25 See Fisher 2001 for a detailed analysis on the role of written histories among the Thakali peoples of Lower Mustang.

published history of Mustang relies on outside texts, namely a wide variety of Tibetan historical chronicles authored by later traveling monks and other visitors to the Kali Gandaki River valley. Because of this, the histories of Mustang are often contentious, containing accounts of scholars searching for community and ethnic identities among groups of people who appear to be searching for that identity just as much themselves (Fisher 2001: 3).

This has also led to a number of competing interpretations of Mustang's history that variably focus on the Sanskritization or Hinduization of local religious practices, the promotion of Tibetan Buddhist practices and cultural identification, and the revival of local cultural touchstones, such as Thakali *ḍhoms* (shamans) or the animistic ritual roots of Bon (which is practiced throughout Mustang). Attempts to clarify the history (and by that agenda, to clarify the cultures) of Mustang also tend to anchor their historical claims in issues of authenticity and claims to traditional pasts, but many of these claims are disparate in form and content, and the validity of each has gone on to be challenged repeatedly by scholars and by the local peoples themselves. However, identity and ethnic consciousness are of special concern in Nepal's current political climate, and the selective use of historical information to construct coherent identities is a widespread practice in relating to caste hierarchies, to state views of "proper" citizenship, and to other claims to political and social legitimacy (see also Guneratne 2002).

Over the years, a number of scholars have sought to reconstruct the history of Mustang by relying primarily on Tibetan texts supplemented with other local sources.[26] Unfortunately, as critics tend to point out, many of these efforts have yet to conclusively demonstrate the links between these early histories of Mustang and the contemporary populations within it. In the past decades, there has been a great deal of animosity between anthropologists and Tibetologists. The former is accused by the latter of lacking sufficient Tibetan language skills to properly address historical issues, and the latter is said to overly privilege Tibetan religion, art, and civilization in the formation of Mustang as a political entity (see Dhungel 2002: 6; Snellgrove 1965; and Oppitz 1968). This is because, generally speaking, Tibetan scholars are particularly keen to study various aspects of culture such as art, history, myths, legends, and religion that pertain specifically to Tibetan civilization. Their research is therefore principally based on literary texts, documents,

26 For historical and archaeological work on the peoples and cultures of the upper Kali Gandaki River valley, see Jackson (1976, 1978, 1980, 1984), Mishra (1994), Pohl and Tripathee (1995), Ramble (1997 and 2008a), Ramble and Seeber (1995), Schuh (1990, 1994, 1995) and Seeber (1994, 1996). Local sources include oral tradition and village chronicles (bemchag) from Panchgaon.

and archaeological studies specifically categorized as Tibetan sources. Anthropologists and ethnographers, in contrast, typically rely on more direct observational techniques for producing subjective and synchronic accounts of given communities and are quick to note where actual socio-cultural practices are at odds with textual accounts (see Ortner 1989).

Today, Himalayan studies have largely come to integrate both Tibetological and anthropological approaches (Oppitz 1968; Dhungel 2002), where the use of literary and documentary evidence is employed for both historical analysis and community observation. Regardless, while Tibetan texts have certainly added a great deal to our understandings of some key events in the upper Kali Gandaki River valley, it still remains extremely difficult to document connections between the current peoples of Mustang and the peoples often referred to in Tibetan texts. Scholars of Tibetan texts have continuously attempted to correlate textual references to the areas called Lo (commonly associated with Upper Mustang) and Serib (commonly associated with Lower Mustang) with what was also referred to respectively as the Kingdom of Lo in the upper Kali Gandaki valley and the present-day Baragaon.[27] The Dunhuang Annals of Tibet, for example, refers to the existence of a Lo and Serib as far back as the seventh century and was said to have come under the influence of the Tibetan Yarlung Dynasty around that time. Conversely, David Jackson has suggested that the fifteenth century Lo kingdom in the upper Kali Gandaki valley in Nepal is, in fact, the Lo mentioned in the Tibetan chronicles. Serib, he then goes on to speculate, must then refer to the area farther south, most likely the Baragaon or the Panchgaon.

Unfortunately, historians of Mustang have not been generally supportive of one another, though they all tend to view Mustang through the lens of a Nepalese-Tibetan borderland, a landscape perpetually caught between competing, and disparate cultures on either side of a contentious dividing line. But while this view of Mustang as a borderland between two conflicting sides is common in the literature, it pays little attention to the realities of contiguous change, migration, and exchange that mark historical narratives within the region itself. Academic reconstruction makes it clear that, while both Upper and Lower Mustang lie in a relatively remote area of the Himalayas, the peoples of the region were never isolated from external influences as other Himalayan groups have been from time to time. Over the

27 For example, Vinding speculates that Serib may be Panchgaon (1998) or a combination of Panchgaon and Baragaon (1998). For an account of the Kingdom of Serib, see Jackson (1976, 1978) and Vinding (1988).

centuries, prior to the Gurkha conquest in the late eighteenth century, the upper Kali Gandaki valley has shifted political affiliation and come under the influence of a number of regional powers over time, such as Ladakh and Jumla in the west, Lo to the north, and Parbat to the south (Fisher 2001: 52).

It is therefore difficult to say with any certainty where any particular community, group of practices, or political and religious system may have originated and certainly flies in the face of many of the modern books and brochures that portray Mustang as a "lost kingdom" of "pure Tibetan culture". The ruins of forts and other military fortifications throughout Mustang attest to a history of outside threats and contentious political relationships, but before the eighteenth century there is little in accounts of Mustang's history that can be taken at face value. To be sure, Tibetologists and other historians of Mustang even admit this contention in their own works, noting that the availability of Tibetan archives and Nepali works based on those archives tends to paint a very specific picture of the region through the primary use of Tibetan cultural elements in historical narratives (see Jackson 1978; Dhungel 2002; Ramble 2008a: 2).

The typical history of Nepal's Mustang (Lo in Tibet) sees the region first appearing as a definable entity in the mid-seventh century as a region of trans-Himalayan trade with a mercantile population in the process of gaining significant transregional visibility. Emerging as a kingdom in AD 1440, Mustang was then considered a stronghold of classic western Tibetan culture until well into the 1700s (Dhungel 2002: 3). However, it was also besieged on multiple sides by stronger neighboring kingdoms, all the while continuing to maintain its status as a trans-Himalayan way station for trade between China, Tibet, and India. By dint of its situation and location, then, Mustang has had a long history of boundary making and unmaking through shifts in power, occasional autonomy, and trade and migration-based economies all the way up to its incorporation into the unified kingdom of Nepal in 1789 and into the formation of the Nepali State in the present day. From the earliest sources pertaining to Lo/Mustang (*La-dvags rgyal-rhabs* – "The Chronicles of Ladakh", Dunhuang/Tunhuang Annals, and the *Deb-ther dkar-po* – "White Annals"), it is assumed that Tibetan influences first arrived in Mustang by the seventh century on the heels of a number of rapid cultural and political changes throughout South Asia. Later, according to these chronicles, as the early Tibetan empire began to disintegrate in the tenth century, the region of Mustang came under the influences of the more local powers, namely the Gung-thang principality; a southwestern Tibetan kingdom (alternatively known as Ngari Me, or Lower Ngari) established under Sa-skya overlordship around AD 1265. Not surprisingly, the name Lo appears in Tibetan literature from the earliest times, but by a number of accounts the region itself remained relatively uncertain and obscure in Tibetan records until it was seized, in the fifteenth century, by Amepal (A-ma-d'pal), a nobleman from Gung-thang. Amepal established himself in a stronghold, which he called Duri Khacho, on a strategic hilltop. The city of Monthang (the current "capital" of Upper Mustang) was founded by Amepal's son, Agonpal, who then later shifted the capital to a plateau near the base of his father's fortifications (Ramble 2008b). This kingdom, whose boundaries are sometimes said to have extended as far south as Kagbeni, then survived through unstable political alliances with neighboring kingdoms for almost four centuries, the wealth of artwork and *gompas* of the northern villages remaining a testament to Lo's cultural and political ascendancy during that time.

According to other Tibetan histories, Tibetan Buddhist and reformed Bon missionaries arrived in the area of Lo and Serib in the twelfth century (Jackson 1978: 200). One of the Bon missionaries was a man called Lubra Tashi Gyaltshan who is said to have founded a monastery in the village of Lubra in Baragaon around AD 1160 (Ramble 1983; Jackson 1978: 204-205). Even today, Lubra's Bon identity is prevalent, and many Mustangis view the village as the locus of Bonpo practices in Mustang. Buddhism and Bon continued to have supporters in the following centuries, though it is difficult to tell from textual readings exactly how and when their influences waxed and waned over time. The oldest and southernmost local evidence of Buddhism in Lower Mustang (Thaksatsae) at present is the temple of Meki Lha Khang, founded some three hundred years later, in the early fifteenth century (Jackson 1978: 218). None of these texts, however, mention the ritual use of Shaligrams or of Shaligram pilgrimage, though both pilgrimage and trade were likely a part of the local economies at this time (much as they are today and still rarely mentioned).

Since Shaligram practices and the Kali Gandaki location of Śālagrāma are mentioned in Puranic texts preceding this time by several centuries, it is interesting to consider that the Tibetan texts make no mention of them, though it may explain why Shaligram pilgrimage is often left out of historical research. The reliance of scholars on Tibetan texts to account for the history of Mustang before the eighteenth century (when Shaligrams begin appearing in Himalayan travel literature) privileges the views of Tibetan Buddhist monks, many of whom were never within a hundred miles of the Kali Gandaki River and have little incentive to discuss religious traditions (especially transient ones) not in line with Tibetan Buddhist orthodoxy. Not surprisingly, their accounts tend to be concerned with the establishment of monasteries and temples, the achievements of missionaries and lamas, and to support the narratives of scholars for whom Buddhism figures prominently. But as William Fisher goes on to note in his ethnography of Mustang's

Thakali peoples, “the lack of other voices should not lead us to assume that the area was uniformly influenced by Buddhism.” (2001: 53). In either case, be it via Tibetan texts or Puranic accounts, the history of Mustang is still largely constructed through outside viewpoints whose choices to include or omit various social aspects or cultural practices continue to render more comprehensive histories of the region unattainable.

While no complete picture emerges of Mustang before the Gurkha conquest in 1789, it is at least clear to some extent that in the centuries preceding the unification of Nepal many different influences on the Kali Gandaki area routinely disrupted local power structures, introduced new religious traditions and ritual systems, and encouraged some degree of population movement and migration in and out of the valley. It is within this context, then, that Shaligram pilgrimage and the exchange of stones with Hindu practitioners in India likely gained its most prominent footholds sometime in the early centuries of the first millennium AD. While no early histories of Mustang directly address the presence of Shaligram pilgrims (or pilgrims in general), these early Tibetan and Nepali texts taken in combination with Indian accounts of Shaligram veneration in the Hindu texts and later Western travelogues noting village Shaligram practices confirm the likelihood that Shaligram pilgrimage and veneration was ongoing at the time, regardless of the agendas and ideologies present in written recording. The coming of the Gurkhas then meant the addition of another external influence, one that also dramatically shifted the orientation of the region from Tibetan influences in the north to burgeoning Nepali influences to the south.

Mustang in the Modern Day

Although Nepal has a history that dates back 2,000 years or more, the sociopolitical picture of modern Nepal was drawn after the Gorkhali conquest of the Kathmandu valley in 1769. The nation-state of Nepal then finally began to crystallize in the second half of the eighteenth century, when Prithvi Narayan Shah, the young king of Gorkha (a town west of the Kathmandu valley), and his army, the Gorkalis, embarked on a series of conquests that would bring a vast portion of the Himalayas under a single autonomous rule. The Gorkhalis (or Gurkhas) were Hindus who claimed high-caste Thakuri status, descendants of the *raja* of Sinja who, in turn, was believed to have descended from the Rajputs who fled India during the Muslim invasions (Fisher 2001: 55; Dhungel 2002: 12). The process of unification, which was later taken up by Prithvi Narayan’s many successors, continued into the early years of the nineteenth century and came to a halt when the boundaries of the new kingdom finally extended from the Sutlej River in the west to the Tista river, between Sikkim and Bhutan, in the east. The Gorkhali campaign of consolidating mountain kingdoms and other territories was then only finally stopped by the Anglo-Nepal War of 1814, and these dimensions were reduced by about a third following the treaty of Segauli in 1815, which concluded a war with the British East India Company. But a substantial area of fertile lowland was later returned to Nepal in recognition of the military help it had provided the hard-pressed British during the Sepoy Mutiny of 1857. Until then, the political borders and territories of Nepal were not clearly defined.

In the years following the conquest, the Gorkhali's ambition to make their kingdom a true Hindustan, a unified homeland of Hinduism, faced a number of both internal and external challenges. Although Lo (Upper Mustang) lies within the border of present-day Nepal, the region is principally inhabited by the Lo-pa (or, in some discussions, Bhotia)[28] peoples and their cultures, which were not incorporated into Nepal until 1789. Until that time, other major high Himalayan settlements such as Dolpo, Manang, Nubri, Nar, and Nyishang were not originally incorporated either, and even after the incorporation of these Himalayan settlements into the Nepali state, they were not considered a part of mainstream Nepali culture and society until very recently. This marginality is what earned the people of Lo/Mustang the term *bhot*, a popular Nepali name for Tibet (*mustanbhot* or *mananbhot*, accordingly). Nepal's Hindu identity was also challenged by the Mughals of India and by the British during the nineteenth and twentieth centuries.

Over the ensuing decades, the political and administrative efforts to integrate the outlying areas, maintain a steady income for the central government in Kathmandu, and forge a common nation out of diverse populations with widely different cultures, languages, and religious practices have had a profound effect on the peoples of Mustang.[29] In more pressing concerns related to Shaligram pilgrimage, this particular historical narrative frames many of the current issues with labeling people or practices in Mustang as "Hindu". Regardless of any actual historical influence, claims to Hindu identity are today often taken as political statements in alignment with the cultural homogenization efforts of the central Nepali government and, therefore,

28 Depending on the context of the discussion, the term Bhot or Bhotia is considered by many Mustangi's to be derogatory.

29 See Fisher 2001 for an in-depth discussion on the effects of Mustang's historical and political situation on the Thakali peoples of Lower Mustang, particularly as it pertains to the Thakali salt trade.

many local Mustangis are circumspect when it comes to claiming any specific religious basis for Shaligram practices or for pilgrimage sites such as Muktinath.

Today, Mustang district comprises a number of administrative enclaves that have been recognized either as the residues of old administrative entities or as the territories of ethnically distinct groups (Ramble 1992, 2008a). The northernmost part is still referred to as Lo (or occasionally as *Glo bo*, *Blo bo*, etc.), while Upper Lo is a designation given to the territory that was once ruled by the King of Mustang at the time of the unification of Nepal (which the Gorkhas recognized as his domain even after conquest). In academic and ethnographic sources, this area is referred to more fully as Loto Tshodun (*Glo bo stod tsho bdun*), "the Seven Sectors of Upper Lo". The Tibetan word *tsho* is often translated "sector" and is an old Tibetan administrative division that might also be rendered as analogous to "county" in present political contexts.

Below Gemi, the southernmost village in the region of Upper Lo, is the large community of Gelung which (with the help of Jumla) broke away from the kingdom in 1754 (Schuh 1994: 85). Immediately to the south of Gelung is Baragaon. South of Baragaon is a group of settlements known as Panchgaon (Nepali: *Panchgaun*, the Tibetan equivalent is Yulkhanga or *Yul kha lnga*). Both terms mean "the Five Villages" which in this case refers to the villages of Thini, Shang, Tsherog, Cimang, and Marpha (Ramble 1992). Mustang's current district headquarters, Jomsom, began life as a little satellite of Thini on the left bank of the Kali Gandaki river but has now acquired the proportions of a mid-sized town along with the region's only airport, a large military barracks, and a great many hotels geared towards the burgeoning tourism and trekking economy. The region between Panchgaon and the southern boundary of Mustang district, comprising thirteen settlements, is known as Thak and the people who inhabit it as Thakalis, an ethnonym that is also sometimes, but not always intentionally, extended to include the inhabitants of Panchgaon (see Fisher 2001). In terms of Shaligram pilgrimage, the Baragaon[30] and Panchgaon regions figure most prominently.

30 While there are apparently two versions of the sectoral groupings of Baragaon's villages, the most commonly referenced one of them is as follows:
1. Purang and Dzar
2. Dzong and Chongkhor
3. Kag and Khyenga
4. Tangbe (two-thirds) and Gyaga (one-third)
5. Tshug
6. Te
7. Tsele, Putrak, Tiri, Samar
8. Lubra, Phelag, Dangkardzong, Pagling

The standard Shaligram pilgrimage trek begins with arrival in Jomsom. From Jomsom, pilgrims can then proceed about an hour's walk north of Jomsom, along the banks of the Kali Gandaki river to where it is joined from the east by the Panda Khola, a small river that traditionally marks the boundary between Panchgaon and Baragaon. The village of Lubra, a traditional center of Bon worship, lies on the southern side (though it is considered to be a part of Baragaon) and is occasionally visited by Shaligram pilgrims interested in some of the more esoteric ritual histories of the stones within the region. The inhabitants of Lubra are also considered to be of a general priestly class (Tibetan: *bla mchod*) as principal followers of the Bon religion and are often referred to as *lama* or *lama guru* by attendant pilgrims (Ramble 1992, 2008a).[31]

The first few decades of the twentieth century, when few Westerners ever traveled beyond Kathmandu, were one of the few times when Mustang was readily accessible to Euro-American travelers. But, soon enough, China's occupation of Tibet would throw Mustang head-first into Tibetan resistance efforts, resulting once again in the political segregation of Mustang from the rest of Nepal and the rest of the world. In the end, the final blow to Mustang's traditional forms of power probably came with the democratic reforms that followed the implementation of the partyless Panchayat System in the 1960s. These reforms then precipitated the decline of many of Mustang's local administrative and reciprocity/exchange systems overall. 1960 was also the year when Tibetan guerillas established a base near Kesang, close to Jomsom. With the help of the US Central Intelligence Agency, close to 6,000 Tibetan rebels, called Khampas, skirmished with Chinese forces across the border. Desperate to appease China but unable to control the Khampas, the Nepali government declared Mustang a restricted zone and censored all news of Tibetan guerrilla activities. In the view of some scholars, this disintegration and segregation have only accelerated following the advent of multiparty democracy in Nepal in 1990 and the opening of Mustang's borders to foreign travel in 1992. Unfortunately for Shaligram pilgrims, this continuing shifting of access, national identity, and political conflict has done more to shape current pilgrimage practices than anything else.

The Tibetan struggle for independence left many of the villagers of Mustang caught between both political and cultural forces that were not theirs. There were Khampa camps in almost every village, and many people

31 When speaking Nepali or English, I often noted that Buddhist monks referred to themselves as *lamas*, even though in Tibetan, *lama* means priest and not monk.

reported that Khampa leaders were not above using intimidation and threats of violence to garner local support. Because of this, Khampa activities posed a political problem for the peoples of Mustang, who could not afford to jeopardize their standing as citizens of Nepal and who also did not want to endanger their cultural and economic ties to Tibet. The CIA ended its support of the Khampa rebels in the early 1970s, when the United States officially recognized the sovereignty of the People's Republic of China (which now included the Tibetan Autonomous Region or TAR). This, however, did not end guerrilla resistance in Mustang, which continued on for some years afterwards and was divided into two factions – one headed by Gyatso Wangdui and the other by Gen Yshe. When the Dalai Lama appealed to the Khampas to surrender, Gen Yshe complied and resettled most of his fighters in Nepal, with a few even in Mustang. But in response to political pressures from China, the Royal Nepal Army moved to Mustang and set up a large military base in Jomsom in 1973. The following year, the army ambushed Gyatso Wangdui's group and killed him in Tinkar in west Nepal near the Indian border.

In the mid-1970s, the Nepali government moved further in Mustang with development plans for schools, health posts, police stations, and water taps. Lower Mustang was opened for tourism and trekking, but Upper Mustang remained closed. The monarch-ruled Panchayat government, then in full force, had a strong ideological base in Nepali nationalism, specifically in resurrecting Prithvi Narayan Shah's image of a unified Nepal. The Panchayat government forwarded its own ideology of a particular pan-Nepali identity that touted its abilities to level ethnic differences and to integrate them into the state's all-encompassing Hindu social model. The king of the Hindu Kingdom of Nepal and the Nepali language became the two most prominent cultural "unifiers", and Nepal was envisioned as a "garden" of four castes and 36 ethnic groups (though the concept of the four Hindu *varnas* – the socioeconomic classes or castes depicted in Hindu books – was largely alien to most of Nepal's Himalayan tribes). Regardless, Hindu norms were advanced through school outreach programs and through Hindu *karmachari* and border guards, all of whom sought to "integrate" and "Nepalise (Sanskritize)" the peoples of Mustang and pull them into a Hindu "mainstream" along with the rest of the nation.

Following the Gorkhali conquests, uniting the country of Nepal politically and administratively proved to be no easy task. There were a series of wars, attempted coups, blockades, occupations, and general political instability at the center of Nepali political life for much of the time up to the present. Politically and geographically marginalized areas, such as Mustang, often had to take advantage of social and political reorientations, migrations, ethnic reinventions, and religious blending between regional and national traditions. This is because state-building in the era following the mid-1800s required the formation of a national ideology, a process that proved to be even more disruptive to social and national cohesion. This attempt to a forge a common nation out of diverse peoples affected the people of Mustang by continuously changing many of the social contexts within which they operated.

With the fall of other Hindu principalities in South Asia to the British, Nepal came to see itself as the only independent Hindu kingdom within the sacred lands of Hindus and began to act accordingly to preserve the sense of "purity" it had developed towards its political culture (Burghart 1984: 106-116). Within the regions consolidated by the Gorkhali conquest, there was a wide array of groups speaking more than forty distinct languages, with three historically and geographically distinct caste hierarchies, loosely defined groupings throughout the middle hills, a variety of culturally distinct Tibeto-Burman-speaking peoples all along the northern borders, and a plethora of ritual and religious systems drawn from local shamanism, Bon, Buddhism, and Hinduism alike (Fisher 1987, 2001: Levine 1987; and Höfer 1979). Over the years, strategies employed to create a Hindu nation out of many religiously and ethnically disparate populations included persuading a number of outlying peoples to adopt some Hindu practices (and broadening the definition of "Hindu" in the process) and outlawing the conversion of Hindus to other religions. In the case of Mustang, this also included the promotion of practices earlier identified as Hindu by their association with pilgrimage from India and other heavily "Sanskritized" regions of South Asia. This promotion did not, however, specifically include Shaligram pilgrimage but rather focused on the more iconic and identifiable aspects of "Hindu" culture such as the preservation of temples, deity statues, and yearly festival traditions to gods of the Brahmanical Hindu pantheon.

Control of the high Himalayas and other marginal hinterlands increased dramatically in the middle of the nineteenth century after the Rana family took control of Nepal's central government, reducing the previous king to no more than a figurehead and beginning a century of rule by a series of hereditary prime ministers. The Ranas initiated a significant step in the process of state formation in 1854 with the codification of a national hierarchy that granted certain rights and political status to each category of social group defined in the legal code. This 1854 code, called the Muluki Ain, was a primarily Hindu model that was superimposed on an otherwise heterogeneous population. It served the dual purpose of distinguishing

Nepal's own national society and culture from that of "foreign" societies and cultures and justified the placement of Rana (and other high-caste Hindu) rulers at the top of the hierarchy. Unsurprisingly, this socio-political composite of ethnicity and caste ranked high Hindu castes at the top, followed by a wide array of non-Hindu hill peoples and *Bhotes* (meaning all Tibetan Buddhists peoples) in the middle, and untouchables at the bottom.[32]

Ambiguities in the middle ranges of the hierarchy established by the Muluki Ain combined with the code's use of categories that did not, for the most part, correspond to precise groups of people has continued to produce a number of creative approaches to identity throughout Nepal. Strategies of genealogical reckoning designed to lay claim to higher caste ancestry became common,[33] as did the tactic of redefining what ethnicity and caste even meant (see Guneratne 2002). In other cases, like those of the Tamang and Chhetri, the legal recognition of new and higher categories improved certain groups' interactions with the state and allowed those who were able to claim membership to achieve better upward mobility and higher positions specifically by emulating Hindu norms (Sharma 1977). The "Hinduization" of the religious practices of Himalayan and other "lower" groups has become one of the most pervasive academic, political, and local concerns since the unification of Nepal.

For scholars, the debate as to the "legitimacy" of Hindu practices among Himalayan peoples (Thakali, Lo-pa, etc.) is a contentious issue because of fears that their work might be taken as an attempt to align academic analysis with state aims. Giuseppe Tucci, for example, argued in 1951 that both Hindu and Buddhist practices were prevalent in the Himalayas but that Buddhism was gaining ground, while just a year later, David Snellgrove argued that Buddhist culture was already in significant decline in Lower Mustang. In 1953, Japanese scholar Jiro Kawakita wrote that the Thakalis of Lower Mustang were following neither Buddhism nor Hinduism in favor of reviving their own local "shamanistic" practices (1957: 92). Then in 1958, Shigeru Iijima saw Hinduism gaining strength in Mustang and argued that the shamanic practices mentioned by previous scholars had already been reduced by Hinduization. By 1962, Fürer-Haimendorf observed that

32 The Muluki Ain also detailed Nepal's new laws on diverse social, religious, economic, and administrative matters in over 163 categories that were meant to ensure a uniform code of punishment to all subjects who violated the law according to their offenses measured against their status (Sharma 1977, Regmi 1976).

33 Up to and including the family of Jang Bahadur Rana, the ruler responsible for the Muluki Ain. After he came to power in 1846, his Chhetri family of Kunwars adopted the name Rana and had a new family genealogy written up to connect the new Rana lineage with the rajputs of Chitor (see Gimlette 1927).

what was viewed as Hinduization was simply secularization camouflaged in the terminology of state-sponsored Hinduism. In 1978, however, Fürer-Haimendorf revised his earlier claims and went on to describe how certain people were intentionally and unilaterally altering their ritual and religious behaviors to conform more closely to high-caste Hindu norms, a view that would be echoed again by Messerschmidt in 1984 (266).

By 2001, this ongoing debate surrounding what appeared to be simultaneous religious revival and religious decline in Mustang had transformed yet again, and scholars now view the religious blending of Mustang as a perpetual kind of "eclectic ritual pluralism", to borrow Fisher's characterization, where the contexts of ritual practices have never been static and religious identities never fixed (Fisher 2001; Guneratne 2002; Walter 2003; Hausner 2007; Craig 2008). Within this context, Shaligram pilgrimage and ritual practices have remained equally difficult to categorize. For some, Shaligram veneration is labeled as a distinctly "Hindu" practice, while others point out that Bon shamans and Buddhists alike also often keep Shaligrams in their homes or use them to mark significant areas of the household or the landscape. Ritual veneration and interpretation of the stones are also not limited to Hindu deities but can also include various Himalayan spirits, such as the Dakini wind spirits and the Buddha himself.

While an assertion of a hegemonic Hindu ideology accompanied the formation of the Nepali state and high-caste Hindus have dominated the political elite ever since, these processes cannot be read simply as the temporal movement of groups undergoing a steady process of Hinduization (or Sanskritization) as they progress towards some kind of ideal Hindu model. While the Muluki Ain may have been designed to create a homogenous society, this aspiration has not been and never will be realized. The Muluki Ain is a projection from political powers above that represents a social order that was, for a long time, little known and even less accepted in the regions of the high Himalayas (Fisher 2001; Höfer 1979). This does not mean, of course, that practices labeled or identified as Hindu did not exist in Mustang prior to 1854 or that Hinduism was unknown in the high Himalayas at any point preceding the unification of Nepal – far from it. Rather, the codification of Hinduism during the era of the Muluki Ain and afterwards to the invasion of Tibet by China in the 1950s, to the end of the Rana government in 1951, and to the ratification of the new secular Nepali Constitution in 2015 has largely been leveraged for a number of political aims that have more to do with foreign policy relations (with India, Tibet, and China) and political maneuvering (support for Tibetan autonomy, aftermath of the Maoist insurgency) than they do with the history of any actual religious practices.

Temples in the Clouds

The history of Mustang, Nepal is a complicated one and can vary in significant ways depending on the textual sources one uses or the narratives one pays particular attention to. This is also how the political history of Mustang tends to blend with mythological narratives of the landscape and the stories that are told about the region and its people become deeply contested and hotly negotiated overall. This has led to an acute awareness on the part of local people and among pilgrims that labeling any particular piece of art, building, or practice as Hindu, Buddhist, or Bon is more often a political statement aimed at securing outside interests in "cultural preservation" than any actual intent to identify origin. Mustang's contested history and ambiguous state of national and ethnic belonging then contextualize the narratives of mobility and place-making central to Shaligram pilgrims' understanding of the ways in which sacred stones are produced, collected, and exchanged.

For both local Mustangis and pilgrims alike, issues of place, space, and time are paramount to the understanding of how claims to "Hindu" or "Buddhist" belonging come to characterize a landscape and peoples who routinely defy such neat taxonomies and whose practices and traditions are never quite so easily quantifiable. Rather, such categories as "Tibetan Buddhist" or "Sanskritized Hindu" form frameworks by which Mustang is continuously politicized and culturally marginalized. This then results in a continuous debate along multiple lines as to where Shaligrams belong and to whom they might belong to.

Beginning with the paleontological history of Mustang's extensive fossil formations and ending with an overview of the political history of the region, this chapter demonstrates how historical narratives have affected patterns of pilgrimage and migration as well as how access to the Kali Gandaki River Valley and to Shaligrams has been shaped and constrained since early days. Claims to specific religious origins or to ethno-national identity have also come under intense scrutiny, as many scholars contest that such attempts to classify any of the Himalaya's continuously shifting and hybridizing traditions tend to be made towards nationalist aims – aims that are just as likely to promote state-sponsored Hinduism as they are the Tibetan fight for independence.

Ultimately, what concerns Shaligram pilgrimage and the practices of Shaligram interpretation and veneration in these cases is not so much whether Shaligrams constitute a historically Hindu, Buddhist, or Bon practice but that claims to Hindu, Buddhist, or Bon belonging say more about the people attempting to control and regulate the region of Mustang, the Kali Gandaki River, and the temple site of Muktinath than they do about the origin of Shaligram ritual traditions. But with such ambiguity and fluidity at work, new possibilities emerge for place-making and belonging that can now take into account both historical and mythological narratives, where physical space and mythic place may or may not even be the same thing but can exist concurrently in the same location. No concern is more highly debated, however, than the question as to whether or not the actual land of Mustang is truly Muktikshetra – the fabled field of salvation – and if it is, might it also be the land of Śālagrāma; the ostensible birthplace of all Shaligrams?

Bibliography

Arkell, W.J., B. Kummel, and C.W. Wright. 1957. *Mesozoic Ammonoidea. Treatise on Invertebrate Paleontology, Part L, Mollusca 4*. Lawrence, KS: Geological Society of America and University of Kansas Press.

Burghart, Richard. 1984. "The Formation of the Concept of Nation-State in Nepal," *Journal of Asian Studies* 44 (1): 101-126.

Craig, Sienna. 2001. "A Tale of Two Temples: Culture, Capital, and Community in Mustang, Nepal." Paper Presented at the New York Conference on Asian Studies, Ithaca, NY.

—. 2008. *Horses Like Lightening: A Story of Passage Through the Himalayas*. Somerville: Wisdom Publications.

Dhital, Megh Raj. 2015. *Geology of the Nepal Himalaya: Regional Perspective of the Classic Collided Orogen* (Regional Geology Reviews). New York: Springer.

Dhungel, Ramesh K. 2002. *The Kingdom of Lo (Mustang): A Historical Study*. Kathmandu: Tashi Gephel Foundation.

Enay, Raymond, and Elie Cariou. 1997. "Ammonite faunas and palaeobiogeography of the Himalayan belt during the Jurassic: Initiation of a Late Jurassic austral ammonite fauna," *Palaeogeography, Palaeoclimatology, Palaeoecology* 134: 1-38.

—. 1999. "Jurassic ammonite faunas from Nepal and their bearing on the palaeobiogeography of the Himalayan belt," *Journal of Asian Earth Sciences* 17: 829-848.

Fisher, William F. 1987. *The Re-Creation of Tradition: Ethnicity, Migration, and Social Change Among the Thakali of Central Nepal*. PhD diss., Columbia University.

—. 2001. *Fluid Boundaries: Forming and Transforming Identity in Nepal*. New York: Columbia University Press.

Fürer-Haimendorf, Christoph von. 1957. "The Interrelations of Caste and Ethnic Groups in Nepal," *Bulletin of the School of Oriental and African Studies* 20: 243-260.

—. 1966. "Caste Concepts and Status Distinctions in Buddhist Communities of Western Nepal." In *Caste and Kin in Nepal, India, and Ceylon: Anthropological Studies in Hindu-Buddhist Context*, edited by Chrisoph von Fürer-Haimendorf, pp. 140-160. London: Asia Publishing House.

—. 1975. *Himalayan Traders*. New York: St. Martin's.

—. 1978. "Trans-Himalayan Traders in Transition." In *Himalayan Anthropology*, edited by James Fisher, pp. 339-357. The Hague: Mouton.

—. 1981. "Social Change in a Himalayan Region." In *The Himalaya*, edited by J.S. Lall, pp. 175-203. Delhi: Oxford University Press.

Garzanti, Eduardo, and Marco Pagni Frette. 1991. "Stratigraphic succession of the Thakkhola region (Central Nepal). Comparison with the north-western Tethys Himalaya," *Rivista Italiana Paleontologia Stratigrafa* 97 (1): 3-26.

Gibling, Martin R., Felix M. Gradstein, Inga L. Kristiansen, James Nagy, Maurizio Sarti, and Jörg Wiedmann. 1994. "Early cretaceous strata of the Nepal Himalayas: Conjugate margins and rift volcanism during Gondwanan breakup," *Journal of the Geological Society, London* 151: 269-290.

Guneratne, Arjun. 2002. *Many Tongues, One People: The Making of Tharu Identity in Nepal*. Ithaca, NY: Cornell University Press.

Hagn, H. 1977. "Shaligram – Cobbles of Malm Limestones with Ammonites as Cult Objects of India," *Communications of the Bavarian State Collection of Paleontology and Historical Geology* 17: 71-102. ("Saligrame – Gerölle von Malm-Kalken mit Ammoniten als Kultgegenstände Indiens." *Mitteilungen der Bayerischen Staatssammlung für Paläontologie und Historische Geologie*).

—. 1988. "On Saligrams – The Holy Stones of India," *Contributions to Himalayan Geology* 4: 53-62.

Hausner, Sondra L. 2007. *Wandering with Sadhus: Ascetics in the Hindu Himalayas*. Bloomington: Indiana University Press.

Hirsch, Eric, and Michael O'Hanlon. 1995. *Anthropology of Landscape: Perspectives on Place and Space*. Clarendon: Oxford.

Höfer, Andras. 1979. *The Caste Hierarchy and State in Nepal: A Study of the Muluki Ain of 1854*. Innsbruck: Universitätsverlag Wagner.

Iijima, Shigeru. 1960. "The Thakali: A Central Himalayan Tribe," *Japanese Journal of Ethnology* 24 (3): 1-22 (Japanese text, English summary).

—. 1963. "The Hinduization of a Himalayan Tribe in Nepal," *Krober Anthropological Society Papers* 29: 43-52.

Jackson, David. 1976. "The Early History of Lo (Mustang) and Ngari," *Contributions to Nepalese Studies* 4: 39-56.

—. 1978. "Notes on the History of Se-Rib and Nearby Places in the Upper Kali Gandaki Valley," *Kailash* 6, no. 3: 195-227.

—. 1980. "A Genealogy of the Kings of Lo (Mustang)." In *Tibetan Studies in Honour of Hugh Richardson: Proceedings of the International Seminar on Tibetan Studies, Oxford 1979*, edited by Michael Aris and Aung San Suu Kyi, pp. 133-137. New Delhi: Vikas.

—. 1984. *The Mollas of Mustang: Historical, Religious, and Oratorial Traditions of the Nepalese-Tibetan Borderland*. Dharamsala: Library of Tibetan Works and Archives.

Kawakita, Jiro. 1955. "Some Ethnographic Observations in the Nepal Himalaya," *The Japanese Journal of Ethnology* 19 (1): 1-57.

—. 1957. "Ethno-geographical Observations on the Nepal Himalaya." In *Peoples of the Nepal Himalaya*, edited by H. Kihara, pp. 1-362. Kyoto: Flora and Fauna Society.

Kohl-Bonn, Joseph. 1936. "To the Indian Stone Cult," *Journal of the German Oriental Society* vol. 90, no. 2: 432-440. ("Zum indischen Steinkult." *Zeitschrift der Deutschen Morgenländischen Gesellschaft*).

Latour, Bruno. 1988. *The Pasteurization of France*. Translated by Alan Sheridan and John Law. Cambridge, MA: Harvard University Press.

—. 1993. *We have never been modern*. Cambridge, MA: Harvard University Press.

—. 1999. "On Recalling ANT." In *Actor Network Theory and After*, edited by John Law and John Hassard, pp. 15-26. Oxford: Blackwell Publishers.

Lavecchia, Giusy, and Giancarlo Scalera, eds. 2003, "Frontiers in Earth Sciences: New ideas and interpretations," *Annals of Geophysics* vol. 49 (supplement).

Le Fort, Patrick. 1975. "Himalayas: The collided range. Present knowledge of the continental arc," *American Journal of Science* 275-A: 1-44. (1981, Manaslu leucogranite: A collision signature of the Himalaya, a model for its genesis and emplacement. *Journal of Geophysical Research*, 86: 545-568.)

Levine, Nancy. 1987. "Caste, State, and Ethnic Boundaries in Nepal," *Journal of Asian Studies* 46: 71-88.

Messerschmidt, Donald A. 1984. "The Thakali of Nepal: Historical Continuity and Socio-Cultural Change," *Ethnohistory* 29: 265-280.

—. 1989. "The Hindu Pilgrimage to Muktinath, Nepal. Part 1 and Part 2: Natural and Supernatural Attributes of the Sacred Field," *Mountain Research and Development* vol. 9, no. 2 (May): 89-104.

Oppitz, Michael. 1968. *History and Social Order of the Sherpa* (*Geschichte und Sozialordnung der Sherpa*). Innsbruck: Universitasverlag Wagner.

Ortner, Sherry B. 1989. *High Religion*. Princeton: Princeton University Press.

Page, Kevin N. 2008. "The Evolution and Geography of Jurassic Ammonoids," *Proceedings of the Geologists' Association* vol. 119 (1): 35-57.

Palmié, Stephan. 2013. "Mixed Blessings and Sorrowful Mysteries: Second Thoughts about 'Hybridity'," *Current Anthropology* vol. 54, no. 4 (August): 463-482.

Ramble, Charles. 1983. "The Founding of a Tibetan Village: The Popular Transformation of History," *Kailash* 10, no. 1: 267-290.

—. 1990. "How Buddhist are Buddhist Communities? The Construction of Tradition in Two Lamaist Villages," *Journal of the Anthropological Society of Oxford* 21 (2): 185-197.

—. 1992. The Archive of Baragaon. http://www.tibetanhistory.net/documents-from-mustang/baragaou/. Accessed 10 October 2016.

—. 2002. "Temporal disjunction and collectivity in Mustang, Nepal," *Current Anthropology* 43, supplement (August-October).

—. 2008a. *Tibetan Sources for a Social History of Mustang, Nepal. Volume 1: The Archive of Te.* Halle: International Institute for Tibetan and Buddhist Studies.

—. 2008b. *The Navel of the Demoness: Tibetan Buddhism and Civil Religion in the Highlands of Nepal.* Oxford: Oxford University Press.

—. 2014. "The Complexity of Tibetan Pilgrimage." In *Searching for the Dharma, Finding Salvation – Buddhist Pilgrimage in Time and Space.* Proceedings of the Workshop "Buddhist Pilgrimage in History and Present Times" at the Lumbini International Research Institute (LIRI), Lumbini, 11-13 January 2010. Edited by Christoph Cueppers and Max Deeg. Lumbini International Research Institute 2014.

Sharma, Prayag Raj. 1977. "Caste, Social Mobility, and Sanskritization: A Study of Nepal's Old Legal Code," *Kailash* 5, no. 4: 277-299.

Snellgrove, David. 1958. *Buddhist Himalaya.* Oxford: Bruno Cassirer.

—. 1961. *Himalayan Pilgrimage: A Study of Tibetan Religion by a Traveller Through Western Nepal.* Oxford. Cassirer.

—. 1965. "Review of *The Sherpas of Nepal* by C. Furer-Haimendorf," *Asian Review of Art and Letters* New Series 2.1.

Stewart, Pamela J., and Andrew Strathern. 2015. *Landscape, Memory and History: Anthropological Perspectives* (Anthropology, Culture and Society). London: Pluto Press.

Stiller, L.F. 1973. *The Rise of the House of Gorkha: A Study in the Unification of Nepal, 1768-1816.* Kathmandu: Ratna Pustak Bhandar.

Stöcklin Jovan. 2008. "Developments in the geological exploration of Nepal," *Journal of Nepal Geological Society* vol. 38: 49-54.

Walter, Damian. 2003. "Among Spirits and Deities: Diverse Shamanisms in the Nepal Himalayas," *CSQ* 27.2 (Summer) Shamanisms and Survival. Available online at: http://www.culturalsurvival.org/publications/cultural-survival-quarterly/nepal/among-spirits-and-dietiesdiverse-shamanisms-nepal-him. Accessed 10 October 2016.

4 A Mirror to Our Being

Locating Muktinath, Finding Śālagrāma

Abstract

The temple complex of Muktinath is both the primary site of Shaligram veneration within Mustang as well as the endpoint of Shaligram pilgrimage. Though identified in Puranic texts as existing in Himalayan Nepal and mentioned by a series of names, such as Muktikshetra and Śālagrāma, it is often unclear as to whether these passages refer to a place, a region, or an ideal. As a Buddhist complex, Muktinath is more commonly referred to as Chumig-Gyatsa (Hundred Waters). As such, Muktinath itself, much like Mustang generally, is not specifically identifiable through any one religious tradition and incorporates a variety of ritual practices, local and foreign customs, and belief systems – all of which are recapitulated in the Shaligram practices that follow.

Keywords: Muktinath, Shaligram, Salagrama, Purana, Chumig-Gyatsa

> "The moment I have realized God sitting in the temple of every human body, the moment I stand in reverence before every human being and see God in him – that moment I am free from bondage, everything that binds vanishes, and I am free."
>
> – *Swami Vivekananda*

A young Muktinath nun, who gave her name as Sister Pemba Dorje, had awoken very early in the morning to begin preparations for the *darshan* of Vishnu-Chenrezig. "Muktinath is a great example of a place of harmony," she told me, still gathering together a few ritual implements. "Many people come here. Hindus and Buddhist from everywhere. They give rice and money and bring their children and elderly parents. Or they come with photos when people cannot come or who have died. I didn't come because I was

Walters, Holly, *Shaligram Pilgrimage in the Nepal Himalayas*. Taylor & Francis Group, 2020
DOI: 10.5117/9789463721721_CH04

Buddhist; I came because I very much like this way. I pray for all beings, for all suffering to stop. I pray for good health and good karma and I look after all the cultures in Nepal and Tibet and India and everywhere. That's why everyone comes to Muktinath. We are for everyone." "And what about the Shaligrams?" I asked. "Oh, yes. Shaligram is also for everyone. Shaligram is a part of Muktinath and Muktinath is for the world."[1]

Textual evidence of Hindu influences in Mustang is extremely limited, and many accounts of practices labeled as either Hindu or Buddhist have drawn on historical texts selectively while often ignoring evidence that might contradict their contentions. While there are no major Hindu temples directly south of the Baragaon (and probably never have been), it is nonetheless untenable to argue that Mustangis were ever unfamiliar with Hindu ideas, status distinctions, and ritual systems prevalent in other parts of Nepal (Fisher 2001: 182; Fürer-Haimendorf 1966: 140; see also Ehrhard 1999). As multiple scholars and ethnographers of Mustang have argued, it is important to recognize that, despite the dearth of historical documentation, Buddhism and Hinduism in the middle hills and high Himalayas of Nepal have always been mixed with indigenous practices.

Earlier scholars' argument that Hindu influences in the region of the Thak Khola as well as further north are relatively recent ignores two significant aspects of the upper Kali Gandaki Valley that have long mitigated against cultural isolation. The first is trade: while it is common to think of Himalayan populations as isolated and remote, it is especially inappropriate to apply this label to people resident in a major trans-Himalayan trade route that has been in constant operation for at least a millennium. The second aspect is pilgrimage: the significant Hindu pilgrimage site of Muktinath lies just north of Kagbeni. and the practice of Hindu pilgrimage to Muktinath (meaning either "Place of Salvation" or "Lord of Salvation", depending on what sources you consult) by way of the Kali Gandaki River valley is even noted to have predated the arrival of a number of modern ethnic groups to the area (Fisher 2001).

There is also a scarcity of documented and reliable sources that can directly address the antiquity of the Muktinath temple complex. As discussed previously, religious sources do not necessarily speak of the Muktinath temple specifically and generally only go so far as to identify the region or area of "Muktinath" as that which is synonymous with "Muktikshetra" (meaning "Field of Salvation"). For this reason, it is almost certain that the current temple site of Muktinath (and its attendant *murti*) was not

1 Conversation in a combination of English and Nepali. Transcribed from fieldnotes.

present in ancient times and likely didn't comprise a significant Shaligram pilgrimage destination until sometime in the last few centuries. The religious significance of Muktinath, then, is not necessarily in the physical temple site itself but in its role as a site of veneration for the pilgrimage practices that far precede it. This is because the pilgrimage to "Muktinath" plays a role in significant portions of Puranic texts as well as warranting mentions in the Purva Vritanta of Kakbhusundi in the Uttarakanda section of the Ramayaṇa and also in the Ramesvarakanda section of the same epic. By way of these texts, many Hindus thus claim that the site of Muktinath must have been revered as "Mukti Tirtha" ("great bridge") as early as the time of the composition of the Ramayaṇa (or around 2,500 to 3,000 years ago). But it must also be repeated that there are ongoing debates among scholars and theologians as to the dating of any of these Puranic or Epic mentions, including those in the Ramayaṇa. In other words, there is currently no way to know for certain when these ideas and descriptions were added to scriptural texts or if they were ever part of the original compositions.

Given the continued references to Muktinath/Muktikshetra and Shaligrams in various Hindu texts, we can safely say that the mountainous landscapes of the Kali Gandaki River (which is the source of – and meets up with – many other sacred rivers across South Asia) was likely highly revered as a sacred landscape possibly as far back as the Vedic period. It should be noted, however, that if we take the description of "Mukti Tirtha" in the Ramayaṇa to mean the present-day location of the Muktinath temple, we can argue that the site has been a site of pilgrimage for the last two to three millennia (to say nothing of attempting to date Shaligram pilgrimage itself), but if we rely on the perspective of the Himavat Khaṇḍa (Skanda Purana) and Varaha Purana, we would have to cede that the current temple site can only be reliably dated to the early Medieval Period of Nepali history (somewhere around the sixth to tenth centuries). To some degree, this ambiguity owes its troubles to the Gupta period of Indian history (approximately AD 320 to 550). Most of the Hindu Puranas as we know them today were composed and standardized during the Gupta period, and it is through these Puranas that knowledge of places such as Muktinath/Muktikshetra and Śālagrāma spread throughout India and became popularized among burgeoning *bhakti* reformer traditions. This is why discussions of Shaligram pilgrimage and practices largely revolve around Puranic texts and the problems of chronology they present.

The temple complex of Muktinath lies at an altitude of roughly 3,710 meters at the foot of the Thorong La mountain pass and close to the village of Ranipauwa (which is also sometimes simply referred to as Muktinath).

The entire complex contains a number of *mandir* and *gompas* including the central temple, or Vishnu-Chenrezig *mandir*, the Narsingh Gompa, the Sarwa (or Sangdo) Gompa (where the eighteenth-century *satguru* Swaminarayan is said to have performed his famous penances), the Shiva-Parvati Mandir, the Mebar Lhakang Gompa (also called the Salamebar Dolamebar Gompa or Jwala Mai Temple), the Yagyashala (sometimes called the fire sacrifice or *hom* temple or the Shaligram *mandir*), a series of *chorten* or monument stupas to the deceased, and, as of June 2016, the tallest standing statue of the Buddha in Nepal.[2]

Considered to be the 105th pilgrimage destination among the 108 Divya Desams[3] (and the only one in Nepal) as well as one of the 51 Shakti Pīthas,[4] Muktinath has long been a site of pilgrimage as well as the principal shrine for the veneration of Shaligrams. Also sacred to Buddhists, who refer to the temple complex as Chumig-Gyatsa (Tibetan: "Hundred Waters"), Muktinath is believed to be one of the main sacred locations for the 21 Taras (female deities) as well as the Dakini,[5] or Sky Dancer Goddesses, and one of the 24 Tantric places of meditation. The central temple shrine contains the *murti* of Sri Muktinath, the deity of Muktinath temple, who is Vishnu for Hindu pilgrims and the Buddha[6] of Compassion, Avalokiteshvara/Chenrezig, for Buddhist pilgrims. This shrine, housed in the Vishnu Mandir, is also considered by Hindu Vaishnavas to be one of the eight Svayam Vyakta Ksetras, or self-manifest fields of salvation.[7] The ongoing theme of self-manifestation is then carried through to the Shaligrams themselves, meaning that no human agency was involved in their physical appearances on Earth.

2 Nepal's Largest Stone Buddha at 12,600 Feet: http://www.huffingtonpost.com/kishor-panthi/nepals-largest-stone-budd_b_10928876.html (Accessed 1 August 2016). And US-Based Nepali Donates Rs 10 Mln To Install Buddha's Statue in Muktinath: http://www.himalayanglacier.com/blog/tag/buddhas-statue-in-muktinath (Accessed 13 October 2016).

3 A Divya Desam is one of the 108 Vishnu temples mentioned in the works of the Tamil Azhvars (saints). Divya means "premium" and Desam indicates "place" (comparable to *dham* or temple).

4 According to some schools of thought within the Shakti Hindu traditions, there are four Adi Shaki Peethas and 51 centers of Shakti worship throughout South Asia, many of which represent various body parts of the Devi Sati goddess. They can be found in India, Sri Lanka, Nepal, Bangladesh, Tibet, and Pakistan, though the precise list of pilgrimage sites tends to vary by specific tradition.

5 The Dakini (or daka) also appear in the medieval legends of India as demons under the tutelage of Kali, who feeds on human flesh. They are comparable to many other malevolent or vengeful female spirits and deities in later times.

6 While Avalokiteshvara or Padmapani is typically understood to be a bodhisattva who embodies the compassion of all Buddhas, the majority of Mustangis I conversed with simply referred to him as "buddha" or "the buddha of compassion."

7 The other seven being Srirangam, Srimushnam, Tirupati, Naimisharanya, Thotadri, Pushkar, and Badrinath.

While the temple is quite small, it houses a copper *murti* of Sri Muktinath (about a meter in height and around 96 cm around) and his two consorts, the goddesses Lakshmi/Bhumi and Saraswati, which measure about 86 cm high each. This altar also contains three principal Shaligrams that reside permanently at Vishnu-Chenrezig's feet (which represent the three principal deities and are recapitulated again in the Yagyashala), a smaller statue of Garuda (Vishnu's celestial mount), and an assortment of other smaller images and icons spread throughout. The *prakaram*, or outer courtyard sanctum, contains the 108[8] bull-faced waterspouts as well as the two front *kunda* (water pools), within which pilgrims to Muktinath must bathe in order to cleanse away all karmic sins. These sacred waterspouts also represent the sacred waters (Pushkarini waters) from all the other Divya Desams including the Divya Desams considered outside of the earthly realm. In Hindu texts, Muktinath is often praised by Hindu saints, particularly in the Vishnu Purana and in the Gandaki Mahathmya (*Gandaki Mahatmya Parishistha*, part of the Himavat Khaṇḍa section of the Skanda Purana).[9] As a Shakti Peetha, Muktinath is also one of the abodes of the Devi (Shaki), formed by the falling body parts of the corpse of Sati while Shiva carried it about, wandering through the Himalayas. Each Shakti Peetha therefore contains a goddess shrine as well as a Bhairava temple (the fiercely monstrous manifestation of Shiva). In the case of Muktinath, the Shakti shrine is referred to as "Gandaki Chandi" and Bhairava as "Chakrapani" and is said to be the place where Sati's forehead fell. Conversely, in Tibetan traditions, Muktinath is said to be the place where Guru Rinpoche (Padmasambhava) meditated on his way to Tibet.

As Sister Pemba motioned me towards the main temple, she offered a small handful of white candies to give to the deities of Muktinath as *prasadam*. "We are so lucky to be here." Her gestures indicated that she was referring to herself and to me. "Shaligram brings you to Muktinath, my life brings me also and we meet here. Now we are friends and we can always be happy. This is why it is so special here. Everyone comes together. Maybe different reason and maybe different life but they come here and then they leave and take it all with them. We stay connected."

8 The number 108 has a variety of significant meanings in Hinduism and in Buddhism. In Hindu astrology, for example, there are 12 Rashi (zodiacs) and 9 Graha (planets) for a total of 108 combinations. There are also 27 Nakshatras (lunar abodes) which are divided into 4 Padas (quarters) also for a total of 108 combinations of Padas.

9 In some historical narratives, the Gandaki Mahatmya is quoted as giving the etymology of 'Nepal' by way of a king called 'Nepa', after whom the country was supposedly named. This is, however, one of a long list of possible etymologies of 'Nepal'.

The first mention of the site of Muktinath in Tibetan texts dates back to the Tibetan king Lhachen Udpal who ruled Ladakh from AD 1080 to 1110 and who is also said to have ruled up to Chhu-la-me-war in the province of Lo. In this account, both Tibetans and local peoples are described as knowing of a place called "Muktijwala", which is also called *Chhu-la-me-war* (fire on water), *Do-la-me-war* (fire on rock), and *Sa-la-me-war* (fire on earth). Conversely, "Muktikshetra" is occasionally mentioned in other Tibetan sources as well. It appears by name in the famous "Pema Kathang",[10] a text related to the activities of Guru Rinpoche (Padmasambhava) while in Mustang, and it also appears in the Tibetan guidebook of Muktikshetra (here noted with the Tibetan name *Chhu-mig-gya-chha;* called *Chhu-mig-gya-chhagye, Ku-chhab-ter-nga, Mu-le-gang gur-sung-phug-sog-kyi-kar-chhag-sal-bo-me-long ngo-chhar-chen-nam.*) Lastly, as Ramesh Dhungel points out, a wide variety of other local official paperwork, treaties, and records written in Tibetan indicates the presence of a site called Muktinath as *Chhu-migya-chha* and Damodarkunda as *Chhu-chhen*. Later on, after control of Mustang was acquired by Jumla, other local accounts begin describing Muktinath as being primarily administered in reflection of the primary deities of Jumla. It is possible, however, that the rulers of Jumla regarded Muktinath as part of their own religious systems even before acquiring control of the region since they were already incorporating Badrinath in the far east into their deity hierarchies. Much of the evidence for this comes from the seal of the kings of Jumla wherein we find the phrase "*Sri Badrinatho Jayati Sri Muktinatho Jayati*".

It is also possible that the tradition of incorporating these sacred sites into their own ritual systems passed down to the Jumla kings from their Malla predecessors. The kings of Jumla were geopolitically related to the Malla kings and succeeded them chronologically. The first mention of Muktikshetra as "Muktinath" may even be evidenced in the seals of Jumla, which are known as *Syaha Mohar*. When all Tibetan sources around them appeared to address the site as *Chhu-mig-gya-cha* and *Chhu-la-me-war*, and other sources called the land "Muktikshetra", it is of particular interest then that the kings of Jumla began calling the site "Muktinath". In other words, the blending of religious traditions between India, Nepal, and Tibet was already heavily underway. Unfortunately, there is no further documentation that establishes whether or not the Jumlas engaged in any temple or *murti* building at Muktinath or

10 Pema Kathang (Wyl.pad+ma bka' thang), the Chronicle of Padma – a biography of Guru Rinpoche, also known as the Sheldrakma(shel brag ma), revealed by Orgyen Lingpa from the Crystal Cave (Wyl.shel brag). It has 108 chapters.

whether or not a temple already existed there. The present-day four-armed copper statue (*murti*) of Muktinath is dated to around the late sixteenth or early seventeenth century but given that this overlaps with the era of Jumla rule, one might speculate that the word "Muktinath", which seems to appear only from the time of the *Syaha Mohar* of the Jumla kings, was brought into contemporary usage after Jumla established the current *murti* of Muktinath and constructed a separate temple away from the Jwala Mai (Jwalaji) – the "Mother Flame" – a short distance away.

According to many Hindu scholars and theologians, the Puranic mentions of Muktinath, Muktikshetra, Śālagrāma, the Krishna or Kali Gandaki River, and Shaligrams demonstrate that the area of Muktinath was important to Hinduism before periods of recorded Nepali or Tibetan history in Mustang. In one overview account, for example, Puranic textual sources were given as evidence that the creation of and pilgrimage focus on Muktinath closely aligned with the standardization of Hinduism as a global religion. This was largely due, in this case, to the widely unifying aspects of Shaligram pilgrimage and ritual practices across variable Hindu traditions from South India all the way north to West Bengal, Nepal, and Tibet (mentioned in Dahal 1988). Therefore, despite the fact that Shaligrams are commonly thought of as direct manifestations of Vishnu, the disparate texts and traditions were said to actually demonstrate that Muktinath was a blending of the two divine powers of Hari and Hara (Vishnu and Shiva). Additionally, as mentioned in the Skanda Purana and Varaha Purana, the unity of water and flames found in Muktinath could only then be read because the unity of the Shaiva and Vaishnava traditions. For Shaligram pilgrims today, this is largely true as both Vaishnava and Shaiva Hindu lineages and sects undertake Shaligram pilgrimage and many local traditions of reading and interpreting Shaligrams take their name-type lists from a mish-mash of deities important to any number of different South Asian cosmologies.

The oldest known religion in Mustang is Bon (Bonpo), an early version of which was likely followed prior to the arrival of Buddhism. In Bon, natural phenomena and nature are worshipped, and today their most prevalent expressions appear in the head-gods of the four Thakali clans who are often depicted as birds or other animal totems.[11] There are also two types of Bon: Bon dKar (white Bon) and Bon gNak (black Bon). According to some accounts, black Bon was indigenous to Mustang before white Bon, which became mixed with eleventh and twelfth-century Tibetan Buddhism of

11 The four Thakali endogamous clan distinctions are Gauchan, Tulachan, Sherchan and Bhattachan, each of which is considered generally equal in social as well as ritual status.

the Shakya-pa sect arriving from the north and continued mixing with other indigenous practices (such as the *dhami-jhankri* shamanic traditions) throughout the Kali Gandaki River valley up until the nineteenth century (Chetri et al. 2004).

The influences of Bon are commonly cited as the reason for the site of Muktinath to also be revered for its singular combination of the five sacred elements, of which – also according to both Hindu and Buddhist traditions – all material things are made.[12] In these philosophies, the five elements are: fire (from natural gas vents coming up through the rocks), water (flowing out of the mountainside into the 108 spouts), sky, earth (or stone, referring to Shaligrams), and air (the high winds typical to the valley). Some traditions also note the presence of sacred trees at Muktinath, growing at an altitude generally considered outside of their normal range (and another possible reference to the *sal/shal* trees of Śālagrāma). The fire aspect, referred to as the Jwala Mai (Mother Flame), is housed in the Jwala Devi *gompa* and is tended to daily by the resident order of Buddhist nuns who live at Muktinath. In fact, the traditional caretakers of Muktinath-Chumig Gyatsa are an order of Buddhist nuns of the Nyingma (or "red hat") sect, the most prevalent form of Buddhism in Mustang today, currently headed by the abbot of Muktinath and head of the *Gye Lhaki Dung*, Lama Wangyal (Hira Bahadur Thakuri). The *Gye Lhaki Dung* is also popularly known as the Lama Domar family, an unbroken lineage of Tibetan Buddhist (Nyingma) *lamas* originally from the Muktinath Valley who have claimed Muktinath as their religious seat for several centuries (MFI 2016). Lama Wangyal, the current abbot of Muktinath, was born in AD 1956 (Tibetan Buddhist Year of the Monkey) during the tenure of his grandfather, Lama Jampal Rabgyè Rinpoche, the author of the Buddhist pilgrimage guide to Muktinath-Chumig Gyatsa, *The Clear Mirror*.

There are, in fact, a number of pilgrimage guides to and discussions of Muktinath that have been written and published over the years. The most prevalent Hindu one available near Muktinath today is Madhu Sudhan Ramanujadas' *Muktichhetra Mahatmyamam* (2003), which includes large sections of Sanskrit, Hindi, and Nepali text drawn from the various Puranas that outline the multitude of Shaligram origin stories discussed previously.

12 In Bon the five elemental processes of earth, water, fire, air, and space are the essential materials of all existent phenomena or aggregates. The elemental processes form the basis of the calendar, astrology, medicine, and psychology and are the foundation of the spiritual traditions of shamanism, tantra, and Dzogchen.
See: Tenzin Wangyal Rinpoche (2002). *Healing with Form, Energy, and Light*. Ithaca, New York: Snow Lion Publications. p. 1.

Another is Pandit Bhavanishankar Shastri's *Shalgram-Rahasyam* ("The Mystery of Shaligram") published with the support of Subba Mohan Man Sherchan of Tukche village in 1947. This work also makes a case for the importance of Muktinath, the Kali Gandaki, and Shaligrams based on Hindu religious texts. The obscure book *Mustang Digdarshan*[13] is another text that is occasionally referenced in terms of historical material, but given that it contains no citations or references, it is usually ignored in favor of more comprehensive texts, such as Rao's 1996 *Śālagrāma – Kosha*. Apart from religiously oriented pilgrimage materials, some devotees also cite Ramesh Dhungel's "Damodarakunda: Ek Parichaya" (Damodar Kunda: An Introduction)[14] as especially helpful in researching the history of Muktinath and Shaligrams as well as his "Muktinath: Kehi Aitihasik Tathya" (Muktinath: Some Historical Facts)[15] and "Dharmic Sahishnutako Prasanga Muktinath" (Muktinath: A Context of Religious Tolerance).[16]

In the first of these short research essays, Dhungel discusses, in quite spectacular detail, the nature of the Damodar Kund as well as the main pilgrimage route to Muktinath, its current geographical location, its present state of administrative affairs, its religious significance and a few historical facts that are related to the reader in the form of a travel log (a format that many pilgrims find particularly useful). In the other two papers, Dhungel references a series of Tibetan and Nepali sources as a basis for formulating a general historical perspective on Muktinath. One of the main points of interest is that many of these pilgrimage literatures are reasonably circumspect about where the boundaries of "Muktinath/Muktikshetra" actually lie, and there continues to be some debate as to the extent of Muktinath's sacred boundaries (i.e., the extent of its *dham*). Dahal, in his discussion of these literatures, points out for example that the Himavatkhanda mentions that Lord Brahma meditated at the center of Muktikshetra and that this particular spot where Brahma meditated is the place where the present-day Muktinath temple and Jwalaji (Jwala Mai) currently reside (1988: 6). In this case, then, the sacred boundaries of Muktinath may be said to lie somewhere between the central temple and the Jwala Mai *gompa*.

13 *Mustang Digdarshan* (Mustang: A Perspective) by Narayan Prasad Chhetri. Published by Kathmandu, 1987.

14 A research paper in the possession of MFI (Muktinath Foundation International) and at Tribhuvan University.

15 1988: "Muktinath (a Hindu Temple in North-western Nepal): Some Historical Facts" *Ancient Nepal* No. 102.

16 1988: "Context of Religious Tolerance in Mukti-Chhetra" (text in Nepali), *Saiva Bhumi* Vol. 3, No.3 (Journal of Pashupati Socio-religious Service Association, Kathmandu, Nepal).

As for the boundaries of Muktikshetra (the Field of Salvation) as a whole, there is additional contention. In some cases, both pilgrims and theologians delineate the areas directly around the Kali Gandaki River Valley where Shaligrams are found as the boundaries of Muktikshetra. In other cases, devotees are apt to cite the Himavatkhanda Purana again in regards to the Jwala Mai (where Lord Brahma invokes Lord Shanker in the form of fire onto Lord Vishnu who takes the form of water in this account) as the center point of Muktikshetra, which then extends outwards from there. However, as to how far north or south of this center point one should consider the land to be Muktikshetra is not universally agreed upon. For example, the Varaha Purana explains that Muktikshetra extends around 1 *yojanas* (approx. 8.8 km) and Shaligramkshetra extends around 12 *yojanas* (approx. 105 km) from the center point.

The Varaha Purana goes on to say: "two ascetics going by the names of Pulatsya and Pulaha sat down at a spot in Muktikshetra and meditated". The present-day "Pulhasrama" of Myagdi district is then subsequently described by some Nepali Hindus as the place where these two ascetics sat down to meditate. Because it is stated in the Varaha Purana that this place of meditation must also come under Muktikshetra, Muktikshetra must then extend from Myagdi District to the south of Mustang to the Damodar Kund in the north. In Dahal's claims, the distance from Damodar Kund to Pulhasrama roughly corresponds to the distance given in the Varaha Purana (1988: 7) but few scholars elsewhere appear to be entirely in agreement on the exact measurements of Puranic distances as translated into modern units. In any case, based on scriptural references, most Shaligram devotees tend to regard the lands between the Myagdi confluence of the Kali Gandaki river to the south and the Damodar Kund mountain to the north as Muktikshetra and not just the areas immediately surrounding the Kali Gandaki where Shaligrams are collected.[17]

This does not mean, however, that there is not some contention as to whether or not Muktinath is the actual location of Śālagrāma (see Chapter 1). Regardless of its continued use in the veneration of Shaligrams, many Hindu schools of thought doubt the actual connections between Muktinath temple and the *dham* of Vishnu connected to the Kali Gandaki River. In some cases, pilgrims cite the source of the Kali Gandaki as the real place (*nath*) of

17 Yogvashishta's Vairagyaprakarana states that Lord Ram himself once came on pilgrimage to a place called Muktikshetra (Shalgram Kshetra). According to this account, Lord Ram was 15 years old when he began his pilgrimage, and he arrived at a place known as Shalgram Kshetra (Muktikshetra) which was surrounded by the Gandaki river.

salvation (*mukti*), while others note the historical and textual ambiguities around calling any number of places in the high Himalayas of Mustang "Muktinath" or "Muktikshetra". Even the pilgrimage travel book of R.S. Gherwal, published in 1927, locates Muktinath at 17,000 feet (as opposed to 12,300 feet) at the base of the Nora Pass rather than at the current site of the temple at the Thorong La Pass (1927: 82).[18] In practice, however, the primary route of Shaligram pilgrimage in Mustang begins at the Kali Gandaki (usually somewhere near Kagbeni) and ends at Muktinath, where Shaligrams recently taken from the river are bathed in the 108 waterspouts and then left at the feet of Vishnu-Chenrezig for special blessings and rituals before being removed from Mustang and taken home.

The preponderance of pilgrimage guides, written over various time periods, has also led to significant disagreement among Shaligram practitioners as to how the Puranic texts should be read in terms of modern-day pilgrimage. Some devotees choose to rely on texts deemed authoritative to their own religious traditions (which vary from one Hindu sect to another), while others view the ambiguity of the texts as evidence of the ever-shifting nature of the Śālagrāma *dham*. In other words, as the river changes course and the Shaligrams move down the valley, so too does the field of Muktikshetra, and it was this movement that each author was attending to. Even more interestingly, many practitioners have also come to link this movement with narratives of environmental conservation and climate change, noting that should the Kali Gandaki River ever dry up or deviate significantly to the east or west, the sacred *dham* of Śālagrāma would follow.

Finally, the Jwala Mai *gompa* is situated slightly south of and a short distance below the central temple and, aside from the *mandir* of Vishnu-Chenrezig, is one of the most popular pilgrimage destination points within the Muktinath complex. Inside the *gompa* (and accompanied by the strong smell of natural gas) is a small, clay and mesh box that rests over the three continuously burning flames and a natural spring that flows just beneath. These three holy flames, "The Flame of Soil", "The Flame of Rock", and the "Flame of Water", are continuously fed by the deep natural gas vents coming up through the rocks below. However, today only two flames are still burning, the third having gone out unexpectedly some years ago. For Hindus, these flames are either the natural representations of Brahma (the creator of the universe) who has set fire to the water (Vishnu) – Jwalaji – or

18 See also Jackson's work on the locations that may or may not be associated with Muktinath. For example, he identifies present-day Mu-Khun as the actual Muktinath. Ramble and Vinding (1987: 21) concur with this identification.

representations of the Shakti in her capacity as the creative force of the universe – Jwala Mai – the Mother-Goddess of Fire.[19] For Buddhists, the sacred flames represent Guru Rinpoche who is thought to have meditated and achieved enlightenment at that specific place. However, the exact pilgrimage routes through Muktinath have changed significantly over the years.

From the time following the establishment of the Jwala Mai Gompa and the central Vishnu-Chenrezig (or Muktinarayan) temple by the kings of Jumla, many other monuments and buildings have been constructed throughout the site: *gompas*, *stupas* and *chorten*, waterspouts, supporting temples, pilgrimage shelters and housing, and so on. The main pilgrimage *gompa* housing the sacred fire, however, has started to slowly fall out of favor for many Hindu pilgrims. The primary sites of veneration today include the 108 waterspouts (Muktidhara) and *darshan* at the central temple. The Jwala Mai/Jwalaji is now largely tended only by the Muktinath's Buddhist nuns (*jyomo* or *tsun-ma*). The main Hindu priests who attend Muktinath temple are also now principally assisted by the nuns, who perform many of the traditional rites of worship. For those pilgrims who do continue their pilgrimage paths throughout the temple site, after completing prayers and *pujas* before the Jwala Mai, many pilgrims then move on to the Mharme Lha Khang Gompa, which is situated slightly north of the central temple. Mharme Lha Khang, which translates as "A Thousand Holy Lamps", is the main *gompa* dedicated to Guru Rinpoche. Inside, his huge clay image is flanked by two Bon deities, the red deity Trakpo[20] to the right and blue deity Singe Doma to the left. Because Singe is also a lion-headed deity, Hindu pilgrims to Muktinath often consider him to be Narasimha, the lion-headed avatar of Vishnu.

The central deity, Sri Vishnu-Chenrezig or occasionally Sri Murti Mahatmyam or just Sri Muktinath, is currently housed in the pagoda-style Buddhist-Hindu temple of Vishnu and Chenrezig (also often referred to by pilgrims as the Vishnu Mandir), which was built sometime between AD 1814 and 1815 by the Nepali queen Subarna Prabha, the second wife of Shah Rana Bahadur (1775-1806), as an offering to the veneration of Shaligrams

19 It is not clear precisely when Hindu pilgrims to Muktinath started referring to the Jwalaji as Jwala Mai. Some historical accounts, dating back only around 70 or 80 years ago, denote the location as one that was regarded as the representation of Vishnu and Shiva (water and flame) and which was still known as Jwalaji. However, there seems to have been a relationship between the current site and a type of localized mother-goddess worship (Mai = Mother, hence Jwala Mai).

20 *drag* or *drag po* (trakpo), 'powerful, terrible'.

(and then renovated around 1929). By a number of local accounts, the original temple at Muktinath was made of clay and some speculate that the subsequent temple built over the site by the Jumla kings reflected an earlier Nepali-style of temple-building that was still intact at the time of the current temple's construction. These accounts may also explain one of the more pervasive stories about the statue of Vishnu-Chenrezig as well who, according to several Shaligram pilgrims and a few ritual specialists, was once a large Shaligram stone (the first found in the region, as a few will tell it) that was then hidden inside of the current *murti* statue when the temple was built. As an interesting counterpoint, there is, in fact, a very large Narayan Shaligram venerated at Muktinath (along with an additionally gigantic Shaligram, roughly the size of a mid-size car tire, brought out only for special occasions), but it does not reside within the main temple *murti*. Rather, it currently sits on the main altar of the Yagyashala/Shaligram Mandir near the main gate, flanked by two additional Shaligram *murti* representing Lakshmi and Saraswati who also stand beside the main icon.

At issue in both Mustang and Muktinath is often then what it means to be Hindu and what it means to Buddhist in these contexts (believing in and doing "Hindu" or "Buddhist" things) and where these ideas of belonging become exclusive or where they overlap. The history of Mustang and of Muktinath temple demonstrates that religious orientation is almost never a matter of historical continuity but more a matter of when, why, and to whom it is an issue, all the while bearing in mind any number of historical, political, and social influences and their effects on a variety of distinct groups (for a longer assessment of the challenges and contentions of Hindu and Buddhist ideas of belonging at Muktinath, see Dana 2011). Recalling Tucci's assessment of the interactions between the forces of Bon, Buddhism, and Hinduism among the Thakali peoples of Mustang, it is therefore advisable to avoid treating such religious distinctions as reified entities at all:

> A deep study of Thakali religious beliefs would demonstrate imprecision of whatever denominational labels might be given to them. We would see that the former Lamaist still survives within every Hindu, but that the Lamaist in turn reserves a not unworthy place in the depths of his soul for Hinduism. In every soul we would see the two religions, not in conflict but coexisting by mutual consent and happily sustaining each other. And surely if we looked deeper still, we would find the secret fire of the original primitive beliefs is still burning. ([1953] 1982: 50).

Viewing religious practices at Muktinath as processes of alignment within a pluralistic ritual landscape, we can see how neither professed religious beliefs nor authoritative religious and historical texts can be used to predict actual practices accurately. Any specific individual (nun, pilgrim, or resident) may perform or participate in rituals from any range of traditions, be they Bon, Buddhist, Hindu, or related to some other local deity or legend without much concern to specific loci of political or spiritual power. Both the varied ritual practices of individuals and group participation in other rituals (such as *darshan* or festivals) continue to fuel contestations over internal and external boundaries that separate different Mustangi populations from one another, Mustangis from other Nepalis, Nepalis from foreigners, and Hindus from Buddhists (see Dana 2011). Scholarship, however, is bound up in endeavors to discover, delineate, or create boundaries: of groups, of cultures, of fields of knowledge, of contexts of action, and so on. This is because boundaries and the categories that they index help make sense of complex and overlapping cultural variations by drawing specific attention to where ideas and objects are incommensurate or where discontinuities threaten the perceived order of the world. Such boundaries also help to heighten our sense of human diversity and allow us to make sense of contradictions, or, as is all too common in the case of Mustang, to view a place or a culture as something pure and in need of preservation. These essentialisms, then, tend to drive the narratives of conservation and preservation that Shaligram practitioners (and many Mustangis in general) find so difficult to engage with, along with the scholarly literature that purports to define and claim the history of Mustang, the history of Muktinath, and the nature of the Shaligram stones themselves.

Introducing the Ritual Landscape

I sat quietly along the side of one of Muktinath's two *kunda*, patiently waiting as Lalita Thapa (who insisted I call her "Lala"), a Nepali Hindu pilgrim from Chitwan who had recently arrived with her husband and two children in Muktinath, arranged five Shaligrams on a small, silver tray along with a number of other sacred objects she continued to produce from within the folds of her sari. "This is *panchayatana puja*," she explained. "We Nepalis typically worship the five major deities of Vishnu, Shiva, Devi, Surya, and Ganesha. My father used to conduct his *pujas* with only one Shaligram for Vishnu. He used a lingam for Shiva that was a gift from his grandfather and then other stones for the other deities, but I am helping my children to learn the rituals of Shaligram. My husband, you see, is a third son. His eldest brother inherited the family Shaligrams some years ago and so now we come to Muktinath so that we can begin the new tradition for our children. These will be their Shaligrams, to begin their households."[21] It was late in the day already, and only a handful of other pilgrims remained in the outside courtyard of the Vishnu-Chenrezig *mandir*, most of them having already completed their required rituals and headed off down the steep mountain walkways to dinner at the local guesthouses. It was a cold day for August and the monsoon rains continued to threaten additional downpours with each passing wave of cloud and fog drifting over the mountain peaks. With little sun to warm the aquifers, the waters of Muktinath were especially freezing that afternoon, and few of the day's pilgrims had been brave enough to strip down and chance a run through the 108 spouts just beyond where we sat.

I watched carefully as she arranged her Shaligrams, with Vishnu in the center and the other four deities at the four corners of the plate. Three she had brought with her from their home in southern Nepal and the final two she had just purchased from a Shaligram merchant near the gates of Muktinath. "I bought these two," she motioned to the newest Shaligrams representing Durga and Ganesha, respectively, "because I did not think we would have enough time to find them in Kali Gandaki. The ones we did find I gave to my husband, who is over there now waiting for *darshan* so that we can have them blessed at Vishnu's feet and take *prasadam* (holy food). Those, I think, we will give to our home temple. One is Ram Shaligram I think, and the other two I'm not sure. The *pujari* (temple priest) here will be able to tell us later. But I only want these five." I nodded, "How will you worship them here then?" "I conducted *puja* down at the Kali Gandaki yesterday after we found the Shaligrams," Lala replied. "Now I will conduct *puja* again here at the water pools for the five that will remain in our home and then I will walk through the waterspouts. This is very important because these five gods are also the five principal elements. Vishnu is sky, Shiva is earth, the Devi is fire, Surya is air, and Ganesha is water. These elements make up all the world and every human body. My father also taught us that one of these elements is especially dominant in each person's personality. I am far more attuned to sky and to Vishnu, I think. That is why I place him in the center."

Lala then pulled a small plastic bag containing tulsi leaves from her handbag and carefully laid a single leaf on each Shaligram as she chanted the *mantras* she had been reciting since girlhood. Then, she washed each Shaligram in the waters of the Muktinath *kunda* before replacing them in

21 Conversation in English. Transcribed from fieldnotes.

the pack she had brought along specially to carry the stones. "Now," she said as she got to her feet, "I'm going to go undress so that I can bathe in the spouts. Will you watch the Shaligrams for me? I don't want them wandering away while I am changing." "Of course," I said, "but how will I know if they are trying to wander away?" Lala laughed as she began to unwind her sari. "Oh, that's easy. You'll see them floating away in the *kunda* waters, trying to go back to Kali Gandaki. That's why I brought them for *puja* up here to Muktinath. First, they are bathed in Kali Gandaki, then bathed again at Muktinath, and then finally when we return home to Chitwan. That way, they can carry with them the *dham Śālagrāma* and are ready to leave it behind. The river is their mother, you see, and all children want to go back to their mother. But it is time for them to leave and live their lives elsewhere. This will help them."

I watched as Lala scurried off into the small, walled, enclave that served as the women's changing room near the nun's quarters. She emerged shortly afterwards clad only in a yellow wrap, her hair loose around her shoulders as she prayed before the first waterspout. And then, with a deep breath and sudden burst of courage, she bowed her head and took off at a brisk pace around the outer courtyard of the *mandir*, careful to ensure that she passed beneath each one of the 108 spouts in turn. As she emerged from the last waterspout at the far end of the courtyard, she stopped to shout out a long string of *mantra*-like phrases that I took to mean that she was both joyous in having made it through the spouts as well as a commentary on their late summer temperatures. From there, she preceded to the main door of the Vishnu Mandir, where she offered prayers to the *murti* within, all the while shivering in the crisp monsoon breezes. Her husband met her there, handing over the recently blessed Shaligrams, resplendent with their newly applied spots of *kumkum* and sandalwood paste that had been done by the Hindu *pujari* within the temple, and a handful of sweets given to him by the Buddhist nun minding the main *darshan*. Later, as I stood near the gate to the Muktinath temple complex, Lala stopped me on the way down towards the village of Ranipauwa. "I am so excited!" She tugged at my sleeve with barely restrained glee. "I feel as though I am welcoming a new child into the family. Which I am I suppose, you know. Shaligrams are the children of the Devi (goddess) and I am now adopting them. Come with me! Let us share our first meal [with them]. After such a long journey, I'm sure we are all hungry."

In the early 1980s, the term "ritual landscape" gained significant ground in the study of British archaeology as a way to describe the means by which people modified the countryside around them as both a representation of and interaction with sacred or spiritual worlds. This concept departed, however, from conventional studies of monuments and sites because it became concerned with more relational aspects of artifacts and landscapes, such as the classification of icons, sacred writings, ceremonial spaces, funerary monuments, and ritual implements. As archaeologists and later anthropologists noted, religion was everywhere: in deity figurines and temple architecture, in standing stones inscribed with religious images, in burial sites, in the careful arrangement of houses, sewn into clothing designs, and in the recordings of natural omens. This term could then be readily applied to Mustang, of course, with its hundreds of *mani* stone walls, ossuary *stupas*, temples and wayside shrines, crossroad offerings, and sacred mountains and rivers, but here I also add the current Shaligram ritual practices as they are carried out along the Kali Gandaki River, on the road to Ranipauwa, and at the temple of Muktinath itself. As a combination of movement and the specific physical undulations of the land itself, this modern ritual landscape takes into account pilgrimage not just as a kind of exploration of the world but as an introspective, personal, and semiotic journey as well. It also must take into account the practice of *puja* – any number of different types of service rituals (*seva*) performed at the juncture of physicality and spirituality, between embodiment and emotion.

The ritual world of Shaligram practitioners contains a relatively constant core of shared beliefs and practices set within a wider field of religious pluralism. This core of rites – which includes various types of *puja* rituals, festival celebrations, and pilgrimages – uses movement in space and time to reinforce sets of social relations and identities, such as *pujas* to obscure the distinctions between family members and gods or festivals to celebrate a deity's life events (marriage, death, birth, etc.). The practice of these rites then derives from and fuels many of the ongoing tensions between national affiliations or political leanings and the sacred landscapes wherein unfettered mobility remains the principal method by which ultimate religious realization is achieved.

While it may appear, at first, that the participation of any given Shaligram practitioner (as well as his right to call himself that) is contingent on the possession of one or more Shaligrams themselves, it became clearer to me over time that the object of the Shaligram itself was taken more as a facilitator of, or impetus for, sacred movement or as a literal guide and companion on life journeys (of both the physical and metaphorical kind) rather than as the end goal to mark the completion of specific pilgrimages or ritual practices (as say, a souvenir might index the travels of the person who keeps it). Though a comprehensively detailed compendium of Shaligram rituals, pilgrimage practices, and performances is beyond the scope of the current work, it is sufficient to say here that Shaligram veneration provides fora within which devotees might demonstrate an identification not with particular religious traditions or lineages (as Shaligrams themselves have no such affiliations) but with broader narratives of belonging that include ties to both ancestors and descendants, relationships to landscapes contingent on the movement between times and places, and continuity with a historical and a mythic past.

"My family's Shaligram has shown us many miracles." Abul Shikdar, a Hindu who had recently arrived from his native Bangladesh, walked slowly through the streets of Kathmandu towards Pashupatinath temple. "I believe in him. I have often wondered if getting another Shaligram will be doing injustice to him, because he has been so integral to my family for such a long time. He is from our ancestral temple in Bangladesh where my grandmother's father had been a *zamindari* (landlord). They had to leave during Partition and when we returned back to Bangladesh many years ago our priest gave it to me. I consider him as my friend first and then a god because he has helped us a lot. As soon as my mother prayed to him, she found her lost diamond earrings, my father's business improved when we prayed again, and my brother passed his school exams with top marks. Now we give him small gifts such as gold crowns, gold necklaces, and a lot of silver items. Every morning I put sandalwood and crimson (*kumkum*) on his forehead here." He held up an image of the large, grey Shaligram on his mobile phone. With his index finger, he then indicated a small depression near the top of the stone. "And I talk to him and bathe him. All the time. So now when I get to Pashupatinath, I intend to pray to Shiva and ask him about getting another Shaligram. I do not know yet if I will go to Kali Gandaki. I want to. But it is a great responsibility. Maybe I will just go to Muktinath and pray with my Shaligram."[22]

"This is my second time to Muktinath." Amit Kumar Kapoor smiled over his tea. We met in a quiet café near Durbar Marg in Kathmandu just as he was preparing to leave for Mustang on a late bus heading towards Pokhara. "My family is Shaiva mostly and a few have converted to Christianity recently, but it doesn't matter. It is not important to be Vaishnava for Shaligrams. What is important is how you treat Shaligram. I have two at home now which I was given by a *pujari* at a temple in Chennai some years ago. Now I am going to try and find a Ganesha, or maybe a Durga or Hanuman. Are

22 Conversation in Hindi. Transcribed from a combination of recorded dialogue and fieldnotes taken during the conversation.

there Hanuman Shaligrams? I don't know. These are the gods of my family and my home village, so I want to bring them home as Shaligram. Then I will do the full *Abishekam puja*. Do you know it? We will gather all our family from all over and do the Ganesh *puja*. This is why I must find Ganesha this time, so he can attend our Ganesha *puja* and come to live with us. Then we will put all the Shaligrams together on the *puja* plate with Shiva lingas and all the other gods in the household. Then we will do *abishek* and bathe them with milk, yogurt, ghee, honey, and plain water. Then Chandan[23] water, then water with flowers, then Rudraksha[24] water, water with gold ornaments in it, and then water with ashes. Then my father will recite the *mantras*, give *aarti* (lamps), and then we will all drink from the Shaligram waters for good health, good family, and good community."[25]

Pilgrimage and *puja* constitute the core of the complex, heterogeneous ritual world of Shaligrams. Understanding the complexities of this ritual landscape is facilitated by bearing in mind several important points. First, while *puja* is often translated as "worship", most Shaligram devotees view ritual *pujas* that involve offerings of food, water, bathing, and *mantras* more as a kind of "service" rendered to the deity that is analogous to the obligations one might have to care for fully functioning and participatory human members of the family or of the community at large. Second, rituals wherein the Shaligram is not necessarily offered service directly but is instead used to offer services to divine others (ancestors, gods, spirits, etc.) often contain complex layers where the Shaligram traverses multiple interlaced fields of mythology, family histories, community events, and personal needs and desires, linking them together through various movements and associations in time and space. Third, different practitioners disagree on the precise details of "proper" Shaligram veneration, and many of these differences are revelatory as to the specific identities, intents, and agendas that exist within and between devotees. Fourth, rituals of Shaligram veneration are derived from a number of culturally plural communities. For example, many devotees are widely versed in concepts and terminology of more than one major religious tradition as well as differences with the various sects and divisions within their own tradition. It is also not uncommon for Shaligram practitioners to use the vocabularies of Hinduism, Buddhism, and other

23 Another name for sandalwood powder.

24 Seeds of the sacred rudraksha ("Shiva's teardrops") tree. These seeds are often used as prayer beads in Hinduism. They are produced throughout South Asia by several species of large, evergreen, broad-leafed trees of the genus *Elaeocarpus*, with *Elaeocarpus ganitrus* being the principal species used in making jewelry (*mala*).

25 Conversation in Hindi and English. Transcribed from recorded dialogue and fieldnotes.

traditions to explain their ritual practices. Fifth, Shaligram rituals may occasionally be used as complements to other ritual complexes – deity *darshan*, festivals in honor of specific deities, ancestor *pujas*, marriage and funeral rites – or Shaligram rituals may be honored and conducted in their own right for specific purposes as determined by the practitioners in the moment. Sixth, within the parameters of these rituals there is a fair degree of latitude for variation. Seventh, pilgrimage is far more politicized than other rituals and, as a result, is a far more contentious subject than the specifics of *puja*.

In its most fundamental form, *puja* is a prayer ritual that is performed in order to host, worship, and to interact with one or more deities. *Pujas* are also performed to spiritually celebrate specific events, such as major festivals, weddings, or funerals.[26] In other cases, *pujas* are conducted to honor the presence of special guests, elderly relatives, and religious teachers, or their memories after they pass away. The word *puja* (Devanagari: पूजा) is originally derived from Sanskrit and can be alternately translated as reverence, homage, adoration, and worship.[27] As one of the most common ritual events throughout South Asia, *pujas* are an integral part of the ritual complexes practiced by Hindus, Buddhists, Jains and Sikhs. An essential part of *puja* for the Hindu devotee is in making a spiritual connection with the divine and in inviting the attention of the divine to matters of the physical world. As mentioned previously, these interactions are most often facilitated through an object: an element of nature, a sculpture or icon, a vessel, a painting, or some other image of the deity in question (*darshan*).

Puja can also be done on a variety of occasions and in a number of different settings. Any given Hindu, Buddhist, Jain, or Sikh practitioner may include daily *pujas* done in their homes or may only participate in *puja* at the occasional temple ceremony, annual festival, or even just a few lifetime events. *Pujas* are, in fact, quite often associated with dramatic shifts in a person's life course, such as the birth of a baby, a wedding, or the beginning of a new business or school venture.[28] The settings of many *pujas* also often reinforce their correlations with certain stages of life. *Pujas* done in the home are typically geared towards domestic concerns (healthy children, plentiful food, a strong household, securing a good marriage, etc.) and *pujas* conducted in the temple towards community concerns (marriages and funerals, births,

26 See Paul Courtright in *Gods of Flesh/Gods of Stone*. Joanne Punzo Waghorne, Norman Cutler, and Vasudha Narayanan, eds. Columbia University Press, see Chapter 2.

27 पूजा Sanskrit Dictionary, Germany (2009).

28 Lindsay Jones, ed. (2005). *Gale Encyclopedia of Religion*. 11. Thompson UGale. pp. 7493-7495.

festivals, agricultural concerns, visitations, etc.). While there is certainly some overlap on the settings within which certain *pujas* are conducted, there remains a continued focus on incorporating the participation of divine persons in the day-to-day life of the family and of the community through ritual. The variability of *puja* specifics[29] also makes it an ideal ritual for both venerating special events and conducting day-to-day ritual tasks, a characteristic that often blurs the boundaries between the human and the divine, between the sacred and the everyday.

The specific steps involved in *puja* varies according to the particular religious school of thought and may also vary by region, occasion, and the deity or deities present. In this way, Shaligram *puja* is no different. Shaligram *pujas* are conducted in a variety of contexts, from the riverbanks of the Kali Gandaki and the temple courtyard of Muktinath during pilgrimage to destination temples the world over and in households where Shaligrams are brought to take up their residence. In both temples and homes, performed alone or with the assistance of a *pujari* or *guru*, Shaligrams are offered food, fruits, sweets, clothing, a bath, a lamp, money, incense and perfume, or even a bed complete with pillows and blankets for them to take rest whenever they might need, all of which, after the prayers are completed, becomes *prasad* – blessed food and other objects that have been ritually "consumed" by the deity and can then be shared by all present at the *puja*.

Within the arrangement of the *puja-darshan* altar (deities, deity accessories, miniature objects, photos, sacred stones, etc.), each piece of the diorama is connected back to the sacred texts, to local events, and to historical narratives that relate to the lives of the people involved. For example, a favorite story of one pilgrim to Muktinath, a Gaudiya Vaishnava from West Bengal by the name of Bhanu Kiran Bawari, involved Krishna's pastimes as a young cow herder. He often spoke of his many Shaligram pilgrimages to Mustang in order to find the specific Krishna Govinda Shaligram he had always wanted, and on one particularly day as we waited at the Jomsom jeep stand, he was keen to explain why. "Back at my home temple near Kolkata we close the temple *darshan* and veil all the deities during the middle of the day. This is because Krishna leaves the temple at this time and engages in his activities within the Vrindavan *dham*.[30] He often goes out and spends

29 Though some festivals and special events certainly have their own dedicated puja rites, such as the Durga Puja and the Lakshmi puja, that are always performed in the same ways at the same times. See Flood, Gavin D. (2002).

30 He is here associating his home village with the town of Vrindavan in the Mathura district of Uttar Pradesh, India. It is here that, according to Vaishnava tradition, Krishna spent his childhood.

his time as a cow herder again with all the cows that wander through the streets and in the fields. When I receive my Govinda (Shaligram), I will dress him as the cow herder and give him small cow figurines, so that the cows might remain with him when he comes into my home."

The objects of both *puja* and *darshan* used in ritual performances then quickly become a kind of sacrifice in as much as they are no longer simply material objects. Fundamentally, they become the very things they signify – a cow figurine is a real cow, and water sprinkled on a Shaligram is water to slake real thirst. This is because, through their relationship to the deity's activities, objects are invested with characteristics beyond 'mere' materiality that are not attributable to substances in the physical world. Object becomes essence or entity. This is how pieces of daily life become linked with mythic events from divine narratives, everything from places where a deity may have once lived and how he or she dresses to intricacies of preferred meals and favorite games all played out ritually across the sacred landscape (Eck 1998: 68). This overall geography is then expanded to encompass all of India and Nepal, and in some cases the entire world, through the construction of a historical, material, and narrative contiguity between a mythological past and other historical and contemporary cultures and events across the globe. For Shaligrams, these kinds of *puja* rituals demonstrate the layering of personhood, agency, and materiality through relationships of "substantive" exchange (see Chapter 5), especially when we note where, when, and with whom Shaligram *puja* is conducted.

Puja is conducted when a Shaligram is born out of Kali Gandaki and may be conducted again, with the same Shaligram, at the birth of a human child, linking new family members with mythic cycles of birth, death, and rebirth located in the origin myths of Shaligrams themselves as well as scriptural narratives related to the specific deity. (For example, Krishna Gopala Shaligrams – Krishna as an infant – are popular in these cases). *Puja* is performed during the festival celebrating the marriage of Tulsi and Shaligram (Vishnu) and then once more during family weddings, linking the newlywed couple to the narratives of reproduction located within the *dham*, or sacred landscape, of Muktikshetra's mountains and rivers. *Puja* is done to welcome the Shaligrams to their new home or temple (and often to care for them every day after) following their long travels, just as *puja* is done for honored guests and great teachers. And finally, *puja* is done to honor the passing of the dead, where an honored Shaligram may accompany the body into the cremation fire and carry them along with it into another life, to be sent as ash into the sacred river and then reborn once again from it. As such, a Shaligram ceases ontologically to be simply stone by its movements between physical landscapes as well as its movements through life events, religious narratives, and communal events. By building relationships, it becomes body and person, replete with all the trappings necessary to carry out a social life among family and community.

The scriptural rules that govern Shaligram veneration therefore do not establish a set of rigid boundaries delineating the correct performance of rituals for Shaligrams but rather function like the banks of a wide riverbed, much like the Kali Gandaki, within which streams of traditional and practical variation may appear, meander, and merge as they move across the landscape. While the river may then slow to a trickle or come flooding out of the mountains, break up or reconverge, or even occasionally overflow its banks, it nevertheless exists within a relationship of myths, narratives, practices, and traditions that locate it within a particular conceptual category – or, as William Fisher might say, a particular riverbed. It may never be the same river twice, but it will always be the Kali Gandaki. The similarity of many of these rituals and the pilgrimages that support them vary in different ways from time to time and from place to place, but their family resemblances remain recognizable as well as the Shaligrams that define them. In other words, though the specifics of time and place may not be especially important, the anchoring of rituals and landscapes within times and places is.

There is no standardized doctrinal consensus among Shaligram practitioners as a whole, but neither can professed religious beliefs be used to predict actual practices accurately. And specific individual practitioners may perform or participate in Shaligram rituals reflective of any range of traditions, be they Buddhist, Hindu, animist, specific to one's guru lineage, or related to a specific local deity or legend. Even secular, agnostic, and scientific atheists have been known to take part in Shaligram ritual veneration. Regardless, both the varied ritual practices of individuals and group performances of rituals fuel the intense fluidity of internal and external boundaries that might otherwise appear to separate devotees in different Hindu traditions from one another, from Hindu traditions and Buddhist traditions, and from those who would mark certain places and concepts as Hindu, Indian, or Nepali from those who would mark them as Buddhist, Tibetan, or foreign. Shaligram ritual practices reflect not so much a religious syncretism as they do a kind of public and private religious pluralism in which the ritual conducted, deities propitiated, and religious experts and authoritative scriptures consulted vary depending on a variety of factors that include age, gender, level of experience and education, ethnicity, nation of origin, nation of current residence, family history, economic standing, and community relationships. But despite these continuous shifts in context and historicity, Shaligram worship remains explicably tied to the land (both origin and destination), to times and places where spirits, ancestors, and deities must be engaged and propitiated, and to concepts of identity, meaning-making, and community building. The specific community may be unique, but the ritual process of its construction remains relatively the same.

The arrival of scholars to Shaligram practice also represents something of a transition point in the history of Shaligram worship. Their focus on standardizing mythologies, locating historical narratives, and contextualizing these narratives within certain times and places tends to predisposition an approach that assumes one "real" story that somewhere, somehow, unfolds in a unilinear and irreversible direction. To revisit the river metaphor of culture; it is far too easy to mistake a single stream for the whole river of Shaligram veneration. When anthropologists began arriving in Nepal in the 1950s, they had already been working with Hindu, Buddhist, and indigenous religious practices in India and elsewhere in South Asia for more than a century. In many ways, the anthropology of Nepal had not only missed entire decades of ethnographic theorizing, it also was immediately contextualized using social theories from religious traditions elsewhere (a common theme, as we have seen).

Much of the early scholarship on Nepal tended to essentialize categories and cultures, with scholars acting as co-artificers of cultural narratives right along with the people themselves. In many cases, this was done by associating certain groups or practices with the historical trajectories of religious traditions from elsewhere in South Asia (i.e., Indian Hinduism or Tibetan Buddhism) and by simplifying and polarizing perceived differences between reductionist dichotomies of Hindu/Buddhist, resident/tourist, and pure/hybrid. The first arrival of anthropologists to Nepal also coincided with dramatic social and political changes more generally, including the closing of Nepal's northern borders by China in 1960 (as a direct result of the invasion of Tibet in the 1950s). During this period, Nepal's internal infrastructure began to grow, as did significant political concerns about Nepal's long-term sovereignty and stability in the face of internal political unrest and pressures from both India and China. As a result, the traditional northern trade routes were closed, merchant groups such as the Thakali began to migrate outwards, and access to Mustang and to the river of Shaligrams was severely curtailed. But this has not meant the end of Shaligrams or the end of Shaligram pilgrimage. Rather, it has been the beginning of something new: a new movement, a new transformation, for people and their sacred *shilas*. Today, a new kind of Śālagrāma is born.

The Gateways of Experience

The temple complex of Muktinath is both the primary site of Shaligram veneration in Mustang as well as the seeming endpoint of Shaligram pilgrimage, where newly acquired Shaligram stones from the Kali Gandaki below are brought before the temple deities to undergo their first *puja* rituals and bathing rites welcoming them to their new families. Muktinath itself, however, is a kind of pilgrimage set in miniature, where multiple shrines, *gompas*, monastic quarters, statues, and sacred buildings form a route by which pilgrims complete their larger physical journey and transition into their ongoing spiritual journey as Shaligram devotees. The history of Muktinath is, however, just as complex and convoluted as that of Mustang and has significant influence on how practitioners view their participation in Shaligram pilgrimage.

As a site of Hindu worship, Muktinath offers a central shrine to the deity Vishnu (who is principally associated with Shaligrams) as well as 108 water-spouts under which pilgrims must pass, along with their Shaligrams, in order to receive life-long blessings and good fortune. Though identified in Puranic texts as existing in what is now Himalayan Nepal and mentioned by a series of names, such as Muktikshetra and Śālagrāma, it is often unclear whether these passages associated with Muktinath specifically refer to a place (such as a village), a region, or an ideal. Through the unity of elements – specifically water, fire (in the form of natural gas vents), stone (Shaligrams), and sky – most modern Hindu scholars take the current location of the Vishnu *mandir* (Vishnu's central temple in the Muktinath complex) to be near enough to the textual references of Śālagrāma to be considered the same thing, even if debate continues as to whether a one-to-one correlation can actually be made between the mythological descriptions of the land of Shaligrams and the actual current pilgrimage route through the Kali Gandaki.

As a Buddhist complex, Muktinath is more commonly referred to as Chumig-Gyatsa (Hundred Waters), though both Muktinath and Muktijwala (water of salvation) also appear in Tibetan texts. Muktinath is also still administered by an order of Buddhist nuns, who are responsible for both the day-to-day care of the temple and the carrying out of the Hindu and Bon rites of veneration throughout the complex. As such, Muktinath itself, much like Mustang generally, is not specifically identifiable through any one religious tradition or label and incorporates a wide variety of ritual practices, local and foreign customs, and belief systems.

The combination of Muktinath and the Kali Gandaki, then, forms the ritual landscape under which Shaligram traditions take shape. First, pilgrims travel through the riverbed to collect new Shaligrams, who are born of sanctified waters through a sacred geological process. Then, a smaller pilgrimage to and within Muktinath begins the Shaligram's life journey with their new families as they prepare to head back home. A shared core of rituals and interpretations then links all Shaligram devotees together, beginning with their collective movement through the landscape of Mustang and into their lives as carers for and worshippers of Shaligram. But if Shaligrams are "born" and they "live" as members of their families in much the same way as their human counterparts do, can we truly call Shaligrams merely objects of veneration? Or are they something else? How could a place transform and then, in essence, become a person?

Bibliography

Chetri, M., L.K. Maskey, N.R. Chapagain, and B.D. Sharma, eds. 2004. *Mustang – The Land of Fascination*. King Mahendra Trust for Nature Conservation, Nepal.

Dahal, Sharada Prasad. 1988. *Muktikshetra: A Historical, Religious and Cultural Explorations* (Original text in Nepali); Muktikshetra Publication Committee. Translated under the direct supervision of Dr. Ramesh K. Dhungel by Rana, Binaya SJB: Post Graduate Diploma in Buddhist Studies, Buddhist Studies Department, Tribhuvan University, Kathmandu, Nepal. Translation sponsored by André Kalden, President of Muktinath Foundation International, 2002.

Dana, Jessamine. 2011. "The Politics of Encounter: Hindu Belonging in a Multi-Faith Pilgrimage Site in Nepal." In *The Politics of Belonging in the Himalayas: Local Attachments and Boundary Dynamics (Governance, Conflict and Civic Action)*, 1st edition, edited by Joanna Pfaff-Czarnecka and Gerard Toffin. New Delhi: SAGE Publications India.

Dhungel, Ramesh K. 2002. *The Kingdom of Lo (Mustang): A Historical Study*. Kathmandu: Tashi Gephel Foundation.

Eck, Diana. 1986. "Darshan of the Image," *India International Centre Quarterly* vol. 13, no. 1, IMAGES (March): 43-53.

—. 1998. *Darśan: Seeing the Divine Image in India*. New York: Columbia University Press.

—. 2012. *India: A Sacred Geography*. New York: Three Rivers Press.

Ehrhard, F.K. 1999 (1993). "Tibetan Sources on Muktinath: Individual Reports and Normative Guides," in *Archaeological, Historical, and Geographical Reports on Research Activities in the Nepal-Tibetan border area of Mustang during the years 1992-1998, Part IV*, H.14, edited by Dieter Schuh, pp. 23-29, Bonn: Archiv für Zentralasiatische Geschichtsforschung (Archive for Central Asian History).

Fisher, William F. 1987. *The Re-Creation of Tradition: Ethnicity, Migration, and Social Change Among the Thakali of Central Nepal.* PhD diss., Columbia University.
—. 2001. *Fluid Boundaries: Forming and Transforming Identity in Nepal.* New York: Columbia University Press.
Fürer-Haimendorf, Christoph von. 1957. "The Interrelations of Caste and Ethnic Groups in Nepal," *Bulletin of the School of Oriental and African Studies* 20: 243-260.
—. 1966. "Caste Concepts and Status Distinctions in Buddhist Communities of Western Nepal." In *Caste and Kin in Nepal, India, and Ceylon: Anthropological Studies in Hindu-Buddhist Context,* edited by Chrisoph von Fürer-Haimendorf, pp. 140-160. London: Asia Publishing House.
—. 1975. *Himalayan Traders.* New York: St. Martin's.
—. 1978. "Trans-Himalayan Traders in Transition." In *Himalayan Anthropology,* edited by James Fisher, pp. 339-357. The Hague: Mouton.
—. 1981. "Social Change in a Himalayan Region." In *The Himalaya,* edited by J.S. Lall, pp. 175-203. Delhi: Oxford University Press.
Muktinath Foundation International (MFI). 2015. Website. http://www.muktinath.org/. Accessed 5 January 2015.
Ramanujadas, Madhu Sudhan. 2003. *Muktichhetra Mahatmyamam.* Kathmandu: Dhaulagiri Osset Press.
Rao, S.K. Ramachandra. 1996. *Shaligram Kosh* [*Śālagrāma – Kosha*]. Sri Satguru Publications. Delhi: Indian Books Center.
Singh, Chandra Shekar et al. 2017. *The Puranas.* Amazon Digital Services.
Tucci, Giuseppe. [1953] 1982. *Journey to Mustang.* Translated by Diana Fussell. Kathmandu: Ratna Pustak Bhandar.

5 A Bridge to Everywhere

The Birth/Place of Shaligrams

Abstract
Shaligrams blur the boundaries between being and object and may be either or both. For this reason, Shaligrams are treated as manifest deities with their own will and agency and as ritual objects who organize community participation around ritual events and social norms. Furthermore, Shaligram practitioners do not typically view juxtapositions of "deity" and "fossil" or "stone" and "body" as incommensurate with one another. Rather, religion and science are leveraged as two narrative frameworks that present possibilities for different ways of knowing. This results in an understanding of Shaligrams as ammonites, Shaligrams as persons, and Shaligrams as deities that views each as one part of a larger reality.

Keywords: Tirtha, darshan, dham, ontology, personhood

Ato'dhisthana Vargesu Suryadisviva Murtisu
Salagrama Silaiva Syad Adhisthanottamam Hareh

"The Lord resides in many places in which he may be worshipped, but of all the places Salagrama is the best."
– *from Garuda Purana, Ch. 9, 1-23*

Early one morning, late in the summer of 2016, I awoke just before sunrise and set out for the Kali Gandaki River. Clad in thick canvas pants and a pair of rubber water shoes, I made it a point to tie my Australian field hat securely to my head with a chinstrap before venturing out into Kagbeni's lively pre-dawn streets. Since the wind was always threatening to steal the hat every time I turned my head, I figured that the discomfort of a spare bit of leather was a small price to pay against an afternoon burnt red in

Walters, Holly, *Shaligram Pilgrimage in the Nepal Himalayas*. Taylor & Francis Group, 2020
DOI: 10.5117/9789463721721_CH05

the glaring Himalayan sun. A mother and daughter in *chubas*, traditional Tibetan dresses, passed me cautiously, hunched over their hand brooms as they swept the previous day's goat droppings from the cobblestones and out into the adjacent fields. An older Mustangi man, passing by with his caravan of mules and donkeys laden with rice and kerosene, shouted out a compliment. "Just like cowboys!" he yelled, touching his own imaginary brim. It was a typical morning in Kagbeni, filled with young women chatting on their way to fetch water from the village taps, small children playing in doorways, and the clink of copper cookware banging out breakfast in nearby guesthouse kitchens. I turned west and headed towards the roar of the water.

The Kali Gandaki riverbed is nearly a quarter mile wide in most places around the village, and as the river slowly meanders back and forth across the valley, breaking up and remerging, undulating from bank to bank over the course of the day, it is continuously revealing a new landscape of stones and silt. The trick to finding Shaligrams, as one veteran pilgrim once taught me, was to first find one of the many small, shallow side streams branching off from the deep central currents. The best streams were the ones in the process of moving off course, easily identified by the tall banks of sediment actively breaking off and sliding down into the water below. Conversely, one could also seek out a stream that had recently petered out in favor of rejoining the main river and walk along its muddy edges slowly upriver, all the while keeping a sharp lookout towards any recently exposed areas.

As one picks one's way carefully along through sun-warmed, clear waters, Shaligrams reveal themselves to the discerning eye. The constant flow of water combined with the settling of the heavy black silt grains that give the Kali Gandaki its name is always exposing new stones, new pathways across the riverbed, and new landscapes. Heraclitus was rather befitting when he said that, "No man ever steps in the same river twice, for it's not the same river and he's not the same man."[1] Apt in this regard, the Kali Gandaki renews its places of pilgrimage as often as it renews its arrival of pilgrims. I too was also discovering a new landscape, stepping out onto the very same riverbed I had visited just the day before, but which now looked completely different – any familiar hills or rocks washed away in the night. Within a few minutes a Hindu pilgrim I had met earlier in the week, a middle-aged Indian man dressed all in white, came up alongside me and asked if any Shaligrams had been revealed to me today. I smiled and replied that they hadn't yet, but that I was ready and the day was still young. He nodded. "*Darshan* will come," he said. "I am waiting too."

1 Fragment 91, "Cratylus".

I was familiar with the practice of *darshan* from my time in India three years earlier. For Hindus, *darshan* is one of the most important aspects of ritual veneration, especially when it comes to the worship of *murti*, the sacred images and statues of Hindu deities present in homes and temples. *Darshan* is a Sanskrit word meaning "to see", but this aspect of "seeing" does not just mean to see the deity physically. *Darshan* means to behold the deity as he or she truly is beyond the material form obvious to the eye and, in return, to be beheld by the deity yourself. In other words, "seeing" is a form of direct contact between persons (human and divine) mediated by an exchange of gazes in the physical world but not limited to the material bodies involved. It is also a kind of knowing (Eck 1998: 2-5); through sight, both deity and devotee are said to participate in the essence of the other.

In the act of *darshan*, the deity is an agent who "gives *darshan*" (*darśan denā* in Hindi), and it is the devotee who "takes *darshan*" (*darśan lenā*). In the views of many Hindus, God presents himself to be seen in material form because humans are, by their nature, limited to the use of their senses in order to apprehend the world they live in. Therefore, when a deity is present to offer *darshan,* devotees arrive to "receive" what is given. What is given then is a kind of physical, bodily, and spiritual interaction through the medium of the senses. Like the physicality of interacting with holy places – the *dhams* (the spiritual abode of the deity) – the reciprocal gift-giving relationship in the *darshan* draws on sense experience to construct a concrete, material appearance of the divine through continuous cycles of relations and obligations exchanged through ritual. Not only does one "see" the deity and be "seen" in turn, one also "touches" the deity with the forehead and hands (*sparsha*) and is "touched" as well. Devotees may also variously touch the limbs of their own bodies to establish the presence of certain aspects of the deity or to invite the deity's attention to a particular physical issue or desire for contact. During the *darshan*, devotees also equally "smell" the incense and lotus flower perfumes and "hear" the sacred sounds of the mantras,[2] the ringing of bells, and the blowing of the conch shell (Eck 1998: 11-12).

This "exchange of gazes" is then what enables a subject/object transformation where it often becomes unclear who is acting upon whom and in what capacity. Similar to Nancy Munn's description of Aboriginal

2 The chanting of the *mahamantra*, for example, requires the repeated chanting of Krishna's names and constitutes another instance in which one "sense aspect" of God is "no different" than another. Put another way, "seeing" God in the form of the deity is no different than "hearing" his name.

"transformations" – where ancestor spirits produce material objects within which they are in some way embodied (1970) – deities in the *darshan* (Shaligrams included) demonstrate their own dynamic subjectivities in an association with an object world (1970: 143-147) that includes human bodies, ritual objects and other sacra, and landscapes. But Hindu deities are not only consubstantial with the objects they produce or inhabit, they are often described as being no different than them – their mythic presence and their material presence as one and the same thing. This is where the exchange or attribution of viewpoints also becomes possible, where the deities' desires and actions are ambiguous, open to interpretation, and communally shared. For Shaligrams, *darshan* constitutes the first vital link merging stone and body as well as between deity and fossil, a link that is initially established beginning with the physical movements and spaces of ritual.

Arrangements of *darshan* altars (deities, deity accessories, miniature animals or people, photographs, sacred stones, etc.) are often carried out with the intention that each piece of the diorama can be connected to sacred texts, local events, household needs, and historical narratives that relate to the place or to the person that the altar currently serves. On an earlier trip to West Bengal, where I was first introduced to Shaligrams at the Radha-Krishna (Sri Sri Radha Madhava) temple in Mayapur, a local *brahmacharya* (celibate monk) once explained that his favorite stories involving Krishna's pastimes were any one of the many tales of his days as a young cow herder. During the middle of the day, when the temple *darshan* altar was closed and veiled, he said that Krishna would then leave the temple at this time and engage in activities within the village *dham*, namely that he would re-enact his time as a cow herder in the nearby *goshala*, where the sacred cows were kept. The *brahmacharya* often liked to represent these activities by placing small cow statues at the Krishna deity's feet before closing the altar. For Shaligram devotees, the altar begins at the Kali Gandaki.

"I think that the river is like the flow of the mother," commented a Hindu woman with a blue sari and a neat white bun sitting near the riverbanks. She held two small Shaligrams in her hand and, as I watched, began preparing a memorial *puja* ritual to mark the first anniversary of her own mother's death[3] and cremation. "It comes from the mountain. Shaligrams come from the mountain first. Then the river. I brought one Shaligram from my home here. It is Krishna Gopala; Krishna the infant with mother Yashoda. And then today another appears to me in Kali Gandaki. Now I have two

3 In Nepal and India, a death anniversary is known as *shraadh*. The first death anniversary is called a *barsy*, from the word *baras*, meaning year in the Nepali and Hindi languages.

Krishna Gopalas. This one you see," she held the slightly larger of the two Shaligrams aloft, "this one is me just like I am with my mother. This one," she now held aloft the other, "this one is my mother, who always worried after her children, letting me know she is with God. She is gone now, but I see her here. Krishna is here. She is here. I see them here, and they see me."[4]

The complex mapping of kinship, deity, time, and distance was common among Shaligram practitioners who often described, as this woman did, a Shaligram as being both a manifestation of God (in this case, Krishna Gopala) as well as evidence of the presence of a deceased loved one. The "birth" of a Shaligram from the mountain and the river could be expressed both as a divine birth and as a representation of the devotee's own birth, the birth of their families, or of specific children. But this layering of time in the context of mythic origin became even more complex within the relationships between Shaligram and devotee where, in the example above, the Shaligram is simultaneously Krishna as an infant in the presence of his mother Yashoda as well as the Hindu woman in the presence of her own mother now deceased. Unsurprisingly, several areas along the banks of the Kali Gandaki River are often used to perform death memorial *pujas* and more often than not, Shaligrams are incorporated. This begins the bridging of birth and death through the flow of the river which mirrors the bridging of birth and death in the familial genealogy (inheritance) of the Shaligram. In this case, an old Shaligram, passed from mother to daughter, was carried and worshipped by a woman who spent her lifetime as a doting mother to her children. Then, a new Shaligram is born out of the river, which becomes that same deceased mother's care beyond death, encapsulated in the story of Krishna Gopala. Through the material linking of myth, ritual, and landscape, both the deity and the dead can then be "seen". This practice of seeing and being seen by the deity (and the dead) is one of the most common and most important parts of ritual practice among observant Hindus and is also one of the major driving forces behind pilgrimage in Mustang and throughout South Asia.

Searching for Shaligrams is its own kind of *darshan*. As I walked with particular care not to disturb too much sediment in the water, I noticed two especially important things about the experience I was undertaking. Firstly, the dark, almost inky black color of a Shaligram is the first thing that tends to catch the seeker's eye (since it stands out against a mix of silty grey and dirt brown); the second was the subtle appearance of ripples or spirals (the tell-tale ridges of the fossil ammonite shell) along an exposed surface that

4 Conversation in Hindi with some English. Transcribed from fieldnotes recorded shortly after the conversation.

might indicate that a stone in question was, in fact, Shaligram. But not every stone that might initially appear this way was really Shaligram. Oftentimes, the refraction of light through the flowing water gave the impression of similar patterns on otherwise smooth stones and the accumulation of silt underneath the current was occasionally responsible for the appearance of analogous ridges in the sand that covered the riverbed. More than once, a burst of excitement and a quick scoop of water to retrieve a Shaligram appearing in the riverbed would end with nothing more than a handful of sand and a plain rock.

Finding a Shaligram often left me with the sense of something truly born from the river, something which was appearing only at the very moment that I happened to see it. Carried down through millennia of time (or 175 million years if we're going by geological counts) by an ancient and sacred *tirtha* (a Sanskrit term meaning "bridge/place of crossing/ford"), revealing itself just at that moment and just for me. Something that I was "seeing", perhaps, that hadn't been there a moment before. *Tirthas* often refer to places where the divine world and the physical world are closer together, and it is not unusual for important pilgrimage sites and sacred rivers throughout South Asia to be labeled as *tirtha*. Additionally, many Shaligram devotees themselves describe their Shaligrams as various kinds of *tirthas* – as links to a wide range of other places, people, and events in their lives. Later on, I also found *tirtha* to be an apt concept for describing Shaligrams and Shaligram practices as a whole. In Western discourses, religion and science are often juxtaposed against one another. But among Shaligram practitioners, "deity" is equally "fossil", and "stone" is also "body". Nor do Shaligram devotees hybridize religion and science as two possible if unrelated points of view regarding the essential nature of the same object but instead use them to draw links between two different ways of knowing. This is to say that, rather than describe a blending of separate, "purer" forms of knowledge (as one might use syncretism to describe the blending of religious traditions), Shaligram practice demonstrates how Shaligrams as ammonites, Shaligrams as persons, and Shaligrams as deities constitute a shared reality.

Tattva Mimamsa, or All Existing Things

The object-person view of Shaligrams is not unprecedented within current anthropological threads regarding the study of ontology (the nature of being). The "ontological turn", however, is a recent movement within cultural anthropology that advocates that representational frameworks move away from an idea of cultures as systems of belief that provide different perspectives on a single objective existence. Rather, theorists who write in this new mood of relativism shift their descriptions of many "cultures" or many "belief systems" to that of many "worlds". Ontological anthropology then seeks to open us to other kinds of possible realities beyond what has typically been taken for granted. While all good ethnography has always been ontological in some fashion (particularly in the ways it strives to get at local "emic" realities), I find only a few select positions in the cultural study of ontology as potentially productive in terms of Shaligrams – which are those focused on the problems of ethnographic misrepresentation and not those more geared towards speculative futurism. This is to say that I do not refer to devotees as "believing" in Shaligrams any more than I might refer to a paleontologist as "believing" in fossils. In other words, just as scientists would not describe invertebrate paleontology as a belief that ammonites are fossils, devotees find it equally incommensurate to describe their practices as a belief that Shaligrams are deities. They *are* fossils. They *are* deities.

For some scholars, the ontological turn offers a way to finally resynthesize anthropology's fractured post-humanist progression (Descola 2013; Kohn 2013), while for others it shifts anthropology's ultimate goal from a critique of the present to the building of better futures (Latour 2013; Holbraad, Pederson, and Viveiros de Castro 2014; cf. White 2013). What is more popular, though, are the three potential positions on ontology laid out by Holbraad, Pedersen, and Viveiros de Castro (2009): ontology as the search for essential truth (*how things are*); ontology as the critique of all possible essences (*how things should be*); and ontology as the exploration and potential realization of other "reals" (*how things could be*). My work here, however, tends to follow along the same lines as that of Eduardo Kohn, for whom ontology means a more thoughtful, critical engagement with different "worlds" rather than with different "worldviews" (2013: 9-10). In this thread of ontological inquiry, the ethnographer can explore how humans relate to other beings (animals, ghosts, spirits, forests, mountains, etc.) and, in turn, how those beings themselves think, act, live, and exert force (Kohn 2014). The aim then is to present "emic" viewpoints carefully and seriously or to become, one could perhaps say, "even more emic than emic" in the analysis and representation of the worlds we study.

Initially, it would appear that the ontology of Shaligrams could also be cogently argued as both multiple perspectives on a singular reality (a stone that is fossil, person, and deity) and as unique ontological beings (see Pederson 2012; Holbraad 2012; Keane 2009; Geismar 2011; and Laidlaw 2012). But by attempting to draw the boundaries of personhood or even entityhood around Shaligrams as a category, we also cannot presuppose that their "beingness" ultimately subsumes the processes by which they are created. The ontology of Shaligrams, by the various intersecting processes through which they are constructed, is neither essential nor absolute. Rather, it shifts continuously as the stones move from one context to another. While the majority of anthropologists would agree that no ontology of an object is ever essential or immutable but is always ultimately socially constructed and shifting, what I argue here is that the ontology of the "Shaligram being" itself is also often open to debate. An ammonite fossil in certain contexts; a deity in others; kin and relative in still others; and yet, all of these simultaneously – where the very essence of who or what might be present could be different depending on the context or situation in the moment even though the physical existence of the object itself remains unchanged.

Shaligrams routinely straddle the boundary between being and object and may be either or both at different times. My point, then, is to foreground the relationships among objects, people, and landscapes as the producer of ontological categories: categories that change depending on the processes in question. Shaligrams are in perpetual motion, constantly reacting to the conditions of possibility and central in the question of their own doing and undoing. In multiple traditions of Hinduism and Buddhism, the boundaries between reality and its representations are often seen as blurry and indistinct, where deities and their icons are both one and the same thing as well as materially separate from one another (Walters 2016). As a result, Shaligrams become as much agents in the making of their own fate as the people who venerate them are. Shaligrams cannot then be theorized as either exclusively objects or as exclusively beings but rather must be analyzed through a hybridized view of agency as object-beings.

I therefore take my cues from Shaligram practitioners themselves and begin my analysis with the view of Shaligrams as *tirthas,* whose continuous forming and breaking of personal and emotional relationships over time constitute them as persons in some of the same ways as human beings are. Over the past several decades, one of the primary themes of sociocultural and ethnographic studies in South Asia has been to question the universality of individual-centered (human) personhood versus other alternatives. In fact, numerous such studies have demonstrated the particularly fluid nature of personhood in South Asian contexts (Marriot 1976; Marriot and Inden 1977; Lamb 1997, 2000), where individuals have relatively permeable and passable boundaries that are continuously shaped and remade through transactions of food, conversation, and interaction; services both economic and intimate; and the exchange of daily and ritual bodily substances.

Bodily Attachments and the Making of Persons

Sometimes called "ties of *maya*" (bodily-emotional attachments that make up the illusion of the material world – Lamb 2000), transactions between persons also include various personal connections to friends and relatives, to places of personal meaning, to favorite household possessions, to village gods and *gurus*, and to the care of household deities. Many scholars have noted that South Asian cultural worlds are made of continuously "flowing substances" (Marriott and Inden 1977) wherein persons are made through networks of interactions and relationships or "dividuals" rather than closed, contained "individuals" of the West (see also Lamb 2000). As both Sarah Lamb and E. Valentine Daniel (1984) emphasize in their work among Indian West Bengalis and Tamils respectively, all things in the material world are perceived to be in constant flux. Persons are then comprised of the inevitable intermixing of substances brought about by their lifetime relationships with other people and with places and things, and which are especially highlighted in significant life events such as birth, marriage, co-habitation, and sharing food (see also Inden and Nicholas 1977).

Through the daily exchange of substances and relationships, Shaligrams become persons in some of the same ways as humans do, even if they are not necessarily perceived as the same kinds of person as humans. They eat and drink, they are concerned with their own cleanliness and ritual purity by accepting or rejecting substances that touch them, they observe the daily activities of the family, they participate in community events, and often travel with their family members on special occasions or sometimes just to the market and back. Shaligram mobility, then, translates to personhood, though a kind of divine personhood where they actively participate in everyday life by sharing meals and engaging in crafts, attending festivals, officiating at weddings and funerals, overseeing the births of children, and where they are bathed, clothed, and cared for as any other member of the family. Many Shaligram devotees went on to describe how their Shaligrams became embedded within their communities such that they could never be unmoored and for that reason could never be given away.

Learning to properly conduct myself in the presence of a Shaligram was an endless source of anxiety in my early days of fieldwork. "Never let your hands touch meat or blood before you touch a *shila* (a Sanskrit term meaning stone and sometimes used interchangeably with Shaligram)" was a constant reminder. Prasad Vipul Yash was one of the first Shaligram ritual specialists I worked with and from whom I initially learned the expected ways of handling sacred stones. From beneath the thick cloths he used to shield his eyes from the glaring summer sun of Kathmandu, he would keep careful watch of each movement I made near the altar [under his direction]. "As a woman you must be especially careful," he noted. "No blood. Not ever. If you do, the *shila* might become angry. Blood is to them a wounding or a threat of sickness. They reject it. You must be without sickness, or it [the Shaligram] will bring you misfortune and pain to your family until you have washed properly and then washed the *shila*. Offer only pure foods with clean hands. Grow tulsi (holy basil) in your home and offer that. Give water from the house or from a holy river every day. If you watch them, you will even see the water go into the *shila* as they drink it up. Then the Shaligram can breathe in all the love and care and will become a part of you. It will live in your house and keep away bad things. It will look for you and wait for you and only you can look after it. This is the responsibility we accept with Shaligram in our homes. You can never give it away once you have done this. Not until death."[5]

Shaligrams as persons links the kinds of "distributed personhood" described by British anthropologist Alfred Gell with concepts of object-agency described by Christopher Tilley and Janet Hoskins and the animate nature of religious icons in the work of Amy Whitehead. Where Gell (1998) describes the dynamic by which any given person's agency or sense of self is said to extend beyond the boundaries of the body in order to relate to and animate other persons, places, or things, Shaligrams act both as bodies themselves and as extensions of the bodies of deities and practitioners. But they are also embedded in material ritual practice.

Christopher Tilley, in his "Ethnography and Material Culture", states that the meaning of an object is produced when that object is used towards a purpose by a group: that "meaning is created out of situated, contextualized social action which is in continuous dialectical relationship with generative rule-based structures forming both a medium for and an outcome of action" (2001: 260.) From this perspective, an object (or a Shaligram) gains agency only when it is used for a specific means by a human. From my own experiences with Shaligram veneration, however, the agency of Shaligrams is at least partially derived from the fact that they are seen to move "of their own accord" through the landscape (via river currents, erosion, wind, etc.), and many devotees regularly relay pilgrimage experiences with Shaligrams "swimming up to them" in the river or "getting away from them" by tumbling into the currents or off a landslide. The processes of fossil formation also

5 Conversation in Nepali with some Hindi for clarification. Transcribed from fieldnotes recorded during and after the conversations.

have agentive meaning in the viewpoint of landscapes as bodies that move, are injured, give birth, and die. It is therefore not human action alone that gives Shaligrams their agency but their capacity as beings in their own right to enjoy movement free of human intervention.

Prasad Vipul Yash did not typically allow women to touch his household Shaligrams. As a devout Sri Vaishnava Brahmin (a high-caste denomination within Hinduism), he was ever concerned with the potential for ritual contamination or for possible insults to the Shaligrams he kept. But on my last day in his company, before I was to travel to the Kali Gandaki in Mustang, he motioned me into the small room where he kept the deity altar. "When you go," he began, "look very carefully into the water. If you are mindful and clean, you will see Shaligram swimming there. If you are quiet and think on spiritual things, they may choose to come up to you. Here, hold Devi Lakshmi [he then placed one of his own Shaligrams into my hands]. Remember how she feels. Understand what she wants. Bring some tulsi with you to offer to the Shaligrams in Kali Gandaki. If they come up to eat it, they accept you."

Janet Hoskins, in "Agency, Biography and Objects", cites Laura Ahearn's understanding that "agency is 'the socio-culturally mediated capacity to act' and is deliberately not restricted to persons, and may include spirits, machines, signs, and collective entities" (2006: 74). Objects, she goes on to say, are then made to act upon the world and on other persons; otherwise, they would not be created (though the non-man-made "creation" of Shaligrams provides something of a complication here, as will become apparent shortly). Hoskins also cites Gell, who explains that "things have agency because they produce effects, because they make us feel happy, angry, fearful, or lustful. They have an impact, and we as artists produce them as ways of distributing elements of our own efficacy in the form of things" (ibid.: 76). To possess true agency in this sense, an object must make some sort of real impact on the mental or physical states of humans, and undoubtedly, Shaligrams readily do so. Shaligram agency is, therefore, rooted in a combination of aspects, from their ability to move about the landscape (as if) on their own accord, from their manifestations as gods whose choices are unrestricted by the material world, and from the feelings and responses they engender within people participating in pilgrimage, *puja*, and the *darshan* (i.e., as ritual objects). What is more, they also carry with them the very nature of the place within which they are formed.

Carl Knappett argues that objects cannot have agency entirely on their own. In his "Animacy, Agency, and Personhood", he states that "if an artifact holds any kind of psychological presence, it is only a secondary effect of its connection with human protagonists, the "real" and primary agents" (2005: 29). He asserts that objects cannot have true agency because they are not alive, whereas when imbued by humans with a purpose, an object may act in a manner that is only similar to that of an actual agent. However, this view of object-agency does not take into account the contexts and mirrored processes by which both humans and Shaligrams are mutually constituted as persons in South Asian ritual practices. It also does not adequately address Shaligrams' link to "placeness" in the landscapes of Mustang nor the ways in which the connections, the *tirthas*, of Shaligram practice "emerge from specific ways of being in the world" (Fowler 2010: 352) or the ways in which they transcend the representational and symbolic (as in Whitehead 2013, "animism" and "fetish") into embodiment and religious performance to interact with people directly in the *darshan*. And perhaps even more importantly, any place where a Shaligram resides is considered to be "no different than" the place where the Shaligram was born (Mustang/the *dham* of Śālagrāma, see Chapters 3 and 4), with all the resulting ties to pilgrimage, landscape, and the karmic life cycle of gods and people.[6]

Shaligrams are also divine persons, different from human persons. As deities, Shaligrams are embedded within broader religious systems that understand the personhood and agency of gods through their material manifestations in *murti* (sacred images) or as *avatars* (deities who descend to earth in physical forms). In fact, people often speak of their *murti*, in homes or in temples, *as* Krishna or Vishnu or God and not merely as representations of them. Shaligrams carry this understanding to the extreme, whereas self-manifest (not man-made), they completely forgo all the usual rituals required to install a deity and call it into the material world (such as the *prana pratistha* – the ritual that establishes or infuses the *murti* with life and divine essence).

Shaligram veneration is also not generally restricted by caste and gender (except in a few notable circumstances, particularly in reference to menstruating women) because, as most ritual specialists explain, there is nothing in the physical world that can possibly violate or unmake a Shaligram – only temporarily insult it. Shaligrams are already God, and nothing can change that. God not only resides within them but *is* them. The fact that divinity

6 Shaligram personhood does not, however, deny the role of discursive understandings. The narratives of Shaligram, of fossil and deity, enmesh each stone in a network of symbolic valences where their meanings and interpretations are not just a superficial veneer layered over a material foundation. Rather, Shaligram origin stories (both scientific and religious) infuse the material objects with the presence of persons with whom they have already been incorporated.

exists in material form and engages in intimate social bonds with humans and have their own agency is, however, quite familiar to and widespread within Hinduism. And as this ethnography shows, Shaligram devotees, as well as Hindus in general, see their deities as quite ordinarily person-like because of how they are cared for, fed, and bathed. These linkages of divinity and everyday life form the bridge between material and immaterial worlds that constitute the *tirtha*, the connection between deity-persons and human-persons in the Shaligram *murti/avatar*.

Tirtha, the Bridge to Everywhere

When I described Shaligrams as *tirthas*, my research participants were also quick to point out that material objects were often used for the purposes of understanding broader realities. As is the case with *murti*, the ubiquitous divine image, the form of the divine is not necessarily an indicator of its nature but the presence of the material object is often required for people to begin understanding the formless complexity of divinity as it truly is. Put more simply, while the material form is necessary in the moment to help in human understanding, the object in question is not actually the divine person in their entirety. The stone is God, but God is not the stone. This division is equally important for understanding how Shaligrams become persons because, while they maintain something of the personality of the divine being who is manifest in their form, they are also often described as having unique personalities of their own – constituted by the specific set of circumstances, attachments, and environments that surround them in the present moment.

This means that, while two different devotees might have Shaligrams who are manifestations of the same deity (two Krishnas, two Kurmas, etc.), the two Shaligrams will differ from one another in their individual personalities because they live in different households and therefore have different relationships. Accordingly, a relationship to a Shaligram acts as both a literal tie to the material world for both deities and people (and for itself) but it also takes on myriad viewpoints related to political, national, or scientific discussions relevant to the immediate present. This is how "fossil" and "deity" often came up in questions and concerns about modern Shaligram personhood because the connections and understandings implied by each category were viewed as potential modifiers to the ultimate "substance" of what made a Shaligram a Shaligram.

As a study of "things", then, this ethnography pivots from the standard anthropological methodologies that view material culture through a kind of experience/analysis divide (Henare et al. 2006; Whitehead 2013) and into an ethnography of people and the divine who are also potential embodied in objects. It wasn't until I had spent a significant amount of time working in the high-altitude fossil beds of Mustang, however, that one traveling *guru*, himself the son and grandson of Shaligram devotees, illustrated the matter even more elegantly.

Having resided in Mustang for some months over the course of two years, I finally visited one of the ammonite fossil beds located a few hundred meters above the Muktinath Valley. This particular fossil layer, the remnants of an ancient sea floor, can be found up around 4,500 meters, just a short distance outside of Chongkhor village at the northernmost point of the valley gorge. As the shale layer slowly erodes out of the mountain, it forms a large wash of broken stones and fossil shells extending some 300 meters down the mountain, tumbling en masse into the Dzong Chu (or sometimes called the Thorong La) river below (which joins up with the Kali Gandaki at Kagbeni a few kilometers to the southwest). My purpose for visiting this particular fossil bed at the time was two-fold: one, it allowed me to observe some of the earlier geological forms that might eventually result in a few of the ammonites becoming Shaligram and two, it gave me a chance to see "raw" unmodified structures in the stones that, given an additional few thousand years rolling through river silts, would become the characteristics of deities as read in the stones' final manifestations. I brought with me one of my favorite Shaligrams, called Krishna Govinda (Krishna the Cow Herder). As is typical of Govinda Shaligrams, mine was a palm-sized, smooth, and perfectly round black stone bearing a white "cow hoof" impression on one side (an effect created by the cross-sectional breakage of a concentric quartz ring inside a belemnite shell). And as luck would have it, I was able to find just such a structure in one of the "raw" fossils in this particular wash-out as well.

When I returned to my lodgings later that day, with both Shaligram and ammonite in hand, I brought them to a man named Sriram Bhavyesh. He had spent decades of his life studying Shaligrams in the temples of South India and was now in Mustang on his seventh personal Shaligram pilgrimage. We had met by chance at Muktinath the year before and had kept in touch as often as possible after I had returned to Boston and he to Chennai and had been happy to meet up again while both of us were visiting Nepal. He took the two stones, touched them to his forehead, and set them on the table before us. He pointed to my Shaligram, "Do you remember this one?" he asked. "Yes," I responded, "Krishna Govinda. The cow hoof makes it clear." "And this one?" He pointed to the ammonite. Though heavily fractured and dark-orange with iron oxidization, the white hoof-like quartz structure was still easily discernable. “It is still the cow hoof,” I answered. “But it is not Shaligram, correct?” He smiled, the wide grin that wrinkled his face in such delightfully characteristic ways immediately putting me at ease. “It is.” He patted my hand. “But it is different. It is still Dasavatara, just changing. Just moving. Not quite there yet. But we can still see what it will be, can’t we?”

The majority of active Shaligram devotees are Vaishnava Hindus, and one of the defining characteristics of Vishnu’s story is the theology of the Dasavatara, or the 10 incarnations of Vishnu. In this particular aspect of Vishnu’s lengthy mythological history, is it said that he has appeared on Earth in some form on 10 particularly notable occasions (or will, given that we are currently only up to 9 in the 10-avatar stretch). This does not mean, however, that each avatar was human (or even human-looking); rather, each avatar took on a specific form and function designed to accomplish some particular set of tasks necessary for the given time in which the avatar appeared. Given the circumstances of his appearance, Vishnu has manifested as a fish (Matsya), a tortoise (Kurma), a boar (Varaha), a half-man/half-lion (Narasimha), a dwarf-man (Vamana), a warrior bearing an axe (Parashurama), Sri Ram the god-king of Ayodhya, Krishna the divine lover and hero of the Mahabharata, the Buddha (depending on what tradition you come from[7]), and finally Kalki, the destroyer of the current age who is yet to come. Given this, it was not difficult for me to imagine that becoming a stone shouldn’t be all that difficult in the grand scheme of Vishnu’s divine omnipotence, but I was not entirely sure at that moment what he meant in saying that the ammonite was “moving.”[8]

“The Dasavatara are in Shaligrams,” he responded. “There are Matsya Shaligrams and Kurma Shaligrams, Ram Shaligrams, and Krishna Shaligrams, each appearing according to the characteristics laid out in the Puranas and in the Epic stories. You call this one ammonite.” He held the fossil in his palm. “This is what science tells us. You think that we reject this, but we do not. Science is right, you see.” I asked him for clarification. “We live in the age of Kali Yuga,”[9] he explained earnestly, his words coming faster with

7 There is some contention on this one. Some traditions place other famous gurus or teachers in this position – such as Chaitanya Mahaprabhu in Gaudiya Vaishnavism – or swap Krishna for his brother Balarama as the eighth and ninth avatars respectively

8 My conversations with Sriram Bhavyesh were typically carried out in a combination of English and Hindi. Many of my recorded conversations are, however, largely in English since he was always especially keen on communicating his understandings of Shaligrams to a Western audience (represented by me).

9 Kali Yuga (literally “age of Kali” or “age of vice”) is the last of the four stages (or ages or yugas) the world goes through as part of a ‘cycle of yugas’ (i.e. Mahayuga) described in the Sanskrit

excitement. “In Kali Yuga, people are very far away from God and it is very hard to understand things we used to understand in ancient times. Science tries to explain it. Religion tries to explain it. But you see,” here he pressed the ammonite into my hand. “This is an age of science. Vishnu comes in the form that is needed most, so this one comes in the form of science. He is God moving as fossil, hiding in fossil, because that is how people are going to come to understand this now.”

It was at that moment that I began to see the conundrum of describing the relationships between people and their Shaligrams. It was not that my approach in this particular case should necessarily be a distinctly scientific one, replacing religious interpretation with geological analysis or “cow hoof” for “quartz erosion”, to look at it another way. But by replacing one method of analysis in favor of the other, I was never going to be able to answer the overarching question: “What is a Shaligram?” As my work continued, I encountered many more fascinating ways in which Shaligram stones joined scientific discourses with religious narratives, as opposed to assuming differing interpretations to be mutually exclusive.

Not everyone agreed, however, and in many cases the association of Shaligrams with ammonites was viewed as a result of the continued influences of European colonialism in South Asia or as evidence of the spread of Western secularism, which, at its core, threatened the legitimacy of religious ways of knowing. As one devotee explained: “We Hindus certainly see divinity in nature, but not everything that occurs in nature is necessarily part of worship. Shaligrams may look like ammonites from other places in the world, but their similarities are only superficial. It should be obvious that just because one thing looks like another thing doesn’t mean that they are the same thing. We recognize the authority of the Shastras and of the sacred texts that tell us about the divine formations of Shaligram and Shiva Linga, and how the Lord comes to dwell within them. Scientists and rationalists do not trust in the authority of the sacred texts and they think they know better. But they do not. This is because scientists can only describe what they see [literally, physically]. But Reality is not something you can see [*darshan*] with just your eyes.”

During my time in India, Rajiv Kamalnayan – a Gaudiya Vaishnava devotee – similarly agreed: “I read online once that scientists name ammonites after themselves or their discoverers and not according to the

scriptures. The other ages are called Satya Yuga, Treta Yuga, and Dvapara Yuga. Kali Yuga is associated with the demon Kali (not to be confused with the goddess Kali). The “Kali” of Kali Yuga means “strife”, “discord”, “quarrel”, or “contention”.

characteristics. I suppose that makes sense if you think it is only a rock. If you knew its personality, if you knew how to relate to it, you wouldn't name it like that." But in dissention, there are links between Shaligrams as persons and deities and Shaligrams as ammonites, with many Shaligram practitioners remarking on the similarities between the two as evidence of the symbolic communication of the divine through naturally occurring materials. "When the Lord takes a body," Rajiv continued, "He does so using ways humans can understand him. The shapes and symbols and characteristics we see in nature are all things He can use to speak to us. It is how we first come to meet Him and know Him. But that doesn't mean that all occurrences of those things are sacred. The Scriptures will tell you how to discern between the two and that is how we will come to understand who Shaligram is for us."

As ammonites, the geological history of Shaligrams spans millions of years and dozens of evolutionary taxonomies. They provide us with a tremendous amount of information about the early ocean environments of ancient Earth and the history of life on this planet. As the direct manifestations of divine movement in the form of deities of the Hindu pantheon, Shaligrams take us through millennia of dedicated history and lore, joining a physical landscape to a sacred landscape and linking individuals and families to traditions and ritual practices that have been in use for at least 4,000 years. In other words, not only have Shaligrams passed down through eons of wind, river currents, and tectonic uplift but they have also equally passed down through inheritance, births, deaths, marriages, and pilgrimage, with no type of movement denying the existence or importance of the other. And most importantly, a Shaligram is not a Shaligram absent either one of these threads. In short, modern Shaligram stones exist at a juncture where scientific and religious discourses are in conversation with one another – in particular, a conversation about what it means "to be" something (or even "to be alive"). This is how Shaligrams can be both ammonite fossil and divine manifestation, person and stone – just as rivers can be both vital economic and social waterways emerging out of the glacial melt as well as *tirthas* into the sacred world. If we are to address *tirthas*, then, we must address where they begin.

The Birthplace of Shaligram

"There are two ways to tell if a Shaligram is real." Anandi devi turned briefly from her *puja* tray to motion me closer. "The first is to place the Shaligram in a bowl of water or milk. Not too much, just enough to cover the bottom.

Then touch the very tip of your finger or your fingernail to the center of the *shila* (stone). If it begins to spin, it is real. This will also happen if you place a pendant or something near it. The pendant will spin around the Shaligram. The second is to place the Shaligram on a small bed of rice. If the rice increases by the next morning, it is Shaligram. This was told to me by a *guru* who said that these were the only reliable ways to test Shaligrams without the help of a great master who would know them by touch."[10]

For a Shaligram to be the deity fully manifest, it must first be authenticated as a real Shaligram. As many devotees note, the marketing and selling of real Shaligrams is enough of a problem without the marketing and selling of fake Shaligrams, which to many represents the worst human tendencies for deception, greed, and mindless souvenir taking. Fake Shaligrams can take a number of forms, from pieces of broken stones glued together, stones that have been cast from cement or epoxy and carved, or real Shaligrams that have been intentionally broken by human hands to reveal their internal *chakras*, most often to increase their value in tourist shops.

Actual Shaligrams that bear intricate religious carvings, on the other hand, present something of a different issue. For some devotees, these painstaking works of art are highly prized and venerated as powerful *murti* representing gods and other celestial beings. For others, carved Shaligrams are viewed with caution: one-part divine manifestation, one-part human intervention. In the end, a Shaligram can never be made by man, who may or may not have performed the proper rituals of cleansing and may not be devoted in practice or possessed of pure intentions. As aniconic and self-manifest, Shaligrams appear just as they are to those worthy of obtaining them, and they appear in only one place in the world: the Kali Gandaki River of Mustang, Nepal.

In most vernacular etymologies, Shaligram is said to have originally been a place-name – supposedly that of a remote Nepali village called Śālagrāma or Shaligraman where the stones were first collected.[11] While it is now also considered an obscure name of Vishnu, Shaligram/ Śālagrāma is typically thought to have once been a village located somewhere along the banks of the Kali Gandaki River but whose precise whereabouts are now essentially unknown. Most published works on Shaligrams then derive the

10 Conversation recorded in Hindi, with occasional English words.

11 The *shila* part of Shaligram Shila simply translates to "stone" and is one of the preferred terms by which Shaligrams are most often referred; as opposed to using the English word "stone" given that the etymology of *shila* carries connotations of "precious stone" rather than simply "rock".

name from the Sanskrit word for "hut" or "house" (*grama* – which in some translations can also mean "village"), referring specifically to the hut of the sage Salankayana who once beheld the form of Vishnu as a tree just outside his door (see Varaha Purana). In this version of the story, this particular hut also sat directly on the banks of the Kali Gandaki where the stones first appeared, which then subsequently bestowed on them the same name. In some cases, the hut of Salankayana is part of a larger village that also shares the indicative name; in other cases, he lives alone and the village appears at the site of his vision much later on in the mythological timeline.

In other variations on the etymology, the region of Mustang which now contains the temple of Muktinath is said to have originally been called *Saligramam* (or in some cases *Thiru Saligramam* is used: *Thiru* is translated as "holy" or "sacred" and is also a Tamil name for Vishnu)[12] before the arrival of Buddhism to the area, but this claim is largely unsubstantiated and tends to be leveraged where claims to the Hindu origin of the region are politically contested. Lastly, in another version of the Shaligram etymology, a number of modern Shaligram devotees tend to reference the sala (or shala) tree (*shorea robusta*) as a possible origin point for the name.

Native to the Indian subcontinent, the sala tree can be found south of the Himalayas, across an area that spans Myanmar in the east and Nepal, India, and Bangladesh and holds significant religious significance for Hindus, Buddhists, and Jains. For Jains, it is stated that the 24th *Tirthankara* (literally "Teaching God", meaning a great *guru*) and founder of modern Jainism, Mahavira, achieved enlightenment under a sala tree. For Buddhists, tradition holds that Queen Maya of Sakya, while en route to her grandfather's kingdom, gave birth to Gautama Buddha while grasping the branch of a sala tree in a garden in the village of Lumbini in southern Nepal.[13]

In many Hindu traditions, the sala tree is especially favored by Vishnu and is praised in Sanskrit literatures for its use as housing timber (perhaps one reason for the similarity in names). In fact, the Kurma Purana

12 *Tiruchhalagramam* is the place-name given in Tamil scriptures.

13 Buswell, Robert Jr; Lopez, Donald S. Jr., eds. (2013). Princeton Dictionary of Buddhism. Princeton, NJ: Princeton University Press, p. 724.
"Then the Blessed One with a large community of monks went to the far shore of the Hiraññavati River and headed for Upavattana, the Mallans' sal-grove near Kusinara. On arrival, he said to Ven. Ananda, 'Ananda, please prepare a bed for me between the twin sal-trees, with its head to the north. I am tired, and will lie down.'" – "Maha-parinibbana Sutta: The Great Discourse on the Total Unbinding" (DN 16), translated from the Pali by Thanissaro Bhikkhu.
The sal tree is also said to have been the tree under which Koṇḍañña and Vessabhū, respectively the fifth and twenty-fourth buddhas preceding Gautama Buddha, attained enlightenment.

identifies "shalgrama" as a village on the banks of the river Gandaki (and the Gandaki as a tributary of the Ganga), so named for its shal trees. In the Kathmandu Valley in Nepal, builders often make use of sala trees in Nepali pagoda-style temple architectures, especially those that contain intricate wooden carvings. A significant number of the temples, such as Nyatapol Temple, are, in fact, ideally thought to be made of bricks and sala wood.

In any case, various Hindu schools of thought continue to have different opinions regarding the exact location of the original village and even of the origin of the name itself, but all agree Śālagrāma is a place that is deeply connected, both mythologically and historically, to the Kali Gandaki River Valley and to the people, plants, and animals that surround it. This has not, however, prevented scholars and explorers from trying to find Śālagrāma and solidify its position on maps and in atlases.

One map, published in a 1975 English translation of the *Mahabharata*, locates Śālagrāma near the source of the Kali Gandaki River.[14] A second map, however, published in *The Geographical Dictionary of Ancient and Medieval India* in 1971,[15] appears to place Śālagrāma along the upper ridge of the Greater Himalayas. It is especially interesting to note here that the course of the Kali Gandaki on both maps does not correspond with its actual course today, for reasons that may be related to both inadequate surveying and a reliance on religious texts, but it is in both cases a central point of orientation for the location of Śālagrāma. While neither map explains precisely what the Śālagrāma label is meant to indicate (i.e., a village versus a natural formation), it is clear that according to the second cartographer, Śālagrāma is located at the source of the Kali Gandaki River near the border with Tibet. The first cartographer, on the other hand, locates Śālagrāma about 60-70 miles below the source of the river.

In his book *Mustang: A Lost Tibetan Kingdom*, author Michel Peissel also mentions that he visited the source of the Kali Gandaki River (1992: 215). In this case, however, he locates the river's source near the border of Tibet at a village called Namdrol, in Mustang.[16] This is somewhat strange considering that, while the source of the Kali Gandaki is indeed near the Tibetan border, the river actually originates at an elevation of 6,268 meters (20,564 ft.) as

14 *The Mahabharata* (Vol. 2). Translated and Edited by J.A.B. van Buitenen. Published by: University of Chicago Press Chicago, USA; London, UK 1975.

15 *The Geographical Dictionary of Ancient and Medieval India*. Nundo Lal Dey, M.A., B.L. Published by: Oriental Books Corporation Delhi, India – 1971 (reprint 1927).

16 Likely referring to Namdrol village/Namdrol Gompa, a roughly 3 to 4-hour walk outside of the city of Lo Monthang in Upper Mustang.

it flows southwards out of Mustang's Nhubine Himal Glacier.[17] Even the Damodar Kund, the glacial lake that lies along the Damodar Himal on the far eastern edge of Mustang and is the point from which the majority of Shaligrams first make their way into the river system, lies further to the south within the rain shadow of the high Himalayas just north of the Annapurna range (4,890 meters) and therefore also does not correspond with the locations of Śālagrāma on the maps in question. While it is sometimes credited as the source of the Kali Gandaki River, the Damodar Kund is actually part of the Damodari Ganga estuary system (springing up along the Shaligram Parvat) which joins up with the Kali Gandaki via the Ghachang Khola further to the west and just south of Surkhang in Upper Mustang.

A true Shaligram can then appear anywhere between the Damodar Kund and Gajendramoksha Divyadham (a temple dedicated to the elephant-king Gajendra located in Triveni, a town in Katari Municipality in Udayapur District in the Sagarmatha Zone of southeastern Nepal, where the Kali Gandaki is called Narayani or the Sapt Gandaki and is known as "Muktinath's Feet"). Triveni is also the confluence where the Kali Gandaki crosses the Indo-Nepal border where it becomes simply the Gandak. Consequently, the regions encompassing the Upper Kali Gandaki, the Muktinath Valley, and portions of Lower Mustang are often referred to in religious literature as Mukti Kshetra (the field of salvation), Mukti-Natha-Kshetra (the field of the lord of salvation), or Saligrama-Kshetra (the field, or place, of Shaligrams). On occasion, sacred sites further afield might also be included in these terms. Depending upon their geographical relationships to either the Kali Gandaki River or any one of the many tributaries and inflows that make up its 46,300-square kilometer catchment on the way to the Ganges near Patna, any sacred site along the way may well also be called Śālagrāma.[18] As for the Puranic scriptures, many of them mention the place called Śālagrāma (such as the Gautamiya Tantra and Varaha Purana), but they're never quite specific on where exactly one might find it.

Locating Śālagrāma within an actual landscape is something of a futile endeavor mainly because its mythological dimensions are actually more instructive than the potential possibilities for its actual physical location (not unlike the religious conceptualization of the *dham* more generally).

17 Garzione, Carmala N. et al. (2000), "Predicting paleoelevation of Tibet and the Himalaya from δ18O vs. altitude gradients in meteoric water across the Nepal Himalaya", Earth and Planetary Science Letters, 183 (1-2): 215-229.

18 Negi, Sharad Singh. *Himalayan Rivers, Lakes and Glaciers*. p. 89. Google Books. Retrieved 19 September 2016.

In other words, determining where Śālagrāma lies on a map is a far less pressing concern to Shaligram practitioners than re-instantiating a sense of its "placeness" through the use of sacred stones wherever they may be undertaking ritual and spiritual practices. Invoking Fisher's metaphor of the river once more, it is the constantly shifting placement of Śālagrāma within the landscape that allows both Shaligram pilgrimage and Shaligram veneration to flow through space and time, breaking up into several meandering streams of tradition and merging again later downstream with each return to Mustang (or other pilgrimage sites, as may be the case). As devotees describe it, wherever Kali Gandaki goes is Śālagrāma, therefore everywhere a Shaligram goes is Śālagrāma. Additionally, each Shaligram is itself a "dwelling place for Lord Vishnu" and acts as the deity's principal home wherever it is kept (another parallel to the etymology of the name). For example, the Skanda Purana says:

> *mlecchadese sucav vapi cakranko yatra tisthati*
> *yajanani tatha trini mama ksetram vasundhare*
> *tanmadhye mriyate yastu pujakah susamahitah*
> *sarva vadhavinirmukto punah so api na jayate*

A Śālagrāma-śila, when duly worshipped at any place inhabited by any class of people, is able to purify an area with a radius of 24 miles. That area should be considered Vishnu-loka, it is non-different from the abode of Vishnu. If someone believing in the sanctity of Śālagrāma-śila, as per the verdict of the shastra, breathes their last within that 24-mile radius, he is sure to attain *mukti*, salvation from material bondage.

For this reason, through the movement of Shaligrams, a particular place of pilgrimage or place of ritual devotion may be described as being Śālagrāma, regardless of said place's actual geographical relationship to Mustang, Nepal. As long as Shaligrams are present, the place in question is Śālagrāma, endowing it with all the sacred characteristics and links to myths and legends possessed by the physical landscapes of the Kali Gandaki River valley (see Chapter 6). For now, what is important to note is that within larger contexts of sacred landscapes throughout South Asia, these kinds of connections between places and landscapes are not unusual. Many famous pilgrimage circuits (such as the Shakti Peethas and the 108 Divya Desam temples – of which the Muktinath temple complex is both) are characterized in this way, where one sacred space is said to be "no different" than another or is linked to other sacred spaces through the mythological movement of deities and the physical movements of people.

The mobile "placeness" of Śālagrāma then directly challenges representations of space, within both political and academic spheres, that are dependent on images of spaces (societies, nations, and cultures) based upon unproblematic divisions and arbitrary borders. But the "placeness" of Śālagrāma also challenges notions of perpetual discontinuity as well. As Gupta and Ferguson (1992) argue, the premise of discontinuity assumes that all spaces are fundamentally discontinuous despite peoples' beliefs and perceptions in the "rootedness" of a culture or society to a particular place (7). In other words, it deconstructs the perception that the borders of a nation on a map also define the location of a distinctive culture or society (i.e., "Indian culture" or "American society"). Discontinuity then forms the starting point from which social scientists can theorize contact, conflict, and contradiction between cultures and societies that blend, overlap, and encounter one another in a variety of ways. For Śālagrāma, however, the perpetual movement and continuous mobility of "placeness" is the key, creating a continuity of practices and beliefs through the exchange and veneration of an object that is both linked to a particular place and carries that place with it. The Shaligram, then, is both an object-person and a place, both continuous and discontinuous – unconcerned with borders and regularly uprooted.

Stones as Bodies

The theology of Shaligrams is based on the concept of *vibhuti*, a philosophy of worship dating back to the composition of the Rig Veda (1, 8, 9 *evāhite vibhūtayaḥ indram āvate* and 6, 21, 1 *ravir vibhūtir īyate vacasā*)[19] that roughly translates, in this usage, to "might" or "power".[20] Specifically, in reference to the material manifestations of this power, subsequent religious works

19 Philological and linguistic evidence indicate that the Rigveda was likely composed in the northwestern region of the Indian subcontinent sometime between c. 1500 and 1200 BC, though a wider approximation of c. 1700-1100 BC has also been given.
Flood, Gavin D. (1996). *An Introduction to Hinduism*, Cambridge University Press.
Witzel, Michael (ed.) (1997). *Inside the Texts, Beyond the Texts. New Approaches to the Study of the Vedas*, Harvard Oriental Series, Opera Minora vol. 2, Cambridge: Harvard University Press.
20 *Vibhuti* (Sanskrit: विभूति; *vibhūti*), also called Bhasma (ash), Thiruneeru and Vibhooti, is a word that has several meanings in Hinduism. In its most common modern-day usage, it denotes the sacred ash which is made of burnt dried wood in Agamic rituals. Hindu devotees apply *vibhuti* traditionally as three horizontal lines across the forehead and other parts of the body as markings in reference to the god Shiva. Vibhuti smeared across the forehead to the end of both eyebrows is called *tripundra*.

provide long listings of such incarnations of the Godhead and of various emanations and characteristics of the gods as suits their intents. In the tenth chapter of the Bhagavad Gita, for example, the thirteenth-century commentator Sayaṇa explains that *vibhuti* means "to possess special powers" and that it is this power which is responsible for all of the variety, creativity, and great expanse of the universe (Rao 1996: 4). The term then connotes a spread of or a great abundance of divine forms within the material world which also forms the foundations for the Hindu concept of omnipotence and, in this analysis, ontology and personhood. In other words, God himself can and does manifest in the world in multiple different forms simultaneously, be they a person, several persons, an animal, a group of animals, man-made works of art, or a series of naturally occurring objects, or some combination of all of them at once. This is another way in which Shaligram persons are not necessarily the same as human persons but rather represent a different kind of personhood – namely, a multi-physical, multi-local divine one.

The direct manifestation of the divine is called *nitavibhuti*, while the presence of the divine in the souls of humans is referred to as *naijavibhuti*; an innate and continuous emanation of the divine. The manifestation of divinity present in man-made icons and statues (those that have been properly installed ritually) is described as *ahita-vibhuti* (and also includes cows, tulsi plants, and the sacred ashvattha tree), which translates as "placed" or "projected". But the manifestation of Shaligrams as well as other sacred stones used in worship such as the Shiva Linga and the consecrated ritual fire (*saṃ skṛtagni*), are called *sahaja-vibhuti*: natural, original, or congenital (Rao 1996: 5). At first glance, this type of theology might seem to imply essentialism in the classical sense, a view where every entity has a set of attributes that are inseparable from its identity and function. But Shaligram ontologies take the simultaneous nature of manifestation quite literally. Not only are there millions of separate, individual Shaligrams in the world but each Shaligram is the same in that it is the same deity (or group of deities) and carries with it the same place (Mustang/Śālagrāma). Each is also a unique person, formed and defined by the specific familial and communal circumstances within which it is worshipped and kept. None of these things unmakes the other.

This hierarchy of manifestations found in *vibhuti* theology is widely accepted in many Hindu traditions and has also found its way into later Vedanta and *bhakti* devotional religious frameworks.[21] Adi Shankara, the

21 Bhakti (Sanskrit: भक्ति) literally means "attachment, participation, devotion to, fondness for, homage, faith or love, worship, piety to (as a religious principle or means of salvation)".

eighth-century philosopher and the codifier of the Advaita Vedanta school of Hindu thought (who is also the theologian most often referenced for his commentary of Shaligrams), famously speaks of *vibhuti* as the manifestations of the one Brahman (the universal divine) that takes varied forms so that devotees on earth might approach a realization of the absolute formless reality in whatever manner most befits their needs.[22] In other words, echoing the Shaligram devotee I quoted at the beginning of this chapter, God appears in whatever form most suitable for what must be accomplished in the moment. Whatever *vibhuti* is, then, it is one with ultimate reality and regarded by Hindus as "non-different" than the godhead itself. Shaligrams, therefore, are not simply regarded as representations of the divine or of some other idea or concept that indexes the presence of the deity, rather, they exist as manifestations of the divine in its entirety.

Each Shaligram, in its broadest sense, is a piece of the undefinable, all-expansive divine that encompasses all of material and non-material reality but that continuously links mundane material concerns with the proper ordering of the cosmos (*tirtha*). It is then Vishnu (or sometimes Shiva), the Supreme Being (the *Svayam Bhagavan*), who is identical to the formless, metaphysical Brahman but who takes on various *avatars* and descends to earth whenever humanity is threated by evil, chaos, and destruction. It is then the specific *avatar*, the exact deity, the precise form necessary for what needs to be accomplished in that moment or in response to the needs of the devotee who has come seeking it. Taking on hundreds of possible forms, from ammonite to *avatara*, from fossil to family, and from stone to deity, Shaligrams look as different from one another as any person does from another person. As a result, extensive traditions of identification have grown up around Shaligram practices over centuries of pilgrimage and ritual veneration, where religious specialists and devotees alike have dedicated their lives to pouring over ancient texts and engaging in local traditions such that the messages written in the characteristics of each stone might be read and the personality of the deity within divined.

These practices of interpretation, or "reading the body" of the Shaligram, are a crucial step in the formation of Shaligram personhood because they will form the foundation by which devotees not only construct the pilgrimage

Bhakti in Hinduism refers to devotion and the love of a personal god or a representational god by a devotee. In ancient texts such as the *Shvetashvatara Upanishad*, the term simply means participation in, devotion to, and a love for any endeavor. It may also refer to one of the possible paths of spirituality and moksha as in the *bhakti marga* mentioned in the Bhagavad Gita.

22 Vedantasutra-bhaṣ ya, 93.2, 23 – *ekam apibrahma vibhūti-bhedair anekadha upasyata iti sthitih*. See also, 2, 4, 10; 3, 3, 23; 1, 4, 4 and 3, 3, 43.

narrative (see Chapter 7) of experience and landscape but will use as the basis for interacting with specific Shaligrams as individuals over their "lifetime" within the family household once they enter the ritual and kinship networks of the community. This is because a new Shaligram, freshly appearing in the river, is often described as relatively unbound, save for its mythological and geological formations. It has yet to form attachments to the people who will ultimately undertake its lifelong care.

In fact, it was not uncommon for pilgrims to Mustang to describe a wide variety of Shaligram "capture" techniques. As one pilgrim, a man by the name of Rajesh Gulati, demonstrated: "I brought with me a large bag for the Shaligrams and I filled it with water from Kali Gandaki. That way, they will not know that they have been born yet and they will not try to run away. I can also take them on the long journey home and not worry that they might escape. They are still sleeping, and they can be born in front of my wife and my parents, who could not come on pilgrimage. We will also keep the water for as long as possible, to put next to them on the altar. They will recognize it and want to stay. They will know they are home and they will share with all of us the cleansing of the river." This control of a Shaligram's "birth" instantiated each Shaligram's place within the family and the household. Rather than being born strictly out of the river, they could also be born at home with mothers, wives, children, and parents in attendance to welcome them and form relationships with them as new members of the family.

The practice of Shaligram interpretation then generally follows their appearance in the river or takes place shortly after the Shaligrams have returned home (depending on the knowledge level of the specific devotee or their access to ritual specialists at home). In a process that recalls the reading of human bodies for evidence of their past lives, the reading of Shaligram characteristics is meant to determine which "person" is present for the sake of knowing his or her proper care. For example, a Krishna Shaligram may be described as more playful and mischievous than other Shaligrams, a Durga Shaligram as more volatile and hard to please, or a Kurma Shaligram as more passive and forgiving. In other cases, specific Shaligrams are sought after for the particular benefits they bring, such as improving wealth or job prospects, healing illness and infirmity, arranging good matches for marriages, and encouraging the birth of healthy children. Such Shaligrams are also typically expected to join a family or temple's already established altar of deities and sacred objects, where they will contribute their qualities and agency to the community deities and human beings as a whole. In short, they will transfer their properties and characteristics to those around them and take in the interactions and exchanges offered in return, intensifying all peripheral intermixings of kinship, friends, community, places, and things by extending the ties of personhood beyond human fellowship and into the world of the dead and divine. But in order to do so, the deity must first be determined, and the person revealed. The final result is the encounter of a person born out of the sacred river – a complex amalgam of bodily pieces, physical and spiritual substances, mythic and geological connections, and pre-existing states of being. Each Shaligram then invokes one or another deity, ancestor, *guru*, place of worship, or familial/community/ethnic lineage made legible by the multivocal readings of the stones and of the landscape to bring forth the first interactions of kinship and identity.

Bridging the Gaps

Shaligrams blur the boundaries between being and object and may be either one or the other or both depending on their circumstances. In previous scholarship on personhood in South Asia, anthropologists have noted the ways in which relationships and ties to other people, places, and things constitute an understanding of persons as existing both within and between bodies. This is a process that is predicated on the "exchange of substances" between entities and which relies on the continued maintenance of personal relationships to the material world. This very same process also applies to Shaligrams, who achieve a kind of divine personhood, in the lives they live and the people with whom they share attachments. As a result, Shaligrams are treated as both manifest deities – with their own wants, needs, desires, and agency – and ritual objects who organize community participation around a series of ritual events and expectations for conduct in their presence. Furthermore, in many academic discourses, religion and science are often perceived as mutually exclusive at best and actively antagonistic at worst. But, as noted previously, among Shaligram practitioners, the juxtapositions of "deity" and "fossil" or "stone" and "body" are not incommensurate with one another.

As object-beings, Shaligrams also act as *tirthas*, or bridges that connect one person to another as well as people to places and those places to histories. Given the contentious history of Mustang, of Muktinath, and of Shaligram pilgrimage overall, it is not surprising that different practitioners view the connections they have with their sacred stones in a variety of ways. From repositories of familial history mobile extensions of sacred places identified as the spiritual location called Śālagrāma to evidence of a paleontological or mythical past and to self-made deities, Shaligrams are both persons and places (*dhams*). Learning to read each Shaligram's unique characteristics in the same way as one might read a text becomes paramount to identifying precisely which deity is manifest in the stone and in what ways they each might prefer to interact with their human devotees. Ultimately, these readings form the foundation for the development of Shaligram interpretive traditions, several of which still exist through combinations of textual and oral custom.

Every Shaligram invokes a number of connections, be it to a deity, an ancestor, a history, a place, or a lineage of kinship. But understanding how exactly Shaligrams link each of these things together calls into question exactly how practitioners understand the relationships between recorded history and folklore and how various accounts of the origins of Shaligram practices have affected those practices today. And no relationship is more important in this case than that of the Shaligram mythic complex, as laid out in Puranic and Shastric texts, and the geological processes that have resulted in the formation of these unique stones in the high Himalayas of Nepal. In other words, once again, fossil and deity are never quite so distinct as one might initially presume.

Bibliography

Daniel, E. Valentine. 1984. *Fluid signs: Being a person the Tamil way.* Berkeley: University of California Press.

Descola, Philippe. 2005. *Par-delà nature et culture.* Paris: Gallimard.

Eck, Diana. 1986. "Darshan of the Image," *India International Centre Quarterly* vol. 13, no. 1, IMAGES (March): 43-53.

—. 1998. *Darśan: Seeing the Divine Image in India.* New York: Columbia University Press.

—. 2012. *India: A Sacred Geography.* New York: Three Rivers Press.

Fowler, Chris. 2010. "From Identity and Material Culture to Personhood and Materiality." In *The Oxford Handbook of Material Culture Studies*, edited by Dan Hicks and Mary C. Beaudry, pp. 352-85. Oxford: Oxford University Press.

Geismar, Haidy. 2011. "Material Culture Studies and other Ways to Theorize Objects: A Primer to a Regional Debate," *Comparative Studies in Society and History* 53 (1): 210-218.

Gell, Alfred. 1992. *The Anthropology of Time.* London: Berg.

—. 1998. *Art and Agency: An Anthropological Theory.* Oxford and New York: Oxford University Press.

Gupta, Akhil, and James Ferguson. 1992. "Beyond 'Culture': Space, Identity, and the Politics of Difference," *Cultural Anthropology* 7 (1): 6-23; reprinted in *Culture, Power, Place* (edited by Gupta and Ferguson, 1997, pp. 33-51) and in *Anthropology of Globalization* (edited by Inda and Rosaldo, 2002, pp. 65-80).

—. 1997. "Discipline and Practice: 'The Field' as Site, Method, and Location in Anthropology." In *Anthropological Locations: Boundaries and Grounds of a Field Science*, edited by A. Gupta and J. Ferguson, pp. 1-46. Berkeley: University of California Press.

Henare, Amiria, Martin Holbraad, and Sari Wastell, eds. 2006. *Thinking Through Things: Theorising Artefacts Ethnographically*. London: Routledge.

Holbraad, Martin. 2012. *Truth in Motion. The Recursive Anthropology of Cuban Divination*. Chicago: Chicago University Press.

—, and Morten Axel Pedersen. 2009. "Planet M: The intense abstraction of Marilyn Strathern," *Anthropological Theory* 9 (4): 371-394.

Hoskins, Janet. 2006. "Agency, biography. And objects." In *Handbook of Material Culture*, edited by C. Tiller et al., pp. 74-84. London: Sage Publications.

Inden, Ronald B., and Ralph Nicholas. 1977. *Kinship in Bengali culture*. Chicago: Chicago University Press.

Keane, Webb. 2009. *On Multiple Ontologies and the Temporality of Things*. Material World blog, 7 July. URL: http://www.materialworldblog.com/2009/07/on-multiple-ontologies-and-the-temporality-of-things/. Accessed 26 October 2017.

—. 2014. "On Spirit Writing: Materialities of Language and the Religious Work of Transduction," *Journal of the Royal Anthropological Institute* vol. 19 (1): 1-17.

Knappett, Carl. 2005. "Animacy, Agency, and Personhood." In *Thinking through material culture: An interdisciplinary perspective*, pp. 11-34. Philadelphia: University of Pennsylvania Press.

Kohn, Eduardo. 2007. "How Dogs Dream: Amazonian Natures and the Politics of Transspecies Engagement," *American Ethnologist* 34 (1): 3-24.

—. 2013. *How Forests Think: Toward an Anthropology Beyond the Human*. Berkeley: University of California Press.

—. 2014. "What an Ontological Anthropology Might Mean." Theorizing the Contemporary, Cultural Anthropology website, January 13, 2014. https://culanth.org/fieldsights/463-what-an-ontological-anthropology-might-mean. Accessed 13 January 2014.

Laidlaw, James. 2012. "Ontologically Challenged." *Anthropology of This Century*, vol. 4, London. http://aotcpress.com/articles/ontologically-challenged/. Accessed 12 October 2016.

Lamb, Sarah. 1997. "The Making and Unmaking of Persons: Notes on Aging and Gender in North India," *Ethos* 25 (3): 279-302.

—. 2000. *White Saris and Sweet Mangoes: Aging, Gender, and Body in North India*. Berkeley: University of California Press.

—. 2004. "Making and Unmaking of Persons: Gender and Body in Northeast India." In *Gender in Cross-Cultural Perspective*. 4th ed., edited by Caroline Brettell and Carolyn Sargent, pp. 230-240. Upper Saddle River, NJ: Prentice Hall.

Latour, Bruno. 1988. *The Pasteurization of France*. Translated by Alan Sheridan and John Law. Cambridge, MA: Harvard University Press.

—. 1993. *We have never been modern*. Cambridge, MA: Harvard University Press.

—. 1999. "On Recalling ANT." In *Actor Network Theory and After*, edited by John Law and John Hassard, pp. 15-26. Oxford: Blackwell Publishers.

Marriot, McKim. 1976. "Hindu transactions: Diversity without dualism." In *India through Hindu categories*, edited by Bruce Kapferer, pp. 109-142. New Delhi: Sage Publications.

—, and Ronald Inden. 1977. "Toward an ethnosociology of South Asian caste systems." In *The new wind: Changing identities in South Asia*, edited by Kenneth David, pp. 227-238. The Hague: Moulton Publishers.

Munn, Nancy D. 1970. "The Transformation of Subjects into Objects in Walbin and Pitjantjatjara Myth." In *Australian Aboriginal Anthropology: Modern Studies in the Social Anthropology of the Australian Aborigines*, edited by Ronald M. Berndt, pp. 100-118. Nedlands, W.A.: University of Western Australia Press.

—. 1992. "The Cultural Anthropology of Time: A Critical Essay," *Annual Review of Anthropology* 21: 93-123.

Pedersen, Morten Axel. 2011. *Not Quite Shamans. Spirit Worlds and Political Lives in Northern Mongolia*. Ithaca, NY: Cornell University Press.

—. 2012. "The Task of Anthropology is to Invent Relations: For the Motion," *Critique of Anthropology* 32 (1): 59-65.

Puranas. 2007. Translated into English by a board of scholars. Motilal Banarsidas Publishers, 74 Volumes.

Rao, S.K. Ramachandra. 1996. *Shaligram Kosh* [*Śālagrāma – Kosha*]. Sri Satguru Publications. Delhi: Indian Books Center.

Singh, Chandra Shekar et al. 2017. *The Puranas*. Amazon Digital Services.

Tilley, Christopher. 2001. "Ethnography and Material Culture." In *Handbook of Ethnography*, edited by P. Atkinson et al., pp. 258-272. London: Sage Publications.

Viveiros de Castro, Eduardo. 2009. *Métaphysiques cannibales: Lignes d'anthropologie post-structurale*. (*Cannibalistic Metaphysics: Lines of Post-structural Anthropology*). Paris: Presses universitaires de France.

—. 1998. "Cosmological deixis and Amerindian perspectivism," *Journal of the Royal Anthropological Institute* 4 (3): 469-488.

Walters, Holly. 2016. "Playing God: Participant Frameworks in the Ras Lilas of Krishna," *Journal of Religion and Popular Culture* vol. 28, issue 2-3 (Fall).

Whitehead, Amy. 2013. *Religious Statues and Personhood: Testing the Role of Materiality*. New York: Bloomsbury Academic Publishers.

6 Turning to Stone

The Shaligram Mythic Complex

Abstract
Shaligram origin stories are as variable as the stones themselves. Whether formed by the *vajra-kita* (thunderbolt worm) whose stone-carving capabilities continue to link religious creation stories with ammonite paleontology or by any number of curses levied at Vishnu for betraying the chastity of the goddess Tulsi, the mountain and river birth of a Shaligram is always preceded by a complex narrative of time, place, and personhood. The core conceptualization of bodies as landscapes, however, remains constant. The birth-death-rebirth processes of the landscape then becomes metonymic for the karmic birth-death-rebirth cycle shared by humans, their deities, and their Shaligrams.

Keywords: Shaligram, tulsi, myth, Himalayas, vajra-kita

"One who thinks the Deity in the temple to be made of wood or stone, who thinks of the spiritual master in the disciplic succession as an ordinary man, who thinks the Vaisnava in the Acyuta-gotra to belong to a certain caste or creed or who thinks of caranamrta or Ganges water as ordinary water is taken to be a resident of hell."
– *SB 4.21.12 from Padma Purana*

As the rickety bus barely rounded another corner, an audible gasp went through the passengers. A recent blizzard had taken out the road between the high Himalayan villages of Ranipauwa and Jharkot, leaving some 800 meters of mountainous mudslides between us and any number of several-hundred-foot dropoffs all the way down to the Kali Gandaki River Valley below. A few feet on our right were the steep walls of the Muktinath Valley and the 8,000+ meter peaks of Nilgiri and Dhaulagiri. To our left was a sheer

Walters, Holly, *Shaligram Pilgrimage in the Nepal Himalayas*. Taylor & Francis Group, 2020
DOI: 10.5117/9789463721721_CH06

vertical drop that began less than a foot away from the trundling wheels of our makeshift vehicle as we wound our way precariously along the peaks. More than once, our bus slid into the treacherous rocks, tilting almost completely sideways over the edge and holding us out over the endless expanse. It would take at least another hour white-knuckling to Kagbeni, the village along the river that would be our stopping point on the trip back to Jomsom, a few kilometers away.

Had I known the road was so poor at the time, I would have made my way down the mountain by my more typical choice of transportation – horseback. But the Himalayas are nothing if not unpredictable, and I hadn't anticipated the late-spring weather to be quite so fickle. My choice in taking the bus was that the trip by horse is somewhat over six hours while the bus is usually only about two, and I had hoped to reach Jomsom before nightfall. Now I, and several other pilgrims to Roof of the World, clutched our seats and each other for dear life, wondering if the half-ton truck would make the trip at all. At least the horse's sense of self-preservation would have been as strong as mine, I remember thinking as we lurched wildly onward.

When I arrived in Jomsom at last, I was more than happy to take up a table at one of the local Thakali tea shops to wait for Binsa Sherchen, a local woman who had been serving as a Shaligram pilgrimage guide for several years. When she finally arrived, delayed by the same late blizzard weather, she slid into a chair and immediately produced a large white shell from her bag. "I thought you'd want to see this first!" she exclaimed. "It's a Lakshmi Conch! One of the very rare ones that spiral to the right."[1] "What do you mean?" I asked, picking up the palm-sized white shell. "That the spiral is clockwise?"[2]

Binsa had always had a particular love of ritual objects and had spent years amassing a collection of *puja* items and festival crafts which she occasionally sold or gave away to pilgrims. "Yes!" Her eyes wrinkled in

1 Shell collectors refer to the reversed Shank as "left-handed" or "sinistral turbinella pyrum"; Hindus call the right-handed version "valampuri" because they orient it with the apical spire downwards and the aperture or siphon (mouth) uppermost and, consequently, on the right side of the shell. Such shells are common in both Hindu and Buddhist rituals of veneration. There are many species in the Conch family, but in South Asia, "shank" always refers to normal smooth white conch shells. However, only a right opening shell is considered to be a real Lakshmi Shank. But in this case, as is likely in other cases throughout India and Nepal, the shell presented here was almost certainly a species of Lightning Whelk (*Sinistrofulgur perversum*) rather than the more favored *Turbinella pyrum*, a species of edible sea snail. As such, it is highly unlikely that the conch shell I was shown was, in fact, a right-turning Lakshmi Conch (*Valampuri Turbinella Pyrum*), given that only a very rare few of these shells are even known.

2 Conversation in Nepali and Trekker's English. Transcribed from fieldnotes.

delight. "The right spiral is the motion of the sun and the moon and all of the stars moving in the sky. It is also the locks of hair on Buddha's head. They spiral to the right just like the curl between his eyebrows and the conch of his navel. I will bring it with me when the pilgrim groups come in a few weeks and we can do *puja* with it on the river. They call it the Lakshmi shell and it is the best kind you can have for bathing *vajra-kita shila*." I had heard this alternative term for Shaligrams only once before. "*Vajra-kita*?" I questioned. "Yes, Shaligram worms," she replied. "The worms [*vajra-kita*] are extinct now, as you know, but they left many Shaligrams [Binsa often conflated *vajra-kitas* with ammonites in our discussions]. That is why there are no new Shaligrams, only very ancient ones."

The mechanisms through which Shaligrams are formed plays into questions of Shaligram agency and ontology. There are a variety of stories that seek to explain precisely why it is that Shaligrams have the form that they do – why their spirals are so clean and precise or why they appear so consistent in shape, even though their sizes may vary drastically. These mechanisms of formation also tend to explain why Shaligrams are not only set apart from other stones and rocks but should be considered bodies with needs and agency rather than as inanimate objects. These narratives foreground particular concepts of intentional making (recall "techne" from Chapter 1), where the iconic spiral shape of the Shaligram is never an accident of geology or the ongoing processes of fossilization but is instead directed by the gods or by the embodied landscape for a purpose – that purpose being to foment interactions between deities and humanity. Or, to put it more succinctly, Shaligrams are made the way they are so that people will see God within them and return with them to their families as kin.

The Formation of Shaligrams by Vajra-Kita or the Thunderbolt Worm

The formation of a Shaligram is generally dependent on two principal entities: the deity who manifests within the *shila* (usually Vishnu or one of his avatars) and the activities of the *vajra-kita* (variously translated as thunderbolt, diamond, or adamantine worm), the celestial worm physically responsible for carving out the holes and coiled chakra formations (recall, once again, the association with "serpent stones" and "worm stones").[3] The

3 In some traditions, the *vajra kita* is replaced by the god Vishwakarma, who presides over art and architecture.

inclusion of the *vajra-kita* in Shaligram mythography is, however, both fascinating and steeped in competing Hindu, colonialist, and Christian missionary perspectives. For this reason, Shaligrams offer something of a fascinating case study for finding new ways to reconcile the study of South Asian folklore with competing voices drawn from English, Sanskrit, and vernacular sources (see Korom 2006) because each of these sources tends to reference one another and either refute or blend their analytical perspectives under a variety of circumstances. For example, many Hindu scholars take the inclusion of the *vajra-kita* to be a later addition to the Puranas, and some theorize that it may represent a particular point in time wherein the peoples of South Asia were gaining greater understanding of the natural processes of the world around them. Additionally, the *vajra-kita* are not always included in the story of Shaligrams, and when they are, they typically appear only briefly in the process of divine manifestation: there to do the work of carving the physical form of the Shaligram and little else. (In some variations of this story, it is the god Vishwakarma, who presides over art and architecture, who physically carves the stone – another possible reference to the meaning of Shaligram as "house stone.") While this may represent an early attempt at reconciling mythic narrative with empirical observation of the natural world, the continued use of the *vajra-kita* today still does the cultural work of linking religious creation stories with modern-day science, and many Shaligram practitioners still reference the adamantine worm as analogous to the ammonite in discussions about the fossil origin of Shaligrams.

By most Hindu accounts, the *vajra-kita* is described as a kind of insect or worm bearing a diamond or adamantine tooth that cuts through the Shaligram in a spiral pattern as the *vajra-kita* burrows inside of it. Once there, the worm remains within the *shila* in perpetuity. Interestingly, this constitutes another way of rethinking the nature of life in terms of describing stones as bodies or when considering the question "Is Shaligram alive?" In some narratives, the Shaligram is acting as a "house" for the life within it, and in other cases it is itself alive. This explanation then tends to lead to a curious mythological blending in that the inclusion of the *vajra-kita* in religious stories never quite results in the same story twice. In one account, from *The Missionary's Vade Mecum* published in 1847,[4] a

4 T. Phillip. 1847. *The Missionary's Vade Mecum, Or, A Condensed Account of the Religious Literature, Sects, Schools, and Customs of the Hindus in the North West of India: With Notices of Missionary Controversial Works, Lines of Argumentation, Etc.* Calcutta. Printed by J. Thomas at the Baptist Mission Press.

missionary instructor, commenting on Vedic Astrology in India, describes the *vajra-kita* thus:

> The Sálagrám, Ammonite-stone found in the river Gunduk and other rivers flowing through Nepal from the Himálaya mountains. Ward says – "the reason why this stone has been deified, is thus given in the Sri Bhágavat: Vishnu created the nine planets to preside over the fates of men. Shani (Saturn) commenced his reign by proposing to Brahmá that he should first come under his influence for twelve years. Brahmá referred him to Vishnu, but this god, equally averse to be brought under the dreaded influence of this inauspicious planet, desired Saturn to call upon him the next day, and immediately assumed the form of a mountain. The next day Saturn was not able to find Vishnu, but discovering that he had united himself to the mountain Gandaká, he entered the mountain in the form of a worm called Vajra-Kita (the thunderbolt worm). He continued thus to afflict the mountain-formed Vishnu for twelve years, when Vishnu assumed his proper shape, and commanded that the stones of this mountain should be worshipped and should become proper representations of himself; adding that each should have twenty-one marks in it, similar to those on his body, and that its name should be Sálagrám. (p. 89)

Another legend similar to this one is recounted in P.K. Prabhu-desai's *Devi-kosa* (vol. III, pp. 158-159, Tilak Maharashtra Vidyapith, Pune, 1968). It omits, however, the actions of the *vajra-kita*. In this version, Vishnu appoints nine principal planetary deities called the *navagraha* whose duty it is to preside over the destinies of all mankind. Having done so, Vishnu asks Shanaishcara (Saturn) to serve a period of training under Brahma where it would be Shanaishcara's duty to cause hardship for a span of twelve years.

After the training was complete, Brahma suggested to Shanaishcara that he should test himself against Vishnu before he set out to trouble mankind with his inauspicious influences. When Vishnu learned of the plan, he transformed himself into a mountain on the banks of the Gandaki river in order to escape. But Shanaishcara was not to be outwitted and he attacked the mountain with all his strength. As a result of Shanaishcara's impact the mountain was shattered into millions of tiny rocks which then fell down into the river. These stones became Shaligram. In these narratives, the formation of Shaligrams is attributed to in-dwelling by Vishnu, who then vacates the stones which are to become his representations on Earth, leaving behind a kind of shell (pun intended) of a divine self. But while these stories are relatively well-known among Shaligram practitioners, they are not typically viewed as authoritative. Rather, many Shaligram devotees take these stories as part of a larger corpus of mythic retellings that attempt to explain why Shaligrams have chosen to take the forms that they do and not how they have come to be self-manifest.

In another account written by Francis Wilford (1761-1822), the German Orientalist and member of the Asiatic Society of Bengal, the *vajra-kita* takes on a bit more of a sinister bent, associating Shaligram veneration with colonialist mentalities of idol worship. In a contribution to the Society's journal *Asiatic Researches*, Wilford includes his own description of Shaligram folklore, stating that:

> Once when Vishnu the Preserver was followed by Shiva the Destroyer he implored the aid of Maya (illusion or Glamour) who turned him to a stone. Through this stone, Shiva, in the form of a worm, bored his way. But Vishnu escaped, and when he had resumed his form he commended that this stone of delusion (*sala-maya*) should be worshipped. As they are found at Salipura or Salagra, they receive their name from the latter. They are generally about the size of an orange, and are really a kind of ammonite. (vol. xiv, p. 413)

A constant contributor to the journal *Asiatic Researches*, Wilford was, however, known for contributing a number of fanciful, sensational, and highly unreliable articles about everything from ancient Hindu geography to mythography and a number of other subjects. For example, between 1799 and 1810, he contributed a series of ten articles about Hindu. geography and mythology for the journal that claimed that all European myths and legends were actually of Hindu origin and that India had produced its own "Christ" (*Salivahana*) whose life and works closely resembled his interpretations of the Biblical Jesus Christ. He also claimed to have discovered a Sanskrit version of Noah (*Satyavrata*) and attempted to confirm the historicity of the Book of Revelation and of the genealogies of Genesis using Hindu and other religious sources. In his essay, *Mount Caucasus – 1801*, he even argued for a Himalayan location of Mt. Ararat (the site on which Noah's Ark comes to rest), incorrectly claiming that *Ararat* was etymologically linked to the Sanskrit name for India, Aryavarta.[5] It is therefore unlikely that most of the colonialist and missionary accounts of the *vajra-kita* are particularly reliable

5 Malik, Jamal (2000). Perspectives of mutual encounters in South Asian history, 1760-1860. Gilroy, Amanda (2000). Romantic geographies: discourses of travel, 1775-1844. "Shelley's Orientalia: Indian Elements in his Poetry" (PDF). atlantisjournal.org. p. 13.

outside of the common mythemes they include. In terms of Shaligram practice, this particular body of colonialist commentary, of which Wilford is a fine example, is generally taken as an attempt by Westerners to blend their own religious traditions into that of Hinduism or Buddhism more broadly or to discredit Vedic beliefs using European Enlightenment logic. As a result, this is seen by many as a misuse of religion as well as a misuse of science. Because Shaligrams are "actually ammonites" just as much as they are "actually deities", Shaligram practitioners tend to view any attempt to leverage the story of the *vajra-kita* as a method for claiming that Shaligram traditions are contrary or inconsistent as an insult to the complexity of Shaligram ontologies and capacities to act.

The *Asura-khanda* section of the Skanda Purana, however, relates the tale of the *vajra-kita* differently, placing it directly in the context of the Gandaki/Tulasi origin story. In this version of the origins of Shaligrams, Gandaki, a pious woman possessed of an insurmountable will, performed severe austerities while residing in the Himalayas over many years. The rather interesting purpose of her penance however, unlike that of most female Hindu ascetics, was to become a mother and to obtain all the gods as her offspring. When her austerities were finally appreciated, the three principle gods – Brahma, Vishnu, and Mahesh (Shiva the Destroyer) – appeared before her and asked her to choose whatever boon it was that she desired most. Gandaki, of course, immediately expressed her wish that each of them should be born out of her womb as her own children. Unfortunately, the gods did not find this request particularly appealing and set out to wonder as to how an immortal deity could be born as child to a human mother. In fact, they considered the request quite unbecoming of the woman, who apparently did not adequately understand the nature of gods (an interesting dilemma given *avatar* theology). Brahma, Vishnu, and Mahesh then pleaded with Gandaki to forget her unreasonable and impossible request and instead to ask for some other desire that they could satisfy in return for her veneration. Regardless, Gandaki remained unmoved and when the gods continued to refuse to grant her wish, she rose up and became indignant. Citing their unwillingness to repay her austerities, she cursed them to become lowly worms. The gods then became angry in response and cursed her in return, this time to become a dark and dangerous river.

Gandaki's curse as well as the counter-curse levied by the three principal gods were soon a matter of great concern for the rest of the gods and celestial beings. Because the pious woman had acquired significant levels of occult power during her penances, her curse could not be avoided. Brahma, Vishnu, and Mahesh were therefore obliged to become the worms they were pronounced to become while she became a river. Distressed, the rest of the deities present took an audience with Brahma and begged him that they should be allowed to intervene and prevent the curse from coming to fruition. Brahma, regrettably, was unable to think of a proper solution and so sent them to Mahesh. Mahesh, however, explained that he was but a destroyer and Brahma a creator. He had no solution either. Vishnu, on the other hand, was the preserver and as the protector of universal order, it was likely that he would have a better idea as to what they might do.

Vishnu did indeed provide the answer: "I have a solution. The curse cannot be undone; they must run their course. But there is a plan whereby the curse and the counter-curse can be pressed for the good of mankind. Our curse on Gandaki has already taken shape. She has become a river, rendered holy by her austerities. Two of my attendant-devotees, the holy *brahmanas*, have previously had to assume the forms of marine monsters (*graha-matangau*) owing to another curse. I will liberate their spirits and enter their cadavers. When their corpses decay and shrivel, you gods can become worms born out of the bone-marrow and fat of the withering cadaver and enter into the stony parts of the cadaver. Although worms, you will have adamantine bodies, and hence you will be known as *vajra-kita*. I will immerse the cadavers of the marine monsters, into which I would have entered, into the river Gandaki. And when you appear as worms inside the cadaverous recesses, you would be regarded as the offspring of the river Gandaki. Thus, Gandaki's curse that you should be born as worms will come true; and she would also have the satisfaction of having you as her children, for this was the boon that she asked for."[6]

By this point, there was a part of the river Gandaki that had become known as a *chakra-tirtha*, a bridge between the physical and sacred worlds which was especially dear to the gods. The cadavers of Vishnu's attendant-devotees were placed at the site of the *tirtha* and the gods then appeared as *vajra-kitas* within these bodies which were now landscapes. Vishnu himself appeared as a discus (*chakra*) in the kingdom of Dvaaravati, where he was also able to mingle with the gods within the river Gandaki. Since that time, it is said that a bath taken in the river at this place, along with the worship of the "fossilized gods" inscribed with the mark of Vishnu's *chakra* (Shaligram), would ensure instant release for the devotee from the karmic cycle. This is because the Shaligram stones were formed out of "cadaverous fossils" (bodies turned to stone, another reference to bodies as landscapes)

6 English translation of the Skanda Purana was published by Motilal Banarsidass, also in Rao's Śālagrāma-Kosha, p. 31.

that were inhabited by the gods as worms (*vajra-kita*) and by Vishnu in the form of his *chakra*.

In contrast to the first story related by the missionaries, I note that in the "divine corpse" version, the *vajra-kita* are not directly responsible for the formation of the *chakra* spirals within the Shaligram stones. Rather, they are manifest simultaneously with the symbol of Vishnu's discus as fulfillment of the sacred river's reproductive mandates. Similarly, the Bhavisya Purana relates a tale wherein Tulasi, a woman who is transformed into the sacred plant that is particularly dear to Vishnu, curses Vishnu to become a stone during one act of their eternal dance (*lila*). (She does this due to his "stone-heartedness".) Vishnu then goes on to say: "To fulfill your curse, I will become a stone (*Salagrama*) and will always live on the banks of the Gandaki River. The millions of *vajra-kita* worms that live at that place will adorn those stones with the signs of my *chakra* by carving them with their sharp teeth."

Finally, one additional legend concerning the formation of Shaligrams through the actions of the *vajra-kita* is recorded in Rao's *Śālagrāma-Kosha*. Though the author is unclear as to where this particular version of the Shaligram creation myth comes from, it leverages the inclusion of the *vajra-kita* in a manner that is yet again different than the previous stories. In this tale, Narayaṇa (Vishnu) chooses to transform himself into a golden insect (who is called a *vajra-kita*) who wandered about the Earth in ancient times. Witnessing his exploits, the other gods also decided to assume the forms of insects and became bees. In short order, the world was apparently filled with these strange, divine insects swarming, humming, and flying about everywhere anyone went. However, seeing his master carousing about in this manner, Garuda (the great golden bird and Vishnu's celestial mount) turned himself into a giant rock that prevented all the gods from flying around. Finding no immediate way around the obstacle, Narayaṇa entered a crack in the rock while all of the other gods (still as bees) followed suit. The insects therefore took up residence in the rock and made homes for themselves in the form of shells shaped like Narayaṇa's chakra. These are now known as Shaligrams.[7]

Shaligram origin stories that include the *vajra-kita* tend to be the outliers in the overall corpus of Shaligram creation narratives, both in terms of volume and in terms of common usage. What I mean by this is that, firstly, the origin stories recounted above constitute only about three legends out of a list of roughly seven to nine creation myths (depending on how

7 Recounted in Rao's Śālagrāma-Kosha, p. 42.

one interprets variants, such as the story of the brothers Jay and Vijay in the Varaha Purana[8]) used to describe the beginnings of Shaligrams and secondly, the majority of Shaligram pilgrims and devotees do not typically reference these particular stories in their own understandings of Shaligram practices. In the remaining four (or six) versions of the origin of Shaligrams, Vishnu and the other deities concerned are directly self-manifest and the appearance of the *chakra* discus in the stone is taken as explicit evidence of the presence of the divine and not as the secondary action of a divine worm.

Among Shaligram devotees, the story of the *vajra-kita* is also taken somewhat piecemeal, or at the very least as a secondary cause. Few devotees subscribe to the presence of the thunderbolt worm in their Shaligrams, and even fewer are familiar with the stories of their manifestations. If they do reference the *vajra-kita*, it tends to be more as a method of detailing the ways in which Shaligrams are made by neither humans nor nature and as evidence of their agency outside of human purviews. For example, one devotee whom I happened to meet at festival in Kathmandu explained, "the *vajra-kita* is just part of the divine formation. Shaligrams are not made by man, and they are not formed in nature either. They are divine, through and through. This is not a shape that can come about through impure intentions, it is made by the machinations of great powers beyond us. That is why they say *vajra-kita*."

It is unclear where precisely the first mentions of the *vajra-kitas* come from or whether or not they were once a part of a localized or indigenous mythological system subsumed by later Hindu influences, but it is interesting to note that their Puranic mentions are comparatively recent in relation to the chronological timeline of Hindu religious texts. The first mention of Shaligrams in architectural inscriptions dates back to around the second century BCE but some Shaligram scholars claim that their origins might go as far back as the Vedic texts around 1500 BCE (Atharvaveda: 1500-500 BCE),[9]

8 In this particular version of Shaligram creation in Chapter 145, the Brahmin brothers Jay and Vijay go to perform a fire offering (*yagya*) for King Marut. Pleased, King Marut gave them a great amount of wealth and money but when the brothers could not decide how to divide it up properly, they began to fight. The fight went on for so long that they eventually cursed one another to become a crocodile and an elephant respectively. Once they had killed each other and the curses came true, they then met again as the animals and continued their fight for a thousand years. This necessitated the then King Bharat to summon Krishna to end the fight because it was disturbing his meditations. Krishna, of course, does so, but as his weapon, the Sudarshan Chakra, strikes the stones of the river in the course of battling the brothers, the markings of the Shaligram are formed.

9 Many practitioners claim that the earliest Vedic reference to Shaligrams is in the Atharva Veda, which states that Shaligrams are supposed to be owned only by Brahmins and treated as

while the textual references to *vajra-kitas* only appear to be highlighted in Puranic texts (such as Bhavisya Purana – probably after the seventh century CE[10] – and Skanda Purana in the ninth century CE) after the sixth century CE. In other words, while the *vajra-kita* cannot be dated specifically, it is possible that its later inclusion may have coincided with a more naturalistic understanding of the world by Puranic writers.

The icon of Vishnu as a bee may date back all the way to the Nad-Bindu Upanishad of the Rig Veda (between 1500 and 1200 BCE) where the deity Dattatreya or Datta (an avatar of the three gods Brahma, Vishnu, and Shiva), one of the oldest deities in the Vedic pantheon, is referred to as a "honey-bee" who collects all of the flowers of Yoga. While Vishnu has, in other circumstances, also been represented as a bee hovering over an inverted triangle (Shiva),[11] the exact relationship between Dattatreya, later Vaishnava worship, and the origins of Shaligrams remains unclear. Regardless of these questions of textual antiquity, however, modern retellings of the *vajra-kita* mythos view the role of the *vajra-kita* or thunderbolt worm in the production of Shaligrams as less a secondary cause of formation and more as a method of explaining the unnatural and uncanny appearance of the *shilas* such as they are. This means that what makes a Shaligram a Shaligram is not just the hierarchy of ideal causes set up in religious scripture but the view of geological processes as Shaligram agency as put forth by the discourses of geological and paleontological science and the transmission of human agency into the agency of deities – a transmission that is maintained when devotees are reminded that "you do not find Shaligrams in the river, they come to you when they are ready".

Before leaving India for Nepal, I visited the home of an elderly *brahmacharya* (a celibate Hindu monk) and his brothers. Mahayogeshvara, a Gaudiya

heirlooms. It goes on to state that a Brahmin's house without a Salagrama sila is as impure as a cremation ground. The water in which a Salagrama sila is washed is considered to be a cleanser of sins. It is also believed that imbibing just one drop of Salagrama water gives the same merit as can be achieved from performing every sacrifice and bathing in every *tirtha*.
The term "Shaligram" (including all other alternate spellings) does not, however, appear in the Vedas. Rather, it is likely that passages discussing Brahmin inheritance have been interpreted to mean Shaligrams at a later date.

10 In records of land grants of the fifth century BCE, verses are quoted that occur only in the *Padma*, *Bhavishya*, and *Brahma Puranas*, and on this basis, Pargiter in 1912 assigned these particular Puranas to an even earlier period. Maurice Winternitz considers it more probable that these verses, both in the inscriptions and in the Puranas, were taken as quotations from earlier dharmashastras, and thus argues that chronological deductions cannot be made on that basis.

11 See: Ransome, Hilda M. 2012. *The Sacred Bee in Ancient Times and Folklore*. Courier Corporation. p. 45

Vaishnava practitioner in West Bengal in India, patiently explained, "The scientists are not completely wrong when they say that Shaligram is a fossil insect. [I did not correct him on this point.] There are places within the sacred scriptures that also say this about Shaligram. But the *vajra-kita* is only a secondary cause because Vishnu himself alone is the principal cause of all of his manifestations, including Shaligram. This is the same as the cursing of Vishnu, which is also only a secondary cause. Another method of the story, not the story. In our (Gaudiya Vaishnava) tradition, Vishnu is the form of Krishna who is the cause of all causes. We say *sarva karana karanam*. This means that the main cause of God's appearance in this world is his own desires and the desires of his devotees, the Vaishnavas, which are the same desires. Since you ask about Shaligrams, you must understand that Vishnu desired to appear in the world of Kali Yuga in a form which could be easily worshipped and maintained by his devotees. This is why we allowed himself to be cursed to become a stone and for the *vajra-kita* to carve out his *chakras*."[12]

The Formation of Shaligrams by River and Mountain

By far the most common narratives surrounding the formation of Shaligrams – and the narratives that most closely bind them to ideals of space, place, and kinship – are the origin stories of Shaligrams as they are born out of the mountain and the river. These narratives also more thoroughly encompass the textual foundations through which pilgrims describe their own pilgrimage and ritual experiences (see Chapter 7). For the most part, these narratives were used as methods of linking biological processes (like birth and death) to geological processes (erosion and river washout) and to reveal the multiple ontologies of Shaligrams through the ways in which they formed. That is to say, all the while a Shaligram was being formed as an ammonite fossil it was also undergoing parallel formation as a deity and would eventually have to be born – the first preceding step to forming as a person.

According to the Varaha Purana (twelfth century CE),[13] some Shaligram stones come from the water (*jalaja*) while others come from the mountainside

12 Conversation in English with some Bangle. Transcribed from audio recordings and fieldnotes.

13 The century in which Varaha Purana was composed is unknown. Wilson suggested twelfth century, during the period of Ramanuja influence. Most scholars concur that this is a relatively late Purana, and a few suggest that the first version of this text was complete by the tenth century.

(*sthalaja*). In common parlance, Shaligram devotees occasionally refer to these two categories as either water-born (*jal*) Shaligrams or mountain-born (*kshetra*) Shaligrams. In practice, "mountain Shaligrams" are the term typically given to the reddish-orange, raw ammonite fossils that can be found slowly sliding down the river valley walls on their way into the Kali Gandaki River below. While many of these fossils could be easily obtained by walking the narrow village paths throughout the Baragaon, few if any Shaligram pilgrims ever actually sought them out and I never encountered any such fossils in the home altars or *puja* trays of active practitioners. Though they often agreed that such stones were holy and acknowledged that *kshetra* Shaligrams were included in the scriptural texts, I did not encounter a single religious use of such stones at any point in the years I worked with devotees. Only the smooth, black formations of Shaligrams born out of the river were ever accepted for ritual use.

As I walked the river with a group of *sadhus* late one morning, one of the Shaiva *babas* explained further: "It is because they are not properly formed yet. This does not mean they are not sacred, but they have not yet flowed through the womb of Himalaya." He motioned down towards the water at our feet. "They have come into the world but are not yet born. They are not ready yet for the home." In practice, then, such stones may be considered holy but they are not yet truly Shaligram; their proper ties of divine personhood and kinship have not yet been solidified. Despite textual ideals that label all aspects of the landscape as sacred, *kshetra* Shaligrams have not yet begun the movements that will ultimately bestow on them the identity of *Shaligram shila*. Because they have not yet entered the life cycle that defines their status as persons, they are not yet ready to be brought into temple, village, or family life.

One Shaligram seller based out of Pokhara remarked that the *jal* Shaligrams were simply of greater spiritual merit due to their contact with both the mountain and the river. *Kshetra* Shaligrams were only of middling merit because they were rough, broken, and "lacked essence", along with a particularly inferior form of Shaligram called *matha* (cell-born): Shaligrams that had been chewed out by insects and were therefore of very poor quality. These particular divisions, however, were rarely expressed by Shaligram devotees themselves, despite their occasional references in Shaligram texts. Many cited the merchant's need to sell the stones for a particular price as their motivations for arranging Shaligrams by level of quality.

Besides the Shaligram origin accounts detailed in Puranic texts (mainly the Brahma Vaivarta, Agni, Padma, Garuda, Nrsimha, Skanda, Brahma, and Brahmanda Puranas), Shaligrams are also mentioned in a wide variety of other Hindu works (many of which are later commentaries or compilations of Puranic texts): the Shalagrama-mahatmya of the Gautamiya Tantra, the Shalagrama-pariksha in the Magh-mahatmya section of the Padma Purana, the Puja-prayoga, the Haribhaktivilasa of the Gopal Bhaṭṭa, the Shalagramarcana-candrika, the Puja-pankaja-bhaskara, the Shalagrama-mimamsa of Somanatha-vyasa, the Shalagrama-lakshaṇa-panjika, the Shalagrama-pariksha of Anupa-simha, the Shalagrama-mula-lakshana-paddhati, the Shalagrama-shila-parikshana-paddhati, and an entire section of the Vaishnavanidhi chapter in Maharaj Krishnaraj Wodeyar III of Mysore's Sri-tattva-nidhi. Many of these later texts advocate the worship of Shaligrams as a method for obtaining material benefits such as great wealth, numerous children, success in business ventures, healthy herds of cattle, and a long and healthy life.

Some Hindu theologians view Shaligram veneration as a "*kamya*", an optional form of ritual worship based on the desires of the practitioners in question and therefore not obligatory for all Hindus. While this concept (that the practice is optional) is largely shared among the attitudes of current Shaligram devotees, few tended to view the ritual worship of Shaligrams as specific to desires for material goods. Rather, the worship of Shaligrams is more commonly associated with religious tradition, family history, and movement across sacred landscapes than the fulfillment of any specific day-to-day desire (though individual wants and needs were certainly not absent from practice). Shaligram devotees therefore tend to follow the approach of the Skanda Purana which advocates Shaligram worship for anyone wishing to perform service or austerities as a way of entering into an intimate relationship with the divine.

Specifically, the Skanda Purana calls for a smooth and shining Shaligram for those who wish to perform mantras (*mantra-siddhi*), a black Shaligram for fame or good renown (*yasas*), a pale-colored Shaligram for liberation from sin (*papa-hara*), a yellow Shaligram for the birth of children and the continuation of the family (*santana*), and a blue Shaligram for exchanging sacrifice for the family's worldly prosperity (*abhyudaya*). Additionally, the Narsimha Purana calls for umbrella-shaped Shaligrams to bring about sovereignty and personal independence or a circular Shaligram for wealth. In this way, what makes a Shaligram a Shaligram in these cases is less about provisioning goods in this life and more about ensuring good outcomes for social life cycles as a whole. As a side note: related to the tensions between river-born and mountain-born Shaligrams, I also encountered similar disagreements about color. In fact, despite a variety of color references in Shaligram descriptions in the Puranas, most devotees described any Shaligram with a color other than black as potentially dangerous, rife with tension and anxiety, and a sure sign of misfortune.

Bouts of Chastity and Other Curses Vishnu has Endured

The origins of Shaligrams espoused by devotees also tended to fall along the lines of the relationships between bodies and landscapes: between the river, the mountain, and the deity. As related in the Padma Purana,[14] there was once a massive and deeply destructive battle that took place between Lord Shiva and the demon Jalandhar. This battle raged on for several days, with neither Shiva nor the demon showing any signs of winning due to the power, in this version of the story, of Jalandhar's pious wife Brinda. In the Vishnu Purana, Shiva then requested help from Lord Vishnu. As the battle between the demon and Shiva continued, Vishnu took on a duplicate form of Jalandhar and went to Brinda's home. Subsequently, as Vishnu broke Brinda's long-held chastity while in the duplicate form of her husband Jalandhar, Brinda's power, her *pativrata* or *sati dharma*, was unable to protect her husband and Shiva was finally able to kill Jalandhar in the battle. As a result of this, Brinda became very angry and cursed Vishnu to take the form of a stone, of grass, and of a tree. It is for this reason, devotees explain, that Vishnu came down to earth to become Shaligram (stone), kush (holy grass), and the Pipal tree.

In the Padma Purana, the events have a slightly different outcome, but the course of the narrative is not particularly divergent. In this account, Vishnu is actually infatuated with Brinda and, because of this, the gods Agni, Brahma, and Shiva decide to approach Maya, the divine manifestation of illusion and concealment. Maya, in turn, directs them to three of her representatives: Gauri (*rajas*), Lakshmi (*sattva*), and Svadha (*tamas*) who give the gods three seeds with instructions to sow them in the place where Vishnu dwells. When the seeds were sown, three plants sprouted: dhatri (*Umblica officialis*), malati (*Linum usitatissimum*), and tulasi (*Ocimum sanctum*). These three plants were then considered aspects (*amshas*) of Svadha, Lakshmi, and Gauri respectively (Rao 1996: 39-40) but it is otherwise unclear precisely what this variation has to do with the origins of Shaligrams other than to emphasize that Shaligram and tulsi plants are strongly associated in worship.

14 Kriya-yoga-sara section.

In the Brahma Vaivarta Purana version[15] of this story (and in ninth *skanda* of the Devibhagavata), the part of Brinda is actually subsumed by the goddess Tulasi (tulsi).[16] This account explains that there was once a daughter of King Dharmadhvaja and his queen Madhavi who was both a beautiful princess and an incarnation of the *hladhini-shakti*, the internal pleasure potency and creative power of the universe (and specifically of the Godhead). When this daughter was born, she was said to have been marked with unusual good fortune and as she matured into an exquisitely beautiful young woman, she never appeared to age beyond sixteen years. As the manifestation of universal divine qualities and blessed with incomparable beauty, she was thus called Tulasi (meaning: matchless). Accordingly, when Vishnu then wanted to perform his *lilas* (sacred pastimes) on earth, he was obliged to do so only in the association of his personal potencies – the potency in this case being that of Vishnu's divine pleasure (*hladhini*) called Tulasi. (In the Gaudiya Vaishnava tradition, this particular manifestation is taken over by Sri Krishna and his *hladhini* who is manifest as his consort Srimati Radharani, who is also the goddess of fortune).

When Vishnu (or Krishna) then descends into the mundane world as *avatara* to perform his pastimes or undertake acts of heroism, his *hladhini* manifests along with him. In many Hindu traditions, these expansions that accompany the avatars of Vishnu are sometimes called Lakshmis, and the princess Tulasi who was born as the daughter of King Dharmadhvaja and Queen Madhavi is also considered an incarnation of Lakshmi, consort of Vishnu and the principal goddess of fortune. Finally, in the Devibhagavata, it is noted that Tulasi's incarnation on earth is actually due to the jealousy of Radha (Krishna's principal consort) who became very angry with Tulasi while in Goloka (the Vaishnava paradise) because Krishna had become overly fond of her. (Non-Puranic accounts sometimes explain that it is Lakshmi who curses Tulasi to become a plant because Tulasi longs to have Vishnu as her husband. Vishnu then joins with Tulasi as a Shaligram stone.)[17]

15 Prakṛtikhaṇḍa Chapter 15 ff.

16 Yet another version of this story is also recounted in the Sthala Purana wherein the king in question is named Kusadwaja and his queen, Madhavi. The girl who born to them is Tulasi, who is married to Jalandhar.

17 In yet another version of this story from the Devi Bhagavata Purana, it was the goddess Saraswati who initially cursed the goddess Lakshmi. In this version, Vishnu had three wives: Sarasvati, Lakshmi, and Ganga. Once Lakshmi and Sarasvati quarreled and cursed each other. Saraswati's curse changed Lakshmi into a tulsi plant and forced her to live on earth forever. Vishnu, however, intervened and modified the curse, saying that Lakshmi would remain on earth as tulsi until the river Gandaki flowed from her body. In the meantime, he would wait by the riverside in the form of a stone to take her back to his abode. This stone is, of course, Shaligram.

As the story continues in the Brahma Vaivarta Purana, it is by the machinations of the karmic cycle that Tulasi is wedded to Sankhacuda, a powerful demon (subsuming here the role of Jalandhar). As fate would have it, Sankhacuda had also received an earlier boon from Lord Brahma to obtain Tulasi as his wife and, having done so, would remain undefeated in battle as long as she remained chaste to him. Taking full advantage of Brahma's boon, Sankhacuda began to terrorize the world and all the demigods as he was wont to. Being severely afflicted by his attacks, the demigods then approached Shiva for protection. Shiva himself then went to fight with Sankhacuda, but due to Tulasi's faithfulness, Shiva was unable to kill him regardless of what he tried. The demigods then fell into despair but Vishnu (naturally) devised a plan to spoil Tulasi and render the demon vulnerable. While Shiva continued to engage Sankhacuda in combat, Vishnu went to the both of them first in the guise of a *brahmana* to beg charity from Sankhacuda. Standing before Sankhacuda, the *brahmana* requested, "My dear Sankhacuda, famous throughout the three worlds as the giver of whatever one desires, please give me your *kavaca* (armor) in charity." Knowing that it was the chastity of his wife, Tulasi, that protected him, Sankhacuda unhesitatingly gave the *brahmana* his armor in charity and resumed his fight with Shiva.

Now dressed in Sankhacuda's armor, Vishnu went immediately to the palace where Tulasi was waiting news of the battle's outcome. Thinking that her husband had returned from the fight to regain his strength, Tulasi welcomed him to the bed chamber for a rest. Thus, the night passed and the faithfulness of Tulasi was broken by Vishnu's deceit, and at that moment Sankhacuda was slain by Shiva in the battle that had also continued throughout the night. When Tulasi realized that the Sankhacuda she had slept with was actually Vishnu and not her husband and that Sankhacuda had been killed by Shiva, Tulasi levied her curse against Vishnu: "By deceiving me, you have broken my chastity and killed my husband. Only one whose heart is like stone could do such a thing. Thus, I curse you to remain on earth as a stone!"

Accepting Tulasi's curse, Vishnu replied, "For many years you underwent very difficult penances to achieve me as your husband. At the same time, Sankhacuda also performed penances to get you as his wife. As a result of a boon from Lord Brahma, the desire of Sankhacuda was fulfilled. Now that Sankhacuda has left this mortal world and gone to the spiritual world, your desire to have me as your husband will be fulfilled [recall the Gandaki/*Vajra-kita* version of the story]. Give up this body, and let your spirit be merged in Lakshmi's, so that I am always with you. This body of yours will be transformed into a river, which will become sacred and celebrated as Gandaki, and from your beautiful hair will grow millions of small trees that will be known as Tulasi. These trees will be held sacred by all my devotees. Furthermore, to fulfill your curse, I will become many stones (*shaligram shilas*) and will always live on the banks of the Gandaki River."[18] Thus Tulasi was transformed and appeared as both the Gandaki River and as the sacred plant tulsi. Vishnu then came into the world as Shaligram, born in the waters and on the banks of the Gandaki. At this point, the Brahma Vaivarta Purana also mentions that Sankhacuda, though a demon in his last manifestation, was also an eternal associate of Krishna by the name of Sudama who manifested in the world as a demon so as to assist in these events coming about.

The tale of Tulasi as recounted in the Brahma Vaivarta Purana (and to some degree the Vishnu Purana) is the most common creation story referenced by Shaligram devotees today. In part, this is due to the availability of the book "Muktichhetra Mahatmyamam", a pilgrimage guide written in 2003 by Madhu Sudhan Ramanujadas which is often available for purchase in the village shops near Muktinath. The book also contains reprints of sections of the Skanda Purana, especially the discussion between Lord Skanda and the sage Agatsya relating to the significance of Shaligrams and their characteristics, and the Varaha and Padma Puranas, where they mention the manifestation of Shaligrams in relation to the region of Muktikshetra.

In other respects, the popularity of this version of the story owes its fame to the Marriage of Tulsi and Shaligram (*Tulsi Vivah*), a festival that takes place throughout India and Nepal on the eleventh lunar day of the Hindu month of Kartik (October/November). What all three variations of this story provide, however, is the links between the chastity-deceit-curse version of the Shaligram story and the literal and metaphorical birth of Shaligrams out of the landscape. To some degree, the variability in the story likely has to do with narrative blending in both Vaishnava and Shaiva traditions of Shaligram veneration where both Vishnu and Shiva are said to play distinctly important roles in the formation of the Kali Gandaki River and of the Shaligrams within it. Furthermore, for many Shaiva and Smarta Shaligram practitioners, the implicit association of Shaligrams directly

18 The final portion of this dialogue also goes on to say, "The *vajra-kitas* will carve out inside these stones my discus – emblem (chakra). I will also dwell in the pot in which tulasi plants are grown." This section is referenced in many commentaries regarding the discussion of Shaligrams in texts but is rarely recounted by devotees when re-telling the story. In other variations of the tale, the *vajra-kitas* are simply omitted entirely.

with Vishnu is not always accepted, noting for example the many instances where Shiva mentions the worship of Shaligrams in the Skanda Purana[19] or the particular quote in the Padma Purana where Shiva himself states:

> My devotees who offer obeisances to the shalagrama even negligently become fearless. Those who adore me while making a distinction between myself and Lord Hari will become free from this offence by offering obeisances to shalagrama. Those who think themselves as my devotees, but who are proud and do not offer obeisances to my Lord Vasudeva, are actually sinful and not my devotees. O my son, I always reside in the shalagrama. Being pleased with my devotion the Lord has given me a residence in His personal abode. Giving a shalagrama is the best form of charity, being equal to the result of donating the entire earth together with its forests, mountains, and all.

For Vaishnavas (particularly Gaudiya Vaishnavas), there is an additional reference to offering Tulasi leaves to Krishna in the Bhagavad Gita (9.26):

> *patram puspam phalam toyam yo me bhaktya prayacchati*
> *tad aham bhakty-upahrtamasnami prayatatmanah*
> (If my devotee offers me with devotion, a leaf, a flower, a fruit, or water, I will accept it.)

According to several current Vaishnava *acaryas* (spiritual masters), the *patram* (leaf) mentioned in this verse particularly refers to the tulsi leaf. Tulsi leaves are also mentioned in the Garuda Purana and in the Brhan-naradiya Purana, which state that the worship of Vishnu without tulsi leaves

19 *Drstva Pranamita Yena Snapita Pujita Tatha* |
Yajna Koti Samam Punyam Gavam Koti Phalam Bhavet ||
Lord Siva speaking to Skanda, "Any person who has seen Salagram Sila, paid obeisances to Him, bathed and worshipped Him, has achieved the results of performing ten million sacrifices and giving ten million cows in charity." --- Skanda Purana – Haribhakti vilas
Pujito'ham Na Tair Martyair Namito'ham Na Tair Narah |
Nakrtam Martya Loke Yaih Salagram Silarcanam ||
Lord Siva speaking to Skanda states, "In this mortal world, if anyone does not worship Salagram Sila, I do not at all accept any of their worship and obeisances."
Lord Shiva also states, "Even if a shila is cracked, split, or broken it will have no harmful effect if it is worshiped with attention and love by a devotee. It further states there that the Supreme Lord Hari, along with His divine consort, Lakshmi, live in the shalagrama that has either only the mark of a cakra, a cakra along with the mark of a footprint, or only a mark resembling a flower garland."- Skanda Purana

is incomplete and is unlikely to be accepted by Vishnu as proper veneration: "Without Tulasi, anything done in the way of worship, bathing, and offering of food and drink to Vishnu (Krishna) cannot be considered real worship, bathing, or offering. Vishnu does not accept any worship or eat or drink anything that is without Tulasi."

The Varaha Purana recalls a Shaligram creation story with similar elements but recounts a somewhat simplified version of the Gandaki/*Vajra-kita* tale as detailed in the Skanda Purana. What is important to note here is, again, the primacy of river-mother birth that precedes the formation of Shaligram deity and divine personhood. In this version, Gandaki, who is already a river-goddess, performs a series of austerities (such as eating only fallen leaves and drinking only air) while meditating on the nature of Vishnu. When Vishnu subsequently appears before her, she begins to sing a series of heart-wrenching verses praising Vishnu and her love for him. Pleased with her devotion, Vishnu then tells Gandaki to choose a boon that he might grant her regardless of how strange or fantastical it might be. Much as in the previous version of this story, Gandaki expresses her desire to give birth to Vishnu as her child: "If indeed you are pleased with me, consent to enter my womb and become my child." Unlike the Skanda Purana, however, Vishnu readily agrees to her request and states that he will enter her womb (here meaning the river's flow) as a *shalagrama* whose worship will therefore confer great prosperity to all mankind (see also Rao 1996: 33-34). Because the stones would then appear out of the flow of the river, they could be said to be its offspring and the river itself to be pure and holy (*mat-sannidhyat nadinam tyam ati-shrestha bhavisyasi*).

The origin of Shaligrams as recounted by the Vishnu Purana (9, 6), the Agni Purana (152), and the Bhagavata Purana (eighth *skanda*) is, however, the story from which the famous references to Shaligrams in the Hindu epics Ramayana and Mahabharata are drawn and also the story that tends to dominate debates regarding the feminine nature of the Kali Gandaki River and the "reproduction" of Shaligrams as persons within the karmic life cycle. In this version of the story, the origin of the Gandaki river is included in the tale of the churning of the milky ocean by both the gods and the *asuras* (demons) to obtain the ambrosia of immortality (*amṛta*). This story begins with a curse levied by the sage Durvasa, which resulted in the loss of all the powers and might of the gods. As the story goes, Durvasa was walking through a forest which was filled with the sweet fragrance of Kalpaka flowers that were being worn in a garland by the celestial maiden Menaka. When the sage met Menaka she offered him the garland and he happily set off with the flowers wound up in his matted hair. Along the way, he then happened to meet Indra, the chief of the gods, who was mounted on his favorite elephant. Thinking the beautiful garland would be more suitable to Indra than himself, the sage presented the flowers to Indra as an offering.

Unfortunately, Indra, who was often arrogant and unresponsive in these matters, took the garland and flung it onto the head of his elephant. The elephant then pulled the garland off with his trunk and threw it on the ground where he trampled it. Infuriated, Durvasa cursed Indra that all his power and glory should instantly vanish. Realizing his error, Indra then begged the sage to forgive him but Durvasa was unmoved by the deity's distress. As time went on, Indra also came to realize that the curse was working not only against him but against all the gods within the sacred realm (the *deva-loka*). As the gods became increasingly powerless, their charms disappeared and their strength was rendered impotent. Even the plants growing in the celestial realms began to wither and die. The world of the gods began to lose its appeal to mankind and the people began to withhold their customary offerings and stop their daily venerations, which rendered the gods even more debilitated.

Finding the assembled gods in such a sorry state, the *asuras* (demons) attacked the celestial realms and humbled the once great deities. No longer immortal or invincible, the gods suffered injuries and some of them died in the ensuing battles. Agni and Brahma then went out and collected all the gods that remained and took them all before Vishnu, seeking his help in overcoming their current crisis. Vishnu counseled them all to partake of *amrita,* the divine nectar of immortality that could only be obtained by churning the milky ocean. Vishnu also explained his strategy in getting the nectar. The gods would need to cooperate with the *asuras* in order to accomplish this arduous task and they would also need the mythical mountain Mandara to use as a churning rod and the dragon (or snake) Vasuki to act as the rope for churning. Vishnu himself decided to take on the form of Kurma, the great tortoise on which they would need to support the mountain so that the churning could remain steady. So here, yet again, it is the land and the water that produce "life" – mortal, immortal, and material alike.

This the gods did, and when at least the bowl of *amrita* emerged from the ocean, the gods and *asuras* immediately began to fight with each other over who would be allowed to drink from the bowl first. Upon witnessing the argument, Vishnu assumed the form of a fetching maiden called Mohini, whose beauty and charm were beyond compare, and it was Mohini who then offered to distribute the *amṛta* to all gods and asuras who had participated in the churning. Fascinated by the extraordinary elegance and refinement of the woman, the gods and *asuras* sat down quietly in two rows. Mohini then took the bowl in her hands and began to serve the gods first but when she came to the end of their row, she suddenly disappeared.

The *asuras* were thus cheated out of the nectar. The gods, having regained their immortality and invincibility after having partaken of *amrita*, easily beat the asuras in the resulting battle and drove them out of the celestial realm. When Mohini (Vishnu) had been serving the drink, however, Shiva became particularly enamored with her and rushed to embrace her before she could finish serving. In the heat of passion, both Shiva and Vishnu perspired copiously, and their combined sweat flowed down as the river Gandaki (a slightly different way of viewing the reproductive qualities of water in this case). This is the reason why the river is sacred: because it contains the essences of both Vishnu and Shiva (Rao 1996: 35-37).

This version of the origins of the Gandaki also reverses the more common narratives wherein the river is expressly female (Tulasi, Brinda, Gandaki, etc.) and transforms the reproduction of the womb (which "births" Shaligram) into the reproduction of semen (sex between two, ostensibly male, deities). It is particularly interesting to note then that, in this story, the Shaligrams themselves are not expressly mentioned, though this origin myth is quite often intertwined with several of the previous origin myths detailed already. In many cases, the story of Durvasa's curse and the churning of the milky ocean mirrors many aspects of the chastity-deceit-curse stories wherein the origin of Shaligrams is the result of both the production of a sacred landscape and the control of female sexuality. But it remains the question of union and reproduction that defines what it is to begin life as a Shaligram. It is within the Shakta texts, however, that we see landscapes as bodily symbolism truly come to fruition.

The Shakta texts speak of fifty-one places, scattered across distant lands, where the dismembered body parts of the goddess Sati fell as a grief-stricken Shiva was carrying her about following her self-immolation on Daksa's sacrificial altar (these places of pilgrimage are called Shakti-piṭhas; of which Muktinath Temple in Mustang is one). The source of the Gandaki river in the Himalayas is one of these places where the texts indicate that Sati's cheeks fell (*ganda-sthala*). Here the goddess takes the form of Gandaki along with her consort Vishnu who appears as the *chakra-pani* (the discus-bearer – i.e., Shaligrams). However, despite the continued association of the Kali Gandaki with divine feminine principles (and its continued reference as a place of profound feminine spiritual power today), many of the legends that recount the transformation of a woman (Tulasi, Brinda, Gandaki) into a river and into a plant are taken as evidence that women are excluded from Shaligram worship.

The Varaha Purana even goes as far as to expressly forbid women from touching Shaligrams. It states that all the merits they have earned by following their *karma* and by praying and performing austerities will be completely nullified if they even touch a *shila*. They are permitted, however, according to this text, to worship Shaligrams from afar or through the men of their families who are duty-bound to perform Shaligram worship. Even Brahmin women are not permitted to worship Shaligrams nor can they inherit one. If their families have produced no male heir, the stone is passed on to another nearby Brahmin. This explanation is, however, almost universally rejected in Vaishnava practice (*Vaishnava viddhi*) because of the numerous other texts wherein it is stated that anyone who is properly initiated can worship Shaligram:

grhita visnu diksako visnu pujaparo narah
vaisnavo'bhihito'bhijnair itaro'smad avaisnavahch
"A person who is initiated in Vishnu mantras, and who is expert in worshiping Lord Vishnu, such a person is known as a Vaishnava. Besides this, everyone else is an avaishnava." (Hari Bhakti Vilasa 1.55, from Padma Purana)

yatha kancanatam yati kasyam rasa-vidhanatah
tatha diksa vidhanena dvijatvam jayate nrnam
"As bell metal is turned into gold when mixed with mercury in an alchemical process, so in that very way, by the process of proper initiation by a true spiritual master, a person becomes a *brahmana*." (Hari Bhakti Vilasa 2.12, from Tattvasagara)

striyo va yadi va sudra brahmanah ksatriyadayah
pujayitva sila cakra labhante sasvatam padam
"Worship of Shalagram shila can be done by women, *sudras* (untouchables), *brahmanas* (twice born), and *ksatriyas* (administrators). Thusly, they can all achieve the eternal abode of Lord Krishna perfectly." (Skanda Purana; conversation between Lord Brahma and Narada Muni)

striyo va yadi va sudra brahmanah ksatriyadayah
pujayitva silacakram labhante sasvatam padam
"If one is initiated as a Vaisnava then whether one is brahmana, kshatriya, vaishya, shudra or a woman, one can worship Shalagrama and attain the Lord's abode." (Skanda Purana)

While there is a plethora of Shaligram origin stories, I have chosen to arrange the Puranic legends by theme rather than by chronology for two reasons. Firstly, though the Puranas can be approximately dated and while one Purana can certainly be said to be older than another, few if any of the Puranic texts owe their compositions to a single author or even a single time period. Rather, most of these texts were written in layers during successive time periods by authors who heavily borrowed from and referenced one another over time. This is why many of the Puranas and other sacred texts contain multiple versions of the same stories or additional commentaries on scriptures that may be shared across several works simultaneously. Secondly, owing to the nature of Hindu worship in general, there is no central scriptural authority followed by all Hindus, and the majority of Hindu traditions do not ascribe to all of the scriptural texts equally. Where one tradition may place more religious authority in one set of Vedas and Puranas, another may disregard them entirely. As a result, the mythic and spiritual dimensions of Shaligrams and Shaligram practices often varies widely from one tradition or sect to another, though they tend to share the majority of key themes related in the extant mythography, especially the themes of birth, death, reproduction, and landscape.

In many of the Puranic tales, the river-goddess Gandaki/Tulsi is sometimes pious and sometimes vengeful, but in the end always seeks to obtain Vishnu and the gods as her children – who will be born out of her waters continuously in fulfillment of karmic order. The water, then, becomes both a passageway between the divine and material worlds (*tirtha*) and a method of producing order out of chaos. In multiple variations on the tale, Gandaki performs a variety of different austerities so as to obtain these divine children, or Vishnu as her husband, but in each story the links to the reproduction of the landscape with birth, death, and rebirth more broadly are often articulated by the nature of the boon she is granted. As a result of her desires and actions, she becomes the mother of gods in material bodies like Shaligrams as well as the producer of human families and communities who rely on her resources, such as water, agriculture, and livestock (the Kali Gandaki is, for example, the only reason why much of anything at all grows in Mustang at such high altitudes). The river, as both literal and figurative fluidity, is a bringer of fortune and a pathway in and out of life. The Shaligram mythic corpus can then be viewed through a network of relationality, extending from the gods – who are united in a variety of male and female forms – to the relations between nature, culture, and humanity.

In each case, the land and the river are characterized as chthonic, primordial, female powers (much like the *sinmo* in Buddhist cosmology) who must be tamed and constrained into the proper cycles of creation by the masculine forces of Vishnu or Shiva (or Guru Rinpoche). It is also not surprising then that images of the Yab Yum (Tibetan lit. "father-mother"), a symbol that represents the primordial union of wisdom and compassion and which is depicted as a male deity in sexual congress with a female consort, and the Satkona, a hexagram *yantra* (six-pointed star) that represents the sexual union of the divine male and female forms as icons of ultimate wisdom,[20] are very common on the walls of *gompas*, temples, libraries, and schools throughout Mustang.

The mythemes of burying and reappearing are also especially salient for the Shaligram corpus of texts and continue to reiterate the agency of the landscape outside of human action. This is because a Shaligram's birth out of the river (and later its return to it through cremation or return pilgrimage) links cycles of life and death with issues of mobility and stasis in both the origin stories of the landscape and with actual human mobility in the present day. Or, as the landscape continues to undergo cycles of concealment and revelation, so too do Shaligrams appear and disappear throughout the course of their lives. Pre-Buddhist and Bon spirits are equally incorporated, where the power and viciousness of the Dakini, the fast-moving female Himalayan wind spirits, for example, are contained by the placement of Shaligrams at key points along roads, over thresholds and doorways, and inside *stupas* due to the Shaligrams' capacity to contain and control movement. The potential for violence in the Himalayas, whether by wind or water or stone, is never far from anyone's mind. Subsequently, the goddesses of the river and the landscape are then much like similar monstrous mothers from other cosmogonies from around the world who have then been repurposed through stasis and mobility within or as landscapes – where her body is controlled and held down while her reproductive energies are redirected towards human endeavors and her more dangerous characteristics are suppressed or pacified so as not to result in destruction.

In related narratives about the marriage of Tulasi, the variations on the story of Tulasi/Brinda and Jalandhar/Sankhacuda continue the theme of a union (again an angry one) between the feminine divine and God Himself – which again results in the production of the landscape as well as its reproductive capabilities. It also preserves the feminine/demon principal who is ultimately tied to the land by the movement and placement of sacred stones (Shaligrams, *mani* stones, foundations of sacred buildings,

20 More specifically, it is supposed to represent Purusha (the supreme being) and Prakriti (mother nature, or causal matter). Often this is represented as Shiva / Shakti.

etc.). In many interpretations of Shaligram origins, these stories then instantiate the existence of each Shaligram as a micro-cosmos within the larger cosmos of the landscape within the largest cosmos of all creation. As my friend and teacher Bikas Shrestha explained, "I once heard someone joke that if everyone keeps taking stones from all of these sacred places, then it will just be a matter of time until the whole mountain or the whole country will be in a village in India. But it's a little different for Shaligrams because each is a cosmos in and of itself already. This is because Shaligram is both made by the forces of the cosmos and contains those forces within it. This is how Shaligram can direct the forces around it, like karma and the spirits you mentioned and people and the land. It is the land and it holds the land."[21] These kinds of interpretations of the manifest mobility of Shaligram also tended to upend the idea that the world was created specifically to accommodate humankind. Rather, in the telling of the landscape as a cosmos, it was humankind who fulfilled the potential of Shaligrams and the world.

Durvasa's curse and the churning of the milky ocean and the transformation of the gods into worms and insects (*vajra-kita*) recapitulate these concerns about life, death, and rebirth as linked to landscapes. In modern contexts, the inclusion of these stories within the Shaligram mythic corpus is particularly revealing through their assembly in the frameworks of Shaligram pilgrimage. The milky ocean provides a progression from the creation of the universe to the creation of the Kali Gandaki and then the Shaligrams within it in an ever increasingly molecular cycle of union, birth, and death from the beginning of time to now. These cycles are then recreated both in pilgrimage and in the social lives of practitioners and their Shaligrams once they return home. The thunderbolt worms or *vajra-kita* are either living entities who carve out the emblems on the static rock or who enter the cadavers of fallen gods in order to turn them into useful landscapes on Earth so that the devout can recognize the right places and objects and find them. In each circumstance, the formation of Shaligrams becomes emblematic for the ordering of natural chaos into set patterns and cycles that will be the same cycles experienced by human individuals and by society as a whole. In other words, the ontology of Shaligrams as divine persons and deities and their agency in ordering the landscape results in the ordering of humanity. It is then the mobility of the river and the mobility of the stones – their flow – out of stationary mountains and established temples and *chorten* that continues to anchor the gods in place and assure that, not only will

21 Conversation in English, with some Hindi. Transcribed from audio recording.

they go on producing goods and resources in the future, but that they will also continue to protect Mustang, its peoples, and its pilgrims from chaos, destruction, and the dangers of natural disaster.

Channels into the Mainstream

Shaligram origin stories are almost as fluid and variable as the stones themselves. Whether formed by the *vajra-kita* (thunderbolt or adamantine worm) whose stone-carving capabilities continue to link religious creation stories with ammonite paleontology or by any number of curses levied at Vishnu for betraying the chastity of the goddess Tulsi, the mountain and river birth of a Shaligram stone is always preceded by a complex narrative of time, place, and personhood. The core conceptualization of bodies as landscapes, however, remains constant.

In one version of the Shaligram origin story, the *vajra-kita,* who cut the distinctive chakra spirals out of the stones, are either themselves cast as living deities who form icons out of inanimate Earth or who enter the bodies of deceased gods in order to remake them into the landscape. But rather than viewing this version of Shaligram birth as antithetical to scientific descriptions of geological processes, Shaligram practitioners tend to note the similarities between taxonomies of "ammonite" and "worm" as a way of blending the two discourses together into a multivalent vision of fossil-who-is-also-deity and mountain-who-is-also-ocean.

Conversely, in the more popular story of Vishnu and Tulsi (alternatively named Brinda/Vrinda or just Gandaki), Shaligram formation is a matter of Vishnu's rebirth as a sacred stone within a sacred river in payment of the debt he owes to the goddess for sexually violating her through deceit, usurping her marriage, and killing her husband. In these variations of Shaligram birth, the links between body and landscape are more straightforwardly reproductive in that Vishnu-as-a-Shaligram must either be literally born into the physical world through the "womb/flow" of the sacred river (the Kali Gandaki) or whose own body must be turned to stone so that he might self-manifest (as an *avatar*) into a landscape made productive again.

It is also for this reason that Shaligram festivals, such as *Tulsi Vivah,* tend to focus on life-cycle events such as marriage, childbirth, and death. The birth-death-rebirth processes of the landscape then becomes metonymic for the karmic birth-death-rebirth cycle shared by humans, their deities, and their Shaligrams. As persons, Shaligrams live the same kinds of karmic lives as people do. They are born, they live lives with their families, they eat and travel and attend to important moments, they have relationships, and eventually they must die and be returned to the land. As deities, they connect mortal human lives with the immortal actions and desires of the gods and with a sense of community that transcends space and place to extend the moments of daily life into the grander order of the natural universe. As fossils, Shaligrams blend Deep Geological Time with mythological spaces to foster greater connections between the unending karmic cycle of the cosmos, the tectonic forces of Earth's creation, and with successive generations of humanity as individuals, as families, and as entire civilizations. And in the end, Shaligram is all of these things simultaneously.

The best way to understand the ways in which these processes and ritual movements overlap and come to stand in for one another is to approach Shaligram pilgrimage ethnographically. By demonstrating how mythological spaces become physical places and how geological time is intertwined with sacred time, attending to Shaligram pilgrimage as an act linked to specific moments and specific places will foreground mobility as the locus of identity in this case, rather than any notion of fixed nationality or religious affiliation. As such, the boundaries of human experience and of Shaligram being continue to blur.

Bibliography

Chand, Devi. 1997 (2002). *Atharvaveda, Sanskrit Text with English Translation.* Delhi: Munshiram.

Korom, Frank. 2001. "Introduction: Fieldwork, Ethnography, and the History of Religions," *Method & Theory in the Study of Religion* vol. 13 (1): 3-11.

—. 2006. *South Asian Folklore: A Handbook.* Westport, CT: Greenwood Press.

Prana-toshini tantra. 1983. Ramatosana Bhaṭṭa and Ramadatta Shukla, translated by Prayaga: Shakta Sadhana Piṭha Publishers.

Puranas. 2007. Translated into English by a board of scholars. Motilal Banarsidas Publishers Pvt. Ltd., 74 Volumes.

Rao, S.K. Ramachandra. 1996. *Shaligram Kosh* [*Śālagrāma – Kosha*]. Sri Satguru Publications. Delhi: Indian Books Center.

Sharma, Ram Charan. 2000. *Shaligram Purana.* S.R.C. Museum of Indology & Universal Institute of Orientology Trust, 199. Digitized 2009. Ann Arbor: University of Michigan.

Singh, Chandra Shekar et al. 2017. *The Puranas.* Amazon Digital Services.

7 River Roads

Mobility, Identity, and Pilgrimage

Abstract
Shaligram pilgrimage is both a macrocosm of Shaligram practice and a microcosm of life itself. Mobility is, however, at the core of all aspects of veneration. Shaligram pilgrimage then offers a glimpse into the methods by which people come to identify with certain places, regardless of whether or not they live in those places or have ever visited them before. It also offers insight into how marginalization, militarization, and economic challenges in Mustang have had significant effects on both Shaligram practices throughout South Asia and the world. Furthermore, due to the plurality of sacred spaces in physical locations, such as happens through the *dham*, Shaligrams become capable of being both from a place and carrying that place with them.

Keywords: pilgrimage, mobility, Mustang, Muktinath, landscape

> "There's a place that I travel, when I want to roam
> And nobody knows it but me.
> The roads don't go there, and the signs stay home
> And nobody knows it but me.
> It's far far away and way way afar, it's over the moon and the sea.
> And wherever you're going, that's wherever you are,
> And nobody knows it but me."
> – *Patrick O'Leary*

Jadav Manjhi carefully lifted the teacup from the woven Tibetan rug where he sat with a small plate of *dal bhat* and a few apples, steadying the tremor in his hands by pressing his elbows onto the tops of his knees. "I first came to Mustang for Shaligram pilgrimage back in 1977," he began, blowing

Walters, Holly, *Shaligram Pilgrimage in the Nepal Himalayas*. Taylor & Francis Group, 2020
DOI: 10.5117/9789463721721_CH07

a strand of his long, grey hair away from the cup. "There weren't many pilgrims back then. It was very difficult to come. You needed special papers and a government official to take you. It's not much better now, sadly. You still need papers, of course, though you no longer need a government man, but it has become so expensive that many can no longer gather the money. Transportation, guides, rooms, food, everything is now focused on Westerners who come to trek, and Westerners always bring a lot of money. Sometimes the guides and drivers here make the price less for pilgrims, but I don't know. I suppose it is better to walk anyway, yes?"

"Was pilgrimage very different back then than what it is like now?" I asked. "Oh yes, very different," he nodded as he carefully placed the teacup back onto its saucer before throwing his sleeve aside to begin mixing the *dal bhat* with his hand. "When I first walked on Kali Gandaki, there were Shaligrams everywhere! Piled high on the banks, rising up out of the waters, there wasn't a single place you could step without seeing Shaligram. You had your pick of the deities then, each one of them watching you as you took *darshan*, waiting for one to call out to you. Now, you must go very high in the mountains to find Shaligrams like this and sometimes they don't appear at all anymore."

"Why is that?" I shifted to sit more comfortably on my feet rather than my knees. "Now there are sellers who come and pick the river clean," he answered. "Take Shaligrams to market in Kathmandu or in India. I am from Hyderabad and I sometimes see them in shops there too. Also, trekkers who come. They find them in the river and take them, even though they don't know what they mean. Just a trinket, I guess. Is that the word? Trinket? Something to remember your trip by, but never someone who also remembers your trip."[1]

I met Jadav Manjhi on my third journey along the pilgrimage route to Muktinath (though I had been living in Mustang for some four months by that time). Sitting in the common room of a Ranipauwa guesthouse, drinking herbal tea over plates of *dal bhat*, he told me of the seven Shaligram pilgrimages he had completed since his late forties. Now in his early seventies, he lamented what he felt would be his last pilgrimage to Mustang. Like almost every Shaligram pilgrim I met, Jadav's concerns about the growing inaccessibility of Mustang were foremost in his mind. Between issues of political unrest, militarization of the Upper Mustang and Tibetan borders, and rising economic hardship leading to the commodification of Shaligrams, both Jadav and the Shaligram practitioner community at large continuously

1 Conversation in Hindi and English. Transcribed from audio recording.

expressed their fears that the mobility of pilgrims and of Shaligrams would remain restricted and might eventually be stopped all together. And should this come to pass, it would mean the death of the practice and of the people.

"The bodies of the Lords do not appear as often as they once did,"[2] Jadav continued. "This is Kali Yuga (the final age), so I think this is how it must be. Maybe one day soon, they will no longer appear at all. And then what will happen? It is not enough to just get Shaligram from the market. It is not enough just to see them in temples or have one given to you by holy men." He threw down his handful of lentils and rice to wipe at the tears starting to spill down his cheeks. "You must come to Kali Gandaki [referring to the river valley in Mustang where Shaligrams can be found]. You must come to where Shaligram is born. You must sit at their feet and listen and learn; this is the yearning of the Vaishnava soul. When they are gone, so we are gone."[3]

Scholars often speak of the images of pilgrimage, but to invoke pilgrimage as an image one must acknowledge what is possibly the most influential text in the anthropology of pilgrimage: Victor and Edith Turner's *Image and Pilgrimage in Christian Culture* (1978). While this work is not strictly about images, despite what its title might imply, it includes a large number of photographs depicting a variety of religious sites from Mexico to Italy, France, Ireland, and England. Though arresting in their presentation, all are revealing in what they do or do not impart about the study of pilgrimage. It would be easy to depict Shaligram pilgrimage in just this fashion and to fill pages with images of sacred stones, pilgrims bathing at Muktinath, or taking *darshan* from the shrines all across Mustang. But these images, though ubiquitous in the coffee table books and travel literatures of Nepal, contain significant, if implicit, stories of movement and embodiment. Just as the photographs of pilgrims kneeling in front of Our Lord of Chalma in Mexico do not depict the journey where pilgrims have approached the shrine by walking on their knees from a distance of more than a mile, images of the pilgrims of Mustang are similarly problematic because they reinforce the snapshot narrative where encountering a Shaligram at any one given point is viewed as a microcosm of its greater process of being.

Shaligram pilgrimage also challenges place-centered and even purely transitional perspectives on the various forms of movement – embodied,

2 See also: "Shaligram Fast Disappearing from Kali Gandaki" *Glocal Khabar*, 13 November 2016: https://glocalkhabar.com/news/national/shaligram-fast-disappearing-kaligandaki/. Accessed 14 November 2016.

3 Concerns over the erosion of the world's largest Shaligram in the Kali Gandaki River near Setibeni: http://english.onlinekhabar.com/2016/07/04/380840. Accessed 4 July 2016.

imagined, metaphorical – that constitute pilgrimage activities in three ways. One, the distinctions between the sacred and the everyday in Shaligram veneration are often extremely fluid and blur the boundaries between temporality and materiality. This is particularly important because this fluidity often extends beyond the Shaligram pilgrimage itself to instantiate the journey as continuously re-enacted throughout an individual's lifetime as well as to frame Shaligram pilgrimage itself as a microcosmic recreation of the entire karmic journey of the soul from lifetime to lifetime to liberation (*moksha*). Two, the physical mobility of Shaligram pilgrimage is often interpreted by pilgrims and devotees as a kind of embodied sovereignty that transcends categories of nationality, ethnicity, economic status, and gender. In other words, Shaligram pilgrimage is perceived as political practice. And three, the goal of Shaligram pilgrimage is both a place and an object, where the place of pilgrimage is significant only in the individual pilgrim's capacity to traverse it and leave the landscape in possession of a sacred stone, which then carries the essence of the place with it while also becoming incorporated as a new member of the family and community. To put it more succinctly: where the material world is the everyday life of the body, religion is the everyday life of the mind.

Shaligram pilgrimage as a political practice refers to a method where belonging and identity are continuously negotiated between multiple frames of national, political, and religious reference. Mustangis, who were historically merchant and migrant populations themselves, often refer to Shaligram mobility and ritual veneration in their conflicts with culturally normative state structures that limit their own mobility and that of the tourists and pilgrims they rely on economically in the name of national unity and security. For Hindu and Buddhist pilgrims, national border controls, militarization, and the economic challenges that result from political conflict are seen as acting in direct opposition to Shaligram pilgrimage and ritual practices that fundamentally rely on the mobility of people and stones to maintain the spiritual links between people and places, between the Himalayan landscape and the *dham* (the spiritual abode of the deities). Ultimately, just as Shaligram pilgrimage reproduces and represents the karmic life cycle and begins the formation of Shaligrams themselves as persons, the political limitations on pilgrimage are translated into fears and concerns about the interference of governments on the progression of life itself.

The final pilgrimage destination, Muktinath temple, is by no means insignificant in all this. Contrary to popular perceptions, however, visiting Muktinath is not the prescribed goal of Shaligram pilgrimage but instead serves as the final stop in venerating the newly acquired Shaligrams within their contexts of origin – or celebrating their "birth" – before they are carried home. This perspective then recalls some of Eade and Sallnow's work (1991) which directly opposes the *communitas* paradigm through a focus on the role of major shrines in hosting and amplifying discrepant discourses among various groups of pilgrims. My work here, however, continues to unmoor the centrality of place in pilgrimage discourses in favor of overcoming a view of sacred movement as simply the arrival to and departure from specific shrines. Through the *dham*, Shaligram pilgrimage disputes the very notion of the fixity of sacred places at all.

The Turnerian notion of pilgrimage as a liminoid phenomenon has, of course, proven to be immensely resonant. In fact, it is something of a trope for anthropologists who study pilgrimage to begin, as I have, with the pronouncements of Victor Turner and then to employ his frameworks as a point of departure for their own work.[4] I do not, however, aim to either defend or deconstruct the notion of *communitas* because Shaligram pilgrimage is neither actually nor ideally divorced from everyday social, political, and cultural processes. Rather, my frameworks echo the work of Coleman and Eade (2004) and Tremlett (2003) who view both Victor Turner and Mircea Eliade as somewhat "romanticist" in their attempt to secure for religion an 'ineffable inner space or realm' that can stand as a critique of modernity and its values. For South Asian scholars, this division between "tradition" and "modernity" is also both equally ubiquitous and equally contentious, noting that the romanticism of tradition often misrepresents modernized (and globalized) India, Nepal, Pakistan, Bangladesh, and others as backward and primitive, or conversely as static societies continuously living out a purer, more authentic past.

What Does It Mean to Move?

Although movement itself has rarely been a major focus for earlier anthropological studies of pilgrimage, when it is discussed, scholars have adopted quite different perspective on it. From the standpoint of cultural geographies, such as Surinder Bhardwaj's *Hindu Places of Pilgrimage in India*, pilgrimage circuits constitute interconnecting links between religious populations and their sacred centers of worship (1973: 7). These routes and landscapes are then viewed through a kind of functional analysis, where the movements

4 See Yamba 1995, p. 9 and Coleman 2002.

of pilgrims integrate geography with religion and are therefore said to help create "pan-Hindu" identities through nationally identified "holy spaces" .In other cases, movement is located within broader semantic fields related to journey-taking, such as Eickelman and Piscatori's (1990) juxtaposition of the Muslim *hajj* (the main pilgrimage to Mecca) with *hijra* (emigration) and *rihla* (travel for learning) or Trapper's demonstration, in the same volume, of how Turkish *ziyet* – voluntary movement for the purposes of paying respects to a person or shrine – establishes the authority of each but in somewhat dissimilar frames of secular or religious reference (1990: 236). Secondary literatures on both Hindu and Tibetan pilgrimage and their sacred geographies, on the other hand, contain abundant examples of particular spaces being perceived and experienced as two or more quite different places.

Recall the *dham* from earlier chapters, where sacred places are said to be no different than another, more distant sacred place. This includes the village of Mayapur in West Bengal, which is often viewed as the Vrindavan *dham* (as being the same place as the town where Krishna spent his childhood), the *Char Dhams* (literally: "four abodes") of Vaishnava pilgrimage (Badrinath, Dwarka, Puri, and Rameswaram) and their four "associated places" (Kedarnath, Rameshwaram, Somnath, and Lingaraja Temple), or where the collection of Shaligrams themselves at Muktinath becomes the same as the *dham* of Muktikshetra and Śālagrāma. Similarly, in Tibetan Buddhist pilgrimages, single mountains may be the abode of a territorial *yul lha* (the "deity of the territory") to whom locals pay a kind of pagan reverence, sometimes even with blood sacrifices, and which pilgrims from all over the world might revere, with specified circumambulations, as the stronghold of a Buddhist or Bonpo *yi dam*[5] (Ramble 2014: 182).

In the contexts of political practice, cultural preservation, and revivalism in Mustang (such as *amchi*/Tibetan Medicine and temple/*gompa* conservation), Shaligram pilgrimage also lies at the juncture of ethno-historical re-creation and national reclamation, where questions of "Hindu" and "Buddhist" belonging remain contentious on a number of levels (see Chapter 3). I therefore locate this ethnography within the context of shifting and competing landscapes as a method for demonstrating how individuals and communities imbue mobility with significance. I am aware, however, that the result of privileging the perspective of Shaligram pilgrimage in Mustang

5 (Tib. ཡི་དམ་, Wyl. *yi dam*; Skt. *iṣṭadevata*) – one of the three roots, the tutelary or chosen meditation deity, which is the root of spiritual accomplishment. Yidams are often classified according to whether they appear in peaceful and wrathful form.

in this case might be seen as overly reductionist of a complex regional phenomenon to a single strand, since time and space constraints mean that I must leave many complementary or rival pilgrimage narratives out of the current account. However, it is precisely the multistrandedness of pilgrimage in Mustang specifically, and in South Asia generally, that contextualizes much of Shaligram pilgrimage itself, given the variety of religious traditions, nationalities, and ethnic groups that participate in the practice.

During Shaligram pilgrimage, movement becomes a locus for transformation and transformation the key to mobility as a political practice. This means that pilgrimage mobility, as well as afterwards, plays a decisive role in devotees' ideas of "going home" and "belonging to a land", where these notions then refer to a routine set of practices in relation to certain people and objects rather than to a specific place (see also Rapport and Dawson 1998). This perspective gets to the heart of one of Shaligram pilgrims' most pressing issues: the continued accessibility of Mustang, the continuation of national and political unrest in the region, and the growing draw-down of Shaligram stones for trade and souvenirs. Just as Shaligram pilgrimage relates to devotees' and residents' views and constructions of locality, landscape, mobility, space, place, the national, and the transnational, it also forms the principal framework for pilgrims' conceptualizations of what is means to be Hindu, Buddhist, or Bon in an era of ever-increasing political identification and mobilization. The continued economic and political restrictions to entering Mustang are then perceived as affronts to individual and collective identity – and meaning-making – where the reduced mobility of pilgrims and Shaligrams comes to represent the current political constraints on the peoples of Nepal and India more broadly as well as the potential life and death of mobility-based religious traditions themselves.

Arriving in Jomsom

The Hotel Majesty was just a short walk from the airport, but the chill morning winds had already begun to gust through the streets of Jomsom and the group of newly arrived pilgrims from India struggled to both carry their small bags and shield their eyes from billows of dust at the same time.[6] Their

6 The account detailed here is actually a combination of two separate groups undertaking two different Shaligram pilgrimages within the same month and year. While I accompanied both groups on their journey, I have chosen to consolidate them in this account, and change their names, for narrative and analytical brevity.

journey up to this point had already taken most of them more than three or four days: from securing tickets to Nepal from a variety of cities in India, applying for their entry permits in Kathmandu, chartering a plane to Pokhara at the base of the Himalayan foothills, and then waiting for any available early-morning mountain flight from Pokhara to Jomsom on the following day. For a few of them, common flight delays had taken up an additional two days; sitting in the departure terminal of Pokhara's small regional airport, standing anxiously by in hopes that the weather might sufficiently clear for the twenty-seat twin-engine Otter planes to make it safely to Jomsom's short high-altitude runway. With little in the way of instrumentation that can safely navigate the steep Himalayan valleys and unexpected mountain down-drafts, pilots to Jomsom must fly by sight, and incidents of plane crashes and other disasters are not unheard of. For those who opted not to wait for the planes, the journey could take even longer: two days by bus or six days on foot. The wealthiest few might even decide to hire a trekking jeep to cover the distance from Pokhara to Jomsom in a single day (about 7 hours), but at nearly $350 one way, few pilgrims could manage the expense.

The majority of the arriving group climbed down from the airplane's modest mobile ladder and immediately set off for their trekking hotels, while others drifted apart to take up rooms at Om's Home, the Dancing Yak Lodge, or the Lo-Monthang Guesthouse. A few stragglers stayed behind to stare in wonder at the surrounding mountains: Nilgiri and Dhaulagiri Himal towering above the town and marking out a space so vast and so boundless and colossal as to humble even the most experienced traveler. Covered in snow even during the height of summer, the two mountains marked out the traditional endpoints of the Kali Gandaki River valley some many miles to the south. The river itself, which flowed just beneath the airport runway, was already white and brown with silt and early spring floodwaters.

It was only around 7am but Jomsom was already bustling with Mustangi villagers riding their horses down from the mountains to restock on supplies of rice, grain, and kerosene. Tibetan women in traditional *chubas* were just coming out to sit along the dusty gravel road to sell apples from Marpha, vegetables from Pokhara, and wool from their home flocks of goats, sheep, and yaks. A young girl darted past, balancing a flat, round basket of apricots, halved and ready to be set out to dry in the relentless sun. Constantly besieged by high Himalayan winds, Jomsom has the feel of an old town despite its somewhat more recent construction: bright blue, pink, and yellow paint less than a year old already peeled and faded, a new cobblestone road already cemented with brown dirt and dust, a road marker painted on large fieldstones barely legible beneath outcroppings of grass and horse dung. Anticipating the morning influx of travelers, small market shops and trekking stands hastily unlocked their doors, and the aroma of hot milk tea and fresh Tibetan bread soon permeated the air.

A weathered Mustangi man towing a line of three horses stopped next to me on the road. “Pilgrims are coming now,” he said, motioning towards the ragged gates of the airport baggage terminal at the end of the gravel street. “It’s a good thing. They will come for Shaligram and for Muktinath. It’s time again. Will you go with them?” “Yes,” I responded. “The pilgrimage is beginning.” He smiled and nodded. “I will wait here then. They will need horses. They will need to get up the mountain and the jeeps are dangerous now, so it is better to take horses. And maybe I will find Shaligrams too. I put them in my house and in the school. Keeps the bad spirits from coming in.” “You are also going to go on pilgrimage then?” I asked. “Oh no.” He turned to calm the first horse as the noise of one of the planes overhead cut through the early morning wind. “I am Bon. I am always here. Always on pilgrimage.” He laughed. “That is how it is in the Himalayas. Nothing ever stops. Not us, not our horses, not our rivers or our cars or our winds. Always going somewhere. It’s because when you stop, you are dead.”

As I waited on the steps of the Lo-Monthang guesthouse, I suddenly heard my name shouted from the direction of the river. Subashna Sharma, a Hindu (though, by her own accounts, also occasionally Buddhist) pilgrim from Kathmandu whom I had known for several months prior to returning to Mustang, raced up to the guesthouse to greet me. “It’s so good to see you again!” She pressed a hand to her forehead to keep the wind from sticking errant strands of her hair to her cheeks as she spoke. “Are you coming with us to the river tomorrow?” When I replied in the affirmative, she smiled broadly and nodded, her round face and delicately crinkled cheeks already becoming red with wind and sun. “We will go out to Kali Gandaki early tomorrow, about 6am I think. We will begin Shaligram *darshan* and do a welcoming *puja* to the river. Then we can go to the jeep stand just near there and get bus tickets to Kagbeni. I want to be sure that we prepare the Devi (Mother-Goddess) *puja* first for the gracious kindnesses of Gandaki and then we will prepare new *puja* depending on the Shaligram *darshan* (meaning which Shaligrams were found that day) before we go to Muktinath. I cannot wait! I even brought tulsi from my home pots, the ones we started to grow after the last Tulsi Vivah (the Marriage of Tulsi and Shaligram festival). I picked them just before we left so that I can offer them fresh to Narayan (Vishnu) when he appears out of the waters.”

Recall the Shaligram origin myths detailed in Chapter 6. These stories, by and large, tend to include three broad themes, each of which is a reference to a life milestone, with a fair degree of exchange and overlap: Landscape as Pious Woman, Divine Death and Rebirth, and The Marriage of Tulasi. In the first category, Landscape as Pious Woman, I include the multiple variations on the tale of the river-goddess Gandaki who performs austerities in order to obtain the gods as her children, or Vishnu as her husband, as well as the story of Sati, whose fallen body parts form the landscape and the river. In the second category, Divine Death and Rebirth, I include the story of the sage Durvasa's curse and the churning of the milky ocean, Vishnu's testing by the planet Saturn, and the transformation of the gods into worms and insects (*vajra-kita*). Lastly, in the third category, The Marriage of Tulasi, I include the variations on the story of Tulasi/Brinda and Jalandhar/Sankhacuda where the union of the deceived woman with God Himself results in the production of the landscape as well as its reproductive capabilities. I have separated the list of Shaligram origin stories into these ostensible categories for two reasons: firstly because these are the primary mythic themes most often referenced and ritually leveraged by Shaligram pilgrims themselves, and secondly because, despite how these origin myths might be presented in religious texts, they are often recalled and retold by devotees as complementary (or competing) narratives related to issues of kinship and relatedness. Though Shaligram veneration has foundations in Puranic, Tantric, and Shastric texts, few Shaligram pilgrims directly consult textual specifics in their practices and, therefore, Shaligram veneration remains a principally oral tradition passed down through the continuous telling and retelling of Shaligram creation stories in relation to an individual's current circumstances.

The relationship of these principal mythemes to the places of pilgrimage is relative to the qualities of a particular space: the constraints of its physical features and the layout of its surrounding landscapes. This means that the particular meanings invested in specific locations and their features and the collectively understood modes of being in them are at least partially a matter of cultural (and political) narrative assignments. Symbolic and mythic representations of landscapes are then canonized and essentialized through narratives such that anyone who has not previously been socialized within the mythic complex (namely, tourists and trekkers) is unlikely to perceive the topography of the place in the same way. For example, without prior knowledge of the stories of Vishnu and Tulsi or even of the *sinmo* whose body forms the mountainous terrain of Mustang, a visitor to the region could never truly comprehend what it means to move through the landscape, to wash away past lives in the waters of Muktinath, or to witness the appearance of a Shaligram as its birth. For Subashna Sharma and her pilgrim group, their relationship to their imminent locality is mutually constitutive, even if their relationship with the precise texts on Shaligram origin may not be. This observation also belies an additional consideration: what Charles Ramble calls "the plural identity of the location" (2014: 181) in which different peoples encounter each other through differing views on the nature of specific places. Furthermore, Ramble's point makes it clear that a specific distinction between two crucial terms related to locality is required: that of "space" and "place".

In standard anthropological formulations, "place" refers to human perceptions and experiences of an abstract "space". "Place" is therefore a construct that holds specific meanings, identities, cultures, and so on, while "space" aligns with the objective, impersonal, and potential categories that structure and constrain experiences of "place". Within contemporary geography, however, the meanings of the two terms are reversed and it can be occasionally confusing as to how these terms are leveraged among different bodies of work. In terms of Shaligram pilgrimage, I adhere in this work to the anthropological convention where space signifies a natural location and place the cultural significance that has been accorded to it. But I also point out that neither "space" nor "place" in the case of Shaligrams is a fixed concept nor is it always rooted to specific physical locations. This is because, as is especially common in South Asian ideas of multi-local (perhaps even multiversal) pilgrimage, physical routes may lead to the same space but not to the same place. And in the case of Shaligrams, for all their physical and spatial dimensions, they traverse categories of belonging continuously and, in the process, mark the people and landscape in a variety of material and immaterial ways.

For many devotees, and quite a few residents of Mustang, the thematic elements of Shaligram practice that specifically revolve around the embodiment of the landscape as female are taken as evidence that the land of the Kali Gandaki was once inhabited by goddess-worshipping peoples and even that the Shaligrams themselves may have one been considered manifestations of the Devi (Mother Goddess) before they were worshipped as incarnations of Vishnu. Though it is difficult to tell, historically, whether or not this is true, the landscape of Mustang is often described as a place of extraordinary feminine divine power (see Oppert 1901), despite the relatively male characterization of most Shaligrams themselves and of the principal deities and *gurus* venerated in shrines throughout the region. According to some local accounts, this is why Muktinath is only ever attended to by nuns and why, according to Shaligram pilgrims, goddess-identified Shaligrams are exceptionally powerful. Read through a Western feminist lens, the transformations of Gandaki, Sati, and Tulasi principally result in the control of female reproduction by male forces, but this viewpoint tends to ignore the multiple contexts of male-female union that permeate both Hindu and Buddhist cosmologies.

Tsering Wangmo, a young Buddhist nun from the village of Thini, once explained: "For many Hindus," she began, "the male god always has a female counterpart who represents his creative energies: Lakshmi and Vishnu, Shiva and Parvati, and all the others. Buddhists have the Yab-Yum, the father-mother, which is painted on many *gompa* walls here in Mustang. It is the primordial union of wisdom and compassion. You will know it easily, since I think many Westerners are surprised when they see it because the Yab-Yum is always a man who is seated but is performing sexual union with a woman in his lap. The male figure is compassion, however, and he represents *upaya*, guidance along the pathway to liberation, while the woman is *panna*, insight into the true nature of reality. I have also heard some Hindus say that Yab-Yum is the same as Satkona, the two triangles that form the six-pointed star (a hexagram).[7] That is Shiva and Shakti also in union, and it represents wisdom and enlightened knowledge – the Supreme Being joined with the Mother of Nature. It is really no surprise that this land is the divine mother, as you require a body to beget another body, don't you? How else could Shaligram be born?" The theme of "life cycle" thus suffuses throughout nearly all Shaligram pilgrimage narratives and is reflected again by Hindu pilgrims as they begin the Shaligram journey in Jomsom.

Subashna Sharma and her group set out from Jomsom just before 6am. Not quite awake and still working on my second cup of strong Nepali tea, I followed them down the main road towards the Kali Gandaki River. Swollen from recent glacial melt, the grey-black waters raced past us in loud, roaring, white currents, crashing up against the narrow valley walls and spraying us with a chilled mist of specks and droplets. Mindful of the dim twilight, we made our way through the cluster of guesthouses near the airport and out onto the road leading past the army base a few hundred meters beyond. This base, established in the wake of the Tibetan resistance in the 1960s, continues to bustle with activity – small regiments of soldiers out for their morning exercises jog in formation or stand around the outside walls, balancing their rifles while trying to sip tea from fragile paper cups. Beyond the army base (and its stark "No Photography" signs), we passed through a few more scattered shops and local Thakali kitchens before finally arriving at the slender and rather precarious-looking wooden bridge that

7 Stylistically, the Satkona *yantra* is virtually identical to the Jewish Star of David and the Japanese Kagome crest.

joined the two sides of Jomsom together – the "old town" on the far side of the Kali Gandaki and the "new town" which includes the airport, the army base, and the most popular trekking hotels.

"I love watching the river from here." Subashna reflexively took my hand as we each began to cross the bridge by twos and only one pair at a time. "It feels angry almost. Raging. I hope that she will impart this into the Shaligrams for me." "What do you mean?" I replied as we stepped out onto the other side of the bridge, making way for an elderly man and a stubby, brown horse piled high with saddle-carpets balanced on a wood frame Tibetan-style saddle. "Every Shaligram has a history, lives that it has lived before this one. Because of this, a Shaligram must match with your household. It must fit in with the family so that you may live in harmony together. When a Shaligram matches with you, then there are no suicides or accidents in your home anymore. It protects your family against all harms and dangers and when someone dies in the household, there will be no rebirth for them. Shaligram will accompany them into oneness with God and then return instead of them. This is Vishnu-Pradyumna Shaligram[8] that we use for this." We paused to wait for the rest of the group to join us before proceeding into the residences and restaurants of old Jomsom visible just up the hill. "I want my Shaligram to rage like the river does, to look after my daughters you see. The world is very bad for girls now – not much education, not much money, much danger. But when Shaligram comes, it will hold back all the bad and the negative. Then they will live good lives. Also, I need a Govinda Shaligram, and a Madhava and a Damodar. These Shaligrams are best for sickness and they will be good for me when I am an old woman."

The walk through old Jomsom followed a cramped and winding fieldstone street threading its way between squat, mud-brick houses and the worn wooden lintels of various shops and market stands, all hoping to catch a traveler's eye just long enough to warrant a few minor transactions for packaged snacks, bottled water, or a handful of fresh steamed *mo mo* (Nepali dumplings). Earthquake damage from the year before was still visible in the walls of many of the houses, and a section of buildings just down the road

8 In the Srimad Bhagavatam, Pradyumna is the son of Krishna and his wife Rukmini. He is also considered to be one of Vishnu's four vyūha avatars (an avatar that bears one of his primary characteristics but is not a complete incarnation) who embodies one of the four-fold (Chaturvyūha) manifestations of the divine: destruction/dissolution of the universe, along with Vishnu's capacity of "knowing". Along with Pradyumna is Vasudeva (as the creator and "feeling"), Samkarshana (as the sustainer and "willing"), and Anirudda (as the purveyor of spiritual knowledge and "acting").

remained mostly still in rubble, revealing the brick, mud, and wood-frame construction common to high Himalayan villages. As we emerged finally from the narrow streets, we came upon an open courtyard. On the furthest edge, across from where we paused to catch our breath, was the Shri Janahit Higher Secondary School, a low, battered building with a white arch at the gate which serves as one of the few secondary schools in the region. Just before the school stood a massive Buddhist monastery, closed off on all sides by a bright, red wall filled with spinning prayer wheels and an ornate gate overlooking a group of small boys dressed in saffron robes and stretching in the early morning sun. And finally, to our right, was the Jomsom jeep stand, a ramshackle ticket counter set back against a high cement wall, behind which were the outdoor public toilets and a few pits for garbage. A few white jeeps emblazoned with red and yellow logos for "Muktinath Darshan" and "Tourists" waited while their drivers sipped tea and smoked cigarettes in the doorways of houses and snack kitchens a short distance away. A massive, tractor-wheeled bus in bright green and blue idled near the *gompa* wall and a pack of shaggy, mud-caked but happily wagging dogs wandered over to us to see if we might have some food to share.

"The river is just ahead." Subashna tugged at my sleeve. "The buses and jeeps won't leave until around 9, so we have some time." We turned and headed for the banks of the Kali Gandaki a few hundred meters up the road, sliding down the rocky embankment until our feet sunk partway into the muddy waters edge. "You probably won't find many Shaligrams here." Subashna called out. "They don't usually come down this far, but it is still possible because pilgrimage begins here. This is where devotees will come first, so sometimes very eager Shaligrams may come to meet you!" Within moments, the group of some six people had dispersed out across the wide riverbed, slogging through the mud and shin-deep waters in search of the first Shaligrams of the journey, buoyed by the enthusiasm of potential discovery. "One day," Subashna sidled up next to me as I searched the shallows for the tell-tale black luster of a Shaligram, "I want to go all the way to Damodar Kund. I have heard that if you leave one tulsi leaf in the *kunda* then the Shaligrams will come to take the tulsi leaf. Then you can catch them right there if you can. Otherwise Shaligram will take that leaf and go right back into the *kunda*."

"I found one! He is here! He is here!" The cry echoed off the steep valley walls. Immediately, the entire group of pilgrims rushed over to where Ranju Thapa, a Nepali woman now living in northern India who was accompanying the group, held up a large, oval stone, still slick and shiny black from the silty water. "Who is he? Do you know?" Ranju immediately handed the stone to

Ranajit Bhusan, the ostensible leader of this particular pilgrimage assembly. Turning the stone over and over in his hands, Ranajit furrowed his brow in contemplation. "It is very large," he said at last. "There are no chakras that I can see, but it has here the white vanamala." He indicated a white quartz line circling the upper portion. "The white vanamala means Sri Ram. Ah yes!" He shouted excitedly as he examined the base of the stone more closely. "Here is his arrow as well." He showed the elongated white marking of a fossil belemnite to the rest of intensely attentive crowd. "This is certainly Sri Ram!" Exclamations of "Jai! Jai!" punctuated the announcement of the Shaligram's identity. "Quickly now, Ranju!" Ranajit handed the Shaligram back to the smiling woman. "Put it in your bag. We will all have *darshan* later today when we get to Kagbeni."

Two more Shaligrams appeared in the Kali Gandaki near Jomsom that morning: the first, a small Krishna Gopala (Krishna as an infant), which brought a great deal of delight to Madhvi Bhusan (Ranajit's wife) who explained that she was eager for grandchildren and that this was now the sign that such an event was imminent (she also looked forward to gifting the small, round, marble-like Shaligram to her daughter-in-law for exactly this reason), and the second a profoundly worn Ananta-Sesha, the serpent of wisdom upon whom Vishnu reclines. As a Shaligram of the spiritual wisdom contained in the Vedas, it was decided by the group that this Shaligram would likely need to become a gift to their local temple. According to a mischievous Subashna later on, apparently the local *brahmacharya* was "in need of some wisdom" and she hoped the Shaligram could help. Along the way, we encountered a pair of local Shaligram sellers making their way up the Kali Gandaki riverbed carrying several large *doko* baskets on their backs filled with stones. The husband and wife team, residents of the village of Ranipauwa to the north, shouted out their greetings as we passed. "Many Lords today! Come and visit us at Muktinath. They are all waiting there."

As the sun finally rose over the mountains, Ranajit gathered the far wandering group back together and we set off back up the embankments and down the road towards the ticket counter. "We will have to go to Kagbeni by bus because the jeeps only seat about six or so. The jeep is also very expensive. It's mostly for trekkers anyway." He nudged me with a side-long smile. "But since you are a Westerner, maybe they would give us a better price." "Why would they do that?" I laughed. "Oh, I think they prefer to take Westerners," he shrugged. "More money."

I turned to Subashna as Ranajit jogged ahead to buy tickets and ensure our seats on the mountain bus. "Is that really true? Do they think pilgrims can't pay?" She nodded, with a terse set to her lips. "Many times, that is true.

I wish they made distinctions between pilgrims and tourists here. There are no services just for religious pilgrims anymore and we have to manage our trips all on our own and at our own risk. Pilgrims are sometimes exploited by agents on websites and by guides and by drivers, all because what they really want is business money (a term she used to mean money from trekkers and tourists). Many of us also must travel alone on pilgrimage, but we have no knowledge of this place or how to travel here. As you can see, this land is wild, and the river can be dangerous. The mountain paths are treacherous and can wash out or fall on you or the weather can turn against you. If we want a local guide, it is also very expensive and how do you know that they know or care about Shaligram? We paid so much money just to get this far, and our pilgrimage has only just started!" Moments later we climbed aboard a towering bus filled to bursting with eager travelers, which included trekkers on the Annapurna circuit and pilgrims anticipating the ride to Kagbeni and Muktinath, and terrifyingly top-loaded with backpacks, personal bags, and boxes of produce and cooking fuel. The driver called out for everyone to hold on and suddenly off we went, swaying madly up a 60-degree, 2000-foot incline, straight up the mountain.

Reaching Kagbeni

The trip from Jomsom to Kagbeni by bus takes just under an hour. Navigating the narrow mountain paths in a bus almost as wide as the road itself was no easy feat, and many times the driver was forced down to less than 10 km/h in order not to accidentally bounce the vehicle right off the edge and send the passengers sliding down the cliffside and back into the river. Through twists and turns we made our way ever higher until the bus finally crested a tall rounded peak and revealed the expanse of the Kali Gandaki below. At the edge of the river, nestled among flat green terraces, was Kagbeni, a village of scattered houses, grain fields, goat herds, trekking lodges, and kitchen-style restaurants. The driver let us off at the edge of the road, stopping only long enough to unload the pilgrim passengers before heading off further up the mountain to the village of Jharkot. Porters immediately began tossing down our bags along with boxes of clothing and supplies that several village men were already waiting to collect. We gathered our things quickly and followed them down the near-vertical indentations in the ground that served as the path into the village. About half-way down, the gravel path turned to uneven fieldstones and then finally to a set of makeshift stairs before we finally stumbled out onto the slate-slab streets near the far edge of village.

"The winds are already coming in." Ranajit waved me closer. "We'll have to find a lodge for the day and go out onto the river tomorrow. I know a pathway over that rise there that leads directly onto the banks. From there, the river is so low we could almost walk back to Jomsom just by way of Kali Gandaki." He chuckled. "But for now, we should have *darshan* and do *puja*. Then we can eat something." We made our way along the streets, weaving past herds of goats and sheep and even, at one point, needing to redirect a small calf from following us too far away from its pen. As we passed beneath the main village *stupa*, a towering structure nearly fifteen feet high, we ducked beneath it into a low chamber filled with intricate paintings of Buddhist, Bonpo, and Hindu deities. For many of the villagers, however, it appeared to just be a handy place to store motorbikes out of the way of the biting winds. There, I noted that a set of four Shaligrams had been placed at each of the four corners beneath the *stupa*, carefully tucked up into the eaves almost out of sight.

"Ranajit?" I called out. "Do you see the Shaligrams here? I did not think that Buddhists worshipped Shaligram." He looked about carefully before replying. "This is Bon practice, more likely. Maybe from Buddhists also. People say here that this land is ruled by Dakini, the female wind spirits who howl and scream through the valleys." He gestured vaguely towards the outside. "Like now, you hear them coming. That's why it is best to be inside at this time. But wherever there is Shaligram, the Dakini will not go. No spirit may enter a place of Shaligram without permission. That is why you see them like this. On *stupas* sometimes, like on roads or near village gates, and in houses. They make the land safe to travel and keep the spirits from entering places where they might cause trouble." I turned to him; my confusion likely clear in my expression. "I don't think I understand. How does Shaligram keep spirits away?" Carefully, he pulled one of the four Shaligrams down from beneath the *stupa*. It was roughly palm-sized with jagged edges and a large, clear chakra imprinted deeply into its top surface. "Do you see the great spiral?" He asked. I nodded. "This is Sudarshan Shaligram, the chakra of Vishnu. But the spiral is also sacred to all religions but most especially here [in Mustang]. It is sacred for Buddhists, who walk in a clockwise circle around sacred places. Because that is the direction of the turning of the universe. The Bonpos revere the counterclockwise circle. The universe turns the other way."

He flipped the Shaligram over and traced the reverse spiral on the opposite side with his index finger to demonstrate. "For us Hindus this is also sacred because it is the karmic wheel and the rotation of the planets. Just like the *shankha* (the conch shell, a symbol of Vishnu and Lakshmi), it is a symbol of water, of life, of the childbearing of women, and of serpents. This is an ancient symbol. It shows the movement of all that is divine and unseen. No spirit can violate that." Gently, he replaced the Shaligram back into the roof of the *stupa*. "It is regeneration and nourishment and continuation. Just like the *naga*[9] serpents coil in a spiral around their treasures to protect them, the movement of the universe coils around us as we traverse through our lives." Subashna gently touched her fingers to a painting of White Tara, a female bodhisattva in Mahayana Buddhism (or a female Buddha in Vajrayana Buddhism). "We must follow this karmic cycle always – being born, living, and dying over and over again. But as a pilgrim, and with Shaligram, you walk the paths like you live your lives. Gandaki is your birth who births Shaligram, and then caring for Shaligram as your family until you reach Muktinath, and there you and Shaligram can be reborn. And then you return home to do it again each day because Shaligram is with you. Each pilgrimage is like a life and Shaligram is a life isn't it? Maybe if we do them enough, we will fulfill many lives all at once."

Within the ritual practices and pilgrimages of Shaligram devotees, a single space often becomes a plurality of places. This is a practice that, in the context of religious traditions in Mustang, is also not unusual (see Ramble 2014). Circumambulations around a sacred space mark the pilgrim's transition from movement through physical spaces into movement through divine spaces as they follow the divine movement of the universe in the proscribed direction. In this case, circumambulation becomes the method by which individuals traverse between a physical holy site and the *dham* superimposed within it: "leaving" Muktinath temple, for example, and "arriving" in Śālagrāma.

In other cases, as I mentioned earlier, this becomes especially poignant when physical routes themselves may lead to the same recognizable space but not to the same places. One particular example of this kind of single-site multilocality happens in Hindu temples throughout Nepal and India where smaller versions of the main temple deities (called *utsav murti*) or their associated Shaligrams are removed from the main altar and carried outside on specified "pilgrimage paths" through the village. These paths often do not follow the main roads of the village, which are easy to traverse and are

9 Naga (IAST: nāgá; Devanāgarī: नाग) is the Sanskrit and Pali word for a deity or class of celestial beings who typically take the form of a giant snake or of a half-human/half-snake creature. Naga are found in mainly in Vedic religions such as Hinduism, Buddhism, and Jainism. A female naga is a nagi or nagiṇi. (Elgood, Heather, p. 2342000. *Hinduism and the Religious Arts*. London: Cassell, 2000).

used for everyday purposes. Instead, devotees carry the deities through older and often less well-tended walking paths or field roads that take them along a specified route through the landscape (and often past specific houses or other shrines). Both roads, however, always bring them back to the temple. But when the deities arrive, the temple is no longer the temple, but a place detailed in religious myth or a historical place located elsewhere wherein the deities may now carry out certain activities particular to certain times or certain places. Sacred shrines and sites map the familiar just as much as they act as signposts to the other world.

For many Shaligram pilgrims, physical pilgrimage is also heavily associated with "inner pilgrimage". In the Tantric traditions, such as in the Kalacakratantra, little attention is paid to the necessity of making physical journeys to the sacred places especially because of the correspondences between the places of sacred sites and the components of one's own body. Therefore, pilgrimage becomes a method of exploring the self through meditation and the performance of austerities.[10] While Shaligram pilgrims, for obvious reasons, almost never reject the necessity of actual physical pilgrimage, the overlay of the physical pilgrimage journey through the landscape with the metaphysical journey of the individual through life, death, and rebirth is especially salient.

I considered Ranajit's words thoughtfully as we continued through the village to our destination, the Hotel Annapurna, a moderate trekking lodge overlooking the river. As we piled through the door, panting for breath from just our short walk, two Mustangi women greeted us from the *baithak* (sitting room). A brief negotiation later and each of us went into our small rooms, unloaded our bags onto the padded bed-pallets, peeled off our shoes, and reconvened in Ranajit and Madhvi's room for *puja*. Subashna sat down next to me as we formed a circle on the floor. In the center, Ranajit had collected the group's Shaligrams and placed them on a small, silver dish he had carried in his pack, covered with a red cloth. As he began to sift through his bag in search of the rest of the *puja* items, Subashna tugged at my sleeve.

"Do you know how to do Shaligram *puja*?" She asked. "I've seen it a few times before," I replied. "But I don't think I could do it myself." She nodded.

10 These various sets of correspondences are detailed in Vesna Wallace's study of the text, the Inner Kalacakratantra. The Kalacakra tradition rejects the inherent sacredness of one place or one human being over another. It suggests that all regions of the world and all human bodies are equally sacred. This view of the human body as containing within itself all the pilgrimage sites is not unique to the Kalacakra tradition. (Wallace, Vesna A. *The Inner Kalacakratantra: A Buddhist Tantric View of the Individual*. New York: Oxford University Press, 2001. See especially pp. 78-86 and tables 5.7-5.11.)

"That's OK. With Shaligram it is easy. There is no calling of God like in other *pujas* because God is always present in Shaligram. It is self-manifest. So, we do not need to do that here. Instead, we will give water and tulsi. At home, we give milk, honey, ghee, sugar, and Ganga water. Then we will use oils to rub on Him, fragrant oils like sandalwood or lotus. I use hibiscus though, because it smells so lovely. We can also offer flowers and fruit too, and then some drinking water. Whatever you have, it is OK. We also don't have lamps with us for *aarti*, but we will make do." Ranajit hushed the group, pulling out his *ghanta* (a hand bell used in *puja*) and setting it aside with a small dish of water and a wrapped leaf carrying *kumkum* and turmeric powder. As he lifted the cloth from the Shaligrams, the pilgrims began to gasp, keen, and pray. Later, as he rang the *ghanta* in a steady and unrelenting rhythm, we all joined together to recite the prescribed *mantras* to Vishnu, Lakshmi, and Shaligram.

That evening, as we sat hunched over our typical meal of *dal bhat*, Ranajit continued to expound on the challenges of Shaligram pilgrimage. "We'll go out onto the river tomorrow and hopefully good Shaligrams will appear." (Good, in this case, typically meant whole, unbroken stones.) "Is it difficult?" I asked. "On this part of the river I mean, to find good Shaligrams?" "It used to be easier," Ranajit replied. "But since they started building (hydroelectric) plants on the river, some areas are now almost impossible to find Shaligrams. Other parts of the river are still restricted because of the Maoists, so there are only specific places you can look. Many people don't even come here anymore. They go to Triveni down by the border (with India) and look for Shaligram there. But you know, those are fossils, I think. They have chakras, yes, and they have many characteristics like Shaligram but I don't think they are. Or they are just not the same. They are not so pleasing to look at. They are covered in holes and they are oddly shaped. They are also not black like Shaligram in Gandaki. Most I have seen are sort of brownish or orange. It is true that *shilas* are still auspicious, even if they are brown or blue or white but Shaligram from Kali Gandaki is unique. They are stronger and you never see pure Shaligrams like Sudarshan or Lakshmi-Narayan at Triveni. The green and white Shaligrams also do not appear in Muktikshetra. No, you must come to Kali Gandaki. This is Śālagrāma."

Overhearing our excited exchanges, the elder son of the guesthouse owner appeared in the doorway. "You have Shaligrams?" He asked. As the group of gathered pilgrims began to show him the Shaligrams from earlier in the day, he grew more fidgety. "I have a Shaligram in my room. I'll bring Him up for you to see Him." Upon producing a very large Harihara (Vishnu and Shiva together) Shaligram and placing it on the table, he leaned in thoughtfully.

"This was a gift from my father-in-law when I got married. My wife is from Pokhara and she came with Him [the Shaligram] at the wedding. I asked him why he would give me Shaligram when I am Mustangi, I can go out and receive Shaligram at any time. He said it was because my family are different Thakalis than they are. So, he sends Shaligram back 'home' to join us back together. Then he said it was because he worried about what the government (in Kathmandu) was going to do to us. Maybe we couldn't come back here, maybe there would be no people coming to Mustang and no jobs and no money. So maybe Shaligram should come back and bring those good things with Him." Ranajit examined the Shaligram carefully. "And has He? Has fortune come back?" The young man shrugged. "No. We are still very poor. But we are together. That is fortune."

In Nepal, many ethno-castes, such as the Thakali, are quite geographically spread apart and it is not uncommon for one particular group to look on another as "not quite belonging" to the ethno-caste in the same way (see Fisher 2001). In the elder son's view of the Harihara Shaligram, then, the binding of Vishnu to Shiva within the stone became analogous to the binding of the two different types of Thakali families in his marriage to the woman from Pokhara. What is more, the theme of "going home" to Mustang was also especially salient to his concerns about keeping the Shaligram properly. It was the hope of his father-in-law that, by returning the Shaligram to Mustang, it would bring with it the prosperity of his wife's family in Pokhara (which is known for its tourist wealth) but in the end, though no money had appeared to follow it back to Mustang, it had brought with it his wife and the wider ties, affections, and support of an affinal family, for which he was grateful.

The following morning, the entire group rose early, bathed, and gathered in the main dining room to sip tea and await the rising of the sun. We were joined by two Australian trekkers, also taking an early breakfast on their rest day before preparing to head northwards towards Lo Monthang in Upper Mustang. Over a course of eggs and coffee, one of the young men produced a large Shaligram from his bag, explaining to his companion that he had found the stone on his walk up from Jomsom and, seeing that it was probably a fossil, wanted to borrow his friend's rock hammer to smash it open and see what was inside. Several of the pilgrims shifted uncomfortably in their seats. Subashna leaned towards me. "You see? They don't know anything. It's just a rock for them. I can't bear to think that they will break it, but I know that's how it goes." Ranajit looked, for a moment, as though he might speak up, but as the two men continued to debate the best way to break the stone so as to get at the shells in its center, I watched as he simply turned back to his tea and scowled.

"Why don't you tell them what it is?" I motioned to Subashna. "Ask them not to break it. Or maybe tell them why it is important and then they won't take it." "This is not how Westerners are," she replied. "They don't understand Shaligram." In a move that was exceedingly rare for me as a participant-ethnographer, I turned in my bench seat and complimented them on their beautiful find. Mentioning that they were incredibly lucky to have found such an exquisite Shaligram, I asked where it had come from. "A Shaligram?" The first of the two looked up at me. "What is a Shaligram?" The pilgrims around the table smiled and by the time the sun had risen, the Shiva Linga Shaligram (as Ranajit identified it) was no longer in danger of destruction (because the men had agreed not to break it) and the two trekkers and the nine pilgrims finally parted ways with good cheer.

The river that morning was cold and fast as we made our way down from a rocky outcropping and onto the soft, muddy river bottom. Immediately, the group began to scatter across the wide expanse of the Kali Gandaki, heads bowed and shoulders hunched in what I always thought of as the characteristic pose of a paleontologist searching the land for bones and other clues. The irony of thinking of Shaligram pilgrims in the same way was not lost on me. Several minutes later, Vijay Pal – Ranajit and Madhvi's nephew – and I met a *sadhu*, carefully poking his walking stick into the riverbed as he too examined the silt for sacred stones. "Jai Baba!" Vijay called out as he approached. The elderly man, with long dreadlocks wrapped tightly around the top of his head, nodded and smiled. "Have you seen Shaligram today?" Vijay asked. The *sadhu* again smiled, placing his hand over his mouth to indicate that he did not speak and rattling his walking stick by way of response. He then pulled a small handful of Shaligrams from the cloth bag hanging from his waist-wrap and gestured towards the river.

Vijay walked over to him and cautiously looked through the Shaligrams clutched in the palm of the *sadhu's* hand. On spying a small Lakshmi-Narayan Shaligram, a characteristic stone marked by partial chakras on both sides, he begged the old man in Hindi if he would be willing to part with it, even offering to trade him the Ananta-Sesha Shaligram that had been found earlier in Jomsom. The *sadhu* glanced over towards me and grinned, handing Vijay the Shaligram without hesitation and shaking his head. He then pointed at me for several seconds before pointing down at a pile of pebbles near my feet. As he walked off, still poking at the riverbed with his stick, I looked down to see another Shaligram, just barely visible within the pile beneath a pool of turgid water. As I bent to pick it up, Vijay held out the new Shaligram he had acquired from the *sadhu* for me to see. "I knew this Shaligram would find His way to me today. Every day I pray that

Vishnu come to me and when I finally got enough money together to come on pilgrimage, I knew He was telling me it was time. I made my journey and He made his. We have crossed the *tirtha* Kali Gandaki together."

Defining the precise journey of Shaligram pilgrimage is problematic given the wide diversity of local and transnational viewpoints and the characteristic merging of geographical and non-geographical categories. Recall, additionally, the Hindu concept of "inner pilgrimage", where individuals visit sacred places within the microcosm of the mind and body and the ways in which inner pilgrimage and physical pilgrimage often co-exist, where one's journey through physical spaces recapitulates one's journey through life. The Hindu conceptualization of the *tirtha* further highlights these correspondences in that a *tirtha* can be shrines or sacred places along the course of a sacred journey but can just as well be applied to a devoted wife, a spiritual teacher, one's parents, or the home in which one is raised (Morinis 1992). Despite cautions concerning reductionism, movement then becomes the key to Shaligram veneration (if not to pilgrimage itself generally). As Morinis notes, pilgrimage is ultimately a term that can be "put to use wherever journeying and some embodiment of an ideal intersect" (1992: 3), and he also writes that the very "essence of journeying is movement" (ibid.: 12). But as Shaligram pilgrimage becomes clearer, the patterns of movement vary significantly within the actual pilgrimage itself: not just to the temple or to the river and back, but a spiral that connects the physical with the metaphysical, bodies with stones, and the everyday troubles of life, death, and rebirth with the movements of both gods and man through time and space. This issue quickly became all the more important when it came time to leave Kagbeni.

We met the same two Australian trekkers the following morning as each of us prepared to disembark: they were headed into Upper Mustang by way of Chusang, Dhi, and Geling villages on their way to Lo Monthang, and we were bound for Muktinath. The pilgrims and the trekkers chatted amicably for several minutes as bills were settled and bags repacked, but as the two young men and their Nepali guide hiked out of the village roads towards the border checkpoint just north of Kagbeni, Ranajit handed me a cup of milk tea and sat down. "Westerners are very lucky. I don't think I will ever get to Damodar. To get guides and permits, it's so expensive." "Has it always been this way?" I asked. "It has been difficult to follow Gandaki past Kagbeni for many years now. Nepal is afraid of China and of India too. Everyone is stuck. People don't belong to those places though; people belong to their people and every time they keep us out, it causes problems and fights. The army patrols the borders and keeps everyone out and the government makes it so hard to go [with fees] that only tourists can go. They want the tourists, though, because of all the money they bring. They don't want pilgrims anymore. They don't care about Shaligram. In fact, you want to know what I just heard? There was a whole group of pilgrims from India just last week who all paid their money, great money, to go to Damodar. They came but they took away their permits at the border (with Upper Mustang). The Army said they could not go anymore." "Why?!" I asked, concerned with Ranajit's soured demeanor. "Simple," he replied. "China was there. Chinese military. That means no one is allowed. They sent them away. No pilgrims. No Shaligram."

Ranajit's concerns about the divisions between tourists and pilgrims was echoed many times over the course of my work in Mustang, framed in particular by the viewpoint that the difference between pilgrim and tourist was that tourists were those who moved through the landscape but not in it. In some cases, pilgrims expressed gratitude at what amenities they could find, such as the building of a pilgrimage *dharmsala* in the village of Ranipauwa near Muktinath, but for many others, the challenges of undertaking Shaligram pilgrimage as well as the near impossibility of reaching the Damodar Kund weighed heavily. In nearly every conversation, the role of national security and the policing of Mustang's contentious northern borders were the focus of concern. For the residents of Mustang, continued political isolation was equally troublesome. "It's a double-sided problem," Karsang Yangchen explained.

Karsang had been born in Mustang before completing her primary education in Kathmandu. She had returned, however, as a young teenager to rejoin her family in the village of Jharkot and had been working as a cook in her uncle's guesthouse for the past several years. "In Upper Mustang it's even worse than it is here though. Everyone around here wants tourists to come because that is how they get money, but Kathmandu keeps all the permit money. It should come here but it doesn't. The government just keeps it. The permits for Upper Mustang are even worse, so few tourists can go there but they don't get the money either and they [the government] won't open the borders because of China and Tibet. They won't give us the money and they won't open the borders. So now it's tourists or nothing. Many people are leaving because of this."

A few weeks later, a group of Hindu pilgrims traveling from the United Kingdom expressed similar sentiments. "I was born in Kolkata, but my family lives in London now," said Ravi Pandey, adjusting his white *dhoti*. "Our foreign passports make us foreign now, I guess. But we didn't used to be. As Indians, it was a little easier to cross the borders because Nepal and India are close friends, but now as English it is very hard to come here. Now we come during monsoon, when the trekkers won't come because it is cheaper, though it is more dangerous. There aren't many drivers, so sometimes there are no buses or jeeps. And sometimes the guesthouses won't take pilgrims because we can't pay so much. But Shaligram pilgrimage is more important now than ever! With Hindus going all over the world, we must come back to these ancient places to keep our traditions alive, to keep connected with our past and our ancestors. But I suppose pilgrimage has always been difficult and it's supposed to be difficult. This is just our karma."

As I and the pilgrimage group set off up the steep and dusty mountain road on the five-hour walk towards Ranipauwa, Subashna jogged ahead to walk next to me. "This land is the land of creation." She smiled. "This is not Mustang. It's not Nepal. It's not India. It's not Tibet and it's not China. This is Muktikshetra, it belongs to no one but those who can walk it."

The trouble of Mustang, and of Shaligram pilgrimage, is the trouble of belonging. They represent a crossroads of identity in virtually every sense. For Mustangis, the trouble lies in historically merchant and migrant populations who continue to struggle against growing state structures that limit the very same in the name of national unity and security. For pilgrims, national divisions and the economic challenges that result from political conflict are viewed as antithetical to pilgrimage sites that are both physically located within nation-states and symbolically and spiritually situated within multi-sited sacred landscapes. Because Shaligram pilgrimage is also considered by devotees to be a continuous recapitulation of the karmic life cycle (as well as a method of living many lives at once) and the Shaligrams themselves are seen as persons, family, and community members, the political limitations on pilgrimage are viewed as the interference of governments on the progression of life itself.

The View from Muktinath

The landscape of Mustang is truly awe-inspiring. Caves dot the mountain-sides, some emerging so high up a sheer cliff as to defy belief that ancient peoples once made them into homes. Monasteries and road-side cairns piled high with *mani* stones are everywhere, some nearly as large as a house and laden with massive slate slabs carved with Tibetan *mantras* or the images of Hindu or Bon deities. Families often commission these stones from carvers as offerings to the gods in memoriam to a recently deceased family member or for luck on a new venture. We also passed trios of earthen *stupas*, painted with mud pigments in red, black, and white – a symbol of the Rigsum Gompo, the Buddhist trinity of Avalokiteshvara, Manjushri, and Vajrapani. As protectors of the sacred geography, these *stupas* are often placed at major crossroads and at the entrances to villages or even atop the roofs of houses. Decorated with sacred juniper sprigs (which the locals call *dhupi salla*), which is also burned as incense, and goat or sheep skulls, these *stupas* often contain small offerings left by pilgrims and villagers in their small square central openings.

These offerings – reliquary *tsa tsa*[11] for Buddhists and more often than not Shaligrams for Hindus and Bonpos – mark the physical passage of people as well as the mystical passage of the dead and the divine. As one local Buddhist monk explained, the colors were additionally meant to symbolize protection (black), compassion (white), and wisdom (red), and the combination of the Bodhi Stupa (white), the Dharma Wheel Stupa (red), and the Serira Stupa (black) were essential to subduing various spirits as they moved across the land. these *stupas* had also been integral in warding off natural disasters. "This is why Mustang survived the great earthquake" he said. Massive oval stones hauled up from the riverbed or taken from washouts higher in the mountains formed fence-like barriers around houses or were used in the foundations of small buildings. Doorways are marked with the images of Shiva and the Buddha along with other celestial beings, or with yarn mandalas and goat skulls to protect the family within from uninvited spirits or *ro lang* (zombies). Groves of poplar trees began to thin out and soon enough, there were no more trees except for a curious spot of green hiding the white-washed walls of Muktinath temple many miles away. This was a landscape that was itself at the crossroads of cultures, religions, and ethnicities. Everywhere, Hindus, Buddhists, and Bonpos lived together and marked the landscape in ever-changing ways.

We reached the village of Ranipauwa nearly six hours later, in the early afternoon of what had turned out to be a clear and sunny day. A painted red archway over the main road leading up to the village greets visitors as they enter, and from there the road splits: we walked to the right down a short pathway to the Ranipauwa bus and horse stand and straight ahead into the village. As we passed the low squat houses dotting the side of the road, we also passed beneath the golden Guru Rinpoche statue a few

11 In the Buddhist tradition, a popular way to memorialize and benefit those who have passed away is to enshrine a share of the person's ashes in a small reliquary, known as a *tsa-tsa*. These reliquaries are created in the form of Shakyamuni Buddha, traditional stupas, and various Buddhist deities.

hundred feet above us. The overlook, containing the new twenty-foot statue and a small visitor's center, was completed in the early months of 2016 and presides over the main stretch of the village from a barren redoubt of granite boulders and slate faults high above. Like many Mustangi villages, the main cluster of residences in Ranipauwa circled a central village square with a government water tap and the primary village *gompa* just a short distance behind it. It was well past midday and a congregation of local women were gathered around the water tap, bathing, washing small children, and attending to the laundry. Their laughter echoed across the open sky, and several called out to welcome us as we slogged by in exhaustion. On the far side of the open roadway, trekking lodges and small shops with bright blue, white, and yellow signs crowded in just beyond the police station and trekking permit checkpoint. Pausing briefly to examine one such restaurant with two mummified yak heads hung over the entrance, we all then dutifully stopped to register at the checkpoint before heading further on.

At over 3,800 meters above sea level, the village of Ranipauwa lies very close to the cold tundra climates of the high Himalayas. Snow falls throughout the winter, but as the village is situated just below the Annapurna rain shadow, monsoons tend to be foggy, grey, and rainy, but without the daily torrential downpours common to other regions of Nepal. With limited possibilities for the cultivation of crops, the majority of Mustangis who do not rely on trekking and tourism tend gardens filled with potatoes, cauliflower, carrots, and lettuce and keep small herds of goats, horses, mules, and yaks. In some areas, *dzo* (a yak-cow crossbreed) also wander down the mountain accompanying small milk cows or miniature donkeys. As we waited to finish our permit registration, a caravan of mules decorated with red and yellow Tibetan saddle blankets began to slowly amble through the village, laden with bags of rice, kerosene, and vegetables just up from Jomsom. The caravan driver waved as he passed.

From the checkpost, we continued on up the compact mud road, passing small stands of woven wool scarves and hats, jewelry makers working in copper and silver, and – as I took quick note of – Shaligram sellers. Advertising Shaligrams directly from the Damodar Kund, I paused at one shop, barely more than a wooden bench, a plastic tarp pinned between two poles for shade, and several hand-woven baskets of Shaligram stones. "These are all from Damodar?" I asked in Nepali. "Yes, Oh yes!" The woman nodded. "Did you collect them?" I inquired. "No, no." She shook her head emphatically, "my husband, he goes to Damodar every year and brings them. He brings them for pilgrims because they cannot go."

Ranajit stopped beside me, listening intently before pulling me away. "You must not buy Shaligram," he said. "This is a karmic sin." I nodded. "And yet there are so many sellers here. Pilgrims must be buying them." "Yes." He paused, worrying about the shoulder straps of his backpack. "Sometimes it is the only way to get Shaligram, especially if you cannot spend the time in Kali Gandaki or if you need a specific Shaligram. Some Shaligrams only come from Damodar, like temple Shaligrams. The very large Shaligrams like that one over there." He pointed to one of the Shaligrams prominent in the woman's baskets, a dinner-plate-sized stone with the ridge of a chakra-shell visible around its entire circumference. "Shaligrams like that only come from Damodar. They do not appear in the river and if there is no other way to get it, then sometimes it is OK to buy. But you shouldn't buy them. If it is the right time and you are worthy of it, Shaligram will come to you on His own." As we talked, Ranajit led the group onwards to our first stop: a small, family-run guesthouse called the Royal Mustang, to secure rooms and wash up before making our way to the temple of Muktinath.

By one o'clock, everyone was eager to make their way up to Muktinath temple and start the ritual *darshan* of the site. Leaving everything but the Shaligram stones in our rooms at the guesthouse, we once again set out up the road, through the village, and towards the temple. As we picked our way carefully through the muddy road, dodging the ever-present cow patty or pile of horse apples, I asked Ranajit how he had first learned about Shaligrams. "My father told me the stories when I was a boy. My grandfather had Shaligrams and his grandfather had Shaligrams. The British never paid much attention to Shaligrams during colonial times, so my family was able to keep theirs even when many of the religious troubles started happening. I think also that's why most Westerners don't know Shaligram, because the British didn't care about them, so they never made rules about them. My father inherited his Shaligrams when I was in school, and then he gave me half of them when I was married. He has given most of the rest of them away now, but he still keeps a few that he wants to take when he dies. They will go with him into the fire he says. This is my duty to make sure. When my children marry, I want to do the same. We have one daughter who is about to go to university, so I will send Shaligram with her. Anirudda, I think, so that she can study well and have great success."

He smiled at the thought. "It is like the story of Chandrahasa. Do you know this one?" I shook my head. "Oh, it is my favorite. Chandrahasa was a boy who was very poor, and he lived on the streets. But one day he found a Shaligram just by chance, though he thought it was nothing more than a normal pebble. He used it to play marbles with his friends and slept with it under his head at night so that no one would steal it. And do you know what happened to him when he grew up? He became king!"[12]

Where the road angles off further on around the Panchgaon villages, the entrance to Muktinath is marked by a white archway (Muktinath-Chumig Gyatsa, it reads) and a short, central pillar set with three spiral Shaligrams embedded within the plaster at the front and a large, round, smooth Shaligram the size of a melon attached to the top. Beyond the archway is the first flight of nearly a hundred stone stairs winding their way through the rocky landscape, ever upwards towards the white walls of the temple complex in the distance. Subashna and her cousin Manish Prasad were the first two to touch their foreheads to the Shaligram on top of the pillar, press their palms together in the *pronam* gesture, and begin ascending the steep stairs one careful step at a time. As each of the pilgrims followed in turn, the group descended into silence; partially in reverence to entering the *dham* of Muktinath and partly due to the constant need to pause and catch our breaths in the thin mountain air. Along the way, hundreds of small stone piles had been erected by various pilgrims along the edges of the stairs and out into the scrubby bushes and patches of grasses, yellow rapeseed, spindly white *Silene moorcroftiana,* and purple amaranths. Many stone piles and bushes were also wrapped in strips of colored cloth, gone threadbare and ragged in the high Himalayan winds. As I stopped to examine one such stone pile, I noted that a circle had been scratched out in the dirt around it with the remnants of red *kumkum* powder still visible on the topmost stone.

Manish stopped next to me. "They're for *puja*," he offered. "To give small offerings as you come and go. Like personal *mandir* or *stupas*. These cloths are prayers to the land. Red is sky, yellow is lake, the red cord is for jungle, the green is land, the white is river, and the thicker white one is mountain, and the blue is clouds. They are also like the Buddhist prayer flags, so everyone leaves them along the trail." As we continued on, we also encountered several *sadhus*, each with a small area staked out with blankets, a small tin to collect donations, several pots of *kumkum* and ash, and a few small deities perched on the rocks next to them. I knew from previous pilgrimages I had accompanied that many of the *sadhus* were Shaiva *sadhus* who overwintered in the *dharmsala* in Ranipauwa, while many of the Vaishnava *sadhus* made

12 Chandrahasa was the king of Kuntala kingdom (roughly corresponding to parts of the north of present-day Karnataka and south of Maharashtra, India, including Gokarna region). The story of Chandrahasa is mentioned in Ashvamedhika Parva in the Mahabharata. He was the son of the king Sudharmika of Kerala. More famously, Chandrahasa befriends Arjuna who was accompanied by Krishna guarding the Ashvamedha horse of Yudhishthira.

the journey from Kathmandu, or even from as far as New Delhi, every summer. One such *sadhu,* who I knew as Naga Baba, had sat with me for many an afternoon at Muktinath, discussing matters of faith, pilgrimage, and Shaligram stones. In fact, one of the Shaligrams I carried with me had been a gift from Naga Baba, who kept a small collection of Shiva Linga and Ananta Shaligrams along with two small brass deities of Shiva and Parvati in his belt pouch. Manish stopped to offer a few rupees to an elderly *sadhu* resting on a stone platform halfway to the temple and receive the old man's blessings and an ash *tikka* on his forehead.

The entrance to Muktinath is guarded by a heavy metal gate, just beyond which is the first of many Buddhist prayer wheels, this one nearly 8 feet high and almost as big around. Another group of *sadhus,* gathered around the gate entrance, looked up as we approached and shouted out "Jai! Jai Sri Muktinath! Welcome! Welcome!" Ranajit smiled and bowed, his hands still folded in front of him. "Jai Babas!" he called out in response before offering a few rupee notes to several of them as he passed. Walking by the Samba Gompa immediately to our left, we began our mini-pilgrimage (as the walk through the complex of Muktinath is often characterized), with our first stop to visit and take *darshan* at the Shaligram mandir (Yagyashala).[13]

Ducking through the low door of the white-washed mud-brick building, we entered to the right of the central ritual fire pit and approached the altar at the north wall. This particular altar was comprised of another large round melon-sized Shaligram at the center, Vishnu-Narayan, and two carved black deities of Bhumi (Lakshmi) and Saraswati on either side of the main Shaligram. To the left of the deities stood a three-tiered table, every inch of which was covered by Shaligram stones. "Many of these Shaligrams have come for retirement," Subashna explained as Ranajit and Vijay began their clockwise circumambulations around the fire pit, reciting *mantras* to the Shaligrams in their hands and on the altar. "Retirement?" I asked. "Yes. You see, if you can no longer keep or care for your Shaligrams, you must either return them to Kali Gandaki or you may bring them to a Vaishnava temple to be looked after. You cannot simply abandon them. This would be a horrible thing. You would not abandon a child or your old parents, would you? No, of course not. Just like elderly parents can go to live at the temple at the end of their lives, Shaligram does this too."

After completing our *darshan* at the Yagyashala, we stepped out onto the stone walkway and turned towards the first rise of the hill. As we did so, two Nyingma Buddhist nuns in their characteristic red and saffron robes

13 In Vaishnava traditions, a yagyashala refers to a place where the fire ceremony is done.

strode past us. "Namaste *aani*!" we called out as they continued further up the path. "Namaste!" They both smiled. "Ranajit?" I asked. "Is it strange to you that Muktinath is attended by Buddhists?" "Not at all." He shook his head. "The nuns might live here all year round to care for Muktinath but there are also Hindu *pujari* and *sadhus* here as well. We worship together, we pray together, we come from the same gods who come from the same places, and so we care for Muktinath together. I think it is actually better this way. No one can say that Muktinath is theirs. Not Shaiva or Vaishnava or Buddhist. It belongs to all of us."

Massive white waves of fast-moving water roared past us, drowning out all but the loudest sounds (acting as a metonymic stand-in for Kali Gandaki). Originating in an aquifer just behind the Vishnu-Chenrezig Mandir, the waters sluiced by through barely contained channels of stone and grass. These waters, which feed the 108 waterspouts, are also the reason that we now stared up in awe at massive green trees and thick, lush vegetation obscuring the Vishnu-Chenrezig Mandir from sight at an altitude where almost nothing else will grow. Ranajit stopped to shake a wide stand of brass temple bells before crossing the walkway to push against another similar stand of bells on the other side. Several more pilgrims coming up behind us announced their approach to the deities in the same way, and together we walked out of the forested gloom and into the bright sun shining down on two 6-foot *mani* wheels spinning madly as the waters rushed over their paddle wheels and spun them around in perpetuity.

"The Scriptures say that if a person can get even a single chance to come to Muktinath, they will be immediately liberated in the afterlife," Ranajit said, motioning the group to hurry along. "From Puranas and the Vedas and I think in Buddhist texts too, we know that Muktinath and Muktikshetra have been a *tirtha* for many religions for a very long time. Sacred places transform what is different back into what is the same. There is no difference here. That is how one can achieve liberation, don't you see? This life is the same as all your other previous lives. You bathe in the waters and cleanse the sins of a million lifetimes and become free from death and rebirth. You can also free your ancestors from their sins and they can then be reborn in Heaven. You can assure that your children are liberated in only one lifetime. When you take Shaligram with you," – he held up the Shaligrams in his hand – "then Muktikshetra is with you, then this life and the life of Shaligram become one."

The Vishnu-Chenrezig Mandir sits atop a high stone platform foundation accessible by two additional flights of stairs beyond the river-run prayer wheels. As we pulled our way up the narrow stairs and into the main courtyard, we were instantly greeted by shouts of activity. About thirty pilgrims already congregated throughout the area, some preparing to plunge beneath the waters of the two *kunda* pools immediately in front of us, many more in various stages of undress along the back benches as they braced themselves for a run beneath the water spouts, and several sitting silently meditating before the *mandir* itself, waiting for the Buddhist nuns to reappear for the afternoon *darshan* of Sri Muktinath himself. Ranajit and Manish immediately started for the waterspouts, carrying their Shaligrams tightly so as not to accidently drop them as they (both men and Shaligrams) were bathed beneath each spout. Vijay and Ranju, on the other hand, opted to begin with the *kunda* pools, promising that they would head for the spouts immediately afterwards.

Subashna darted off with a joyful shout to enter the temple and await *darshan*. Madhvi, however, came to sit beside me as I, in my typical ethnographer's fashion, found an out-of-the-way bench and sat down to observe. "I am far too old for those cold waters," she said, playfully nudging my shoulder. "I will wait for *darshan* and then maybe I will put my hands in the waters later." I smiled as Ranajit and Manish emerged shivering and bouncing from the far-side of the waterspouts, jogging towards the first *kunda* while dripping water down the slick stone incline. "I can see why you think so," I replied. "They look absolutely freezing." "Oh yes," she nodded. "But the body is always cursed to suffer." I did not ask whether she was referring to her own age or to the relative temperature of the water. "But the soul does not suffer. Human bodies are like all other bodies that way. They must eat, come together for children, defend, and die. You see now how Shaligram is not stone but body. They [the gods] come to us like this so that we may know them and so that they may know us. Do you see? This is why government cannot get in the way of pilgrims. We come and go in birth and death, and they can do nothing. We come and go as pilgrims of Śālagrāma, and they can do nothing."

Approaching the Summit

Shaligram pilgrimage is both a macrocosm of Shaligram practice that includes a variety of *puja* and *seva* rituals and a microcosm of life as a Shaligram practitioner and as a Shaligram stone. At the core of both aspects of veneration is, however, mobility: the capacity of both pilgrim and deity to traverse a sacred landscape and to undertake a momentous and meaningful journey together until the end. This challenges largely place-centered analyses of pilgrimage in that, while Muktinath temple is an important aspect of Shaligram pilgrimage, it is not necessarily the destination or reason of travel for people arriving in Mustang. Instead, Shaligram pilgrimage offers a glimpse into the methods by which people come to identify with certain places, regardless of whether or not they live in those places or have ever visited them before. It also offers insight into how marginalization, militarization, and economic challenges in Mustang have had significant effects on both Shaligram practices throughout South Asia and the world and on the ways in which local peoples in the region continue to struggle with political isolation and the imposition of government restrictions on the movement of life itself.

Due to the plurality of sacred spaces in physical locations, however, such as happens through the conceptualization of the *dham*, Shaligrams become capable of being both from a place and carrying that place with them. Through Shaligrams' parallels to their human kin, then, pilgrimage is reframed as the sovereignty of mobile persons within mobile localities that defy national and ethnic labels. This mobility as a political practice also underlies pilgrims' narratives of what it means to be Hindu, Buddhist, or Bon in response to increasingly political restrictions on religious movement. As a result, economic and political red-zoning of Mustang becomes an offense against communal and practitioner identity – where reductions on Shaligram pilgrimage come to represent the current political constraints on people traveling in and between Nepal and India generally as well as the potential life and death of mobility-dependent religious traditions.

"Pilgrimage as a life journey" is therefore a common metaphor for religious practice among Shaligram devotees and is a literal way of referring to the birth, life, and death of Shaligram deities and the families with which they come to live. In other words, Shaligram veneration is both begun by and necessitates continued pilgrimage to Mustang and is a life-long pilgrimage in and of itself. Ultimately, what predicates many of the concerns Shaligram pilgrims are focused on is the discrepancy between being labeled as "foreigners" and the numerous intimate connections they feel with the lands, objects, and histories that are foundational to their identities as Hindus, Buddhists, of Bonpos above and beyond national, ethnic, or caste identity. This, then, is how the Shaligram community of practitioners comes together: as pilgrims who all walk the same paths from Jomson to Muktinath and from birth to death, until the day both they and their deities must die and be returned to the river.

But the "death" of a Shaligram is not its end any more than the physical death of a person is thought to be their end. Rather, Shaligrams not only live on in successive reincarnations through inheritance or new pilgrimage appearances, they also live on in digital representation, where the traditional exchange of Shaligrams from landscape to person and then from person to person has been upended by the advent of the Internet and the relatively new practice of selling stones online.

Bibliography

Bhardwaj, Surinder Mohan. 1973. *Hindu Places of Pilgrimage in India: A Study in Cultural Geography*. Berkeley: University of California Press.

Coleman, Simon, and John Eade, eds. 2004. *Reframing Pilgrimage: Cultures in Motion*. European Association of Social Anthropologists. London: Routledge.

Eade, John, and Michael J. Sallnow. [1991] 2000. *Contesting the Sacred: The Anthropology of Christian Pilgrimage*. Chicago: University of Illinois Press.

Eickelman, D., and Piscatori, J., eds. 1990. *Muslim Travellers: Pilgrimage, Migration, and the Religious Imagination*. Berkeley and Los Angeles: University of California Press.

Fisher, William F. 1987. *The Re-Creation of Tradition: Ethnicity, Migration, and Social Change Among the Thakali of Central Nepal*. PhD diss., Columbia University.

—. 2001. *Fluid Boundaries: Forming and Transforming Identity in Nepal*. New York: Columbia University Press.

Morinis, Alan. 1992. *Sacred Journeys: The Anthropology of Pilgrimage* (Contributions to the Study of Anthropology). Westport: Praeger.

Oppert, Gustav. 1901. "On Shaligrams: Sacred Stones of Indian Aborigines Become Emblems of the God Vishnu," *Review of the History of Religions* vol. 43: 325-332. (Sur Les Sālagrāmas: Pierres Sacrées Des Aborigènes De L'inde Devenuesemblèmes Du Dieu Vishnou. *Revue de l'histoire des religions*).

Ramble, Charles. 1983. "The Founding of a Tibetan Village: The Popular Transformation of History," *Kailash* 10, no. 1: 267-290.

—. 1990. "How Buddhist are Buddhist Communities? The Construction of Tradition in Two Lamaist Villages," *Journal of the Anthropological Society of Oxford* 21 (2): 185-197.

—. 1992. The Archive of Baragaon. http://www.tibetanhistory.net/documents-from-mustang/baragaou/. Accessed 10 October 2016.

—. 2002. "Temporal disjunction and collectivity in Mustang, Nepal," *Current Anthropology* 43, supplement (August-October).

—. 2008a. *Tibetan Sources for a Social History of Mustang, Nepal. Volume 1: The Archive of Te*. Halle: International Institute for Tibetan and Buddhist Studies.

—. 2008b. *The Navel of the Demoness: Tibetan Buddhism and Civil Religion in the Highlands of Nepal*. Oxford: Oxford University Press.

—. 2014. “The Complexity of Tibetan Pilgrimage.” In *Searching for the Dharma, Finding Salvation – Buddhist Pilgrimage in Time and Space.* Proceedings of the Workshop “Buddhist Pilgrimage in History and Present Times” at the Lumbini International Research Institute (LIRI), Lumbini, 11-13 January 2010. Edited by Christoph Cueppers and Max Deeg. Lumbini International Research Institute 2014.

Rapport, Nigel. 1998. “Coming Home to a Dream: A Study of the Immigrant Discourse of ‘Anglo-Saxons’ in Israel.” In *Migrants of Identity: Perceptions of Home in a World of Movement,* edited by N. Rapport and A. Dawson. Oxford: Berg.

Tremlett, Paul-Francois. 2003. “The problem of belief: A response to Matthew Engelke,” *Anthropology Today.* 19 (4): 24.

Turner, Victor. 1969. “Planes of Classification in a Ritual of Life and Death.” From *The Ritual Process: Structure and Anti-Structure.* London and New York: Routledge.

—, and Edith Turner. 1978. *Image and Pilgrimage in Christian Culture.* New York: Columbia University Press.

8 Ashes and Immortality

Death and the Digital (After)Life

Abstract
When a Shaligram dies, it begins a new journey: either cremated in the hands of the dead or handed down to the next generation. As such, the ever-present issue of mobility foregrounds both the life of a Shaligram and the lives of practitioners. In an age of ever-increasing diaspora and digital life, however, keeping up with changes in communities and families living abroad has been a significant challenge for those maintaining Shaligram practices. As a result, many Shaligram devotees are growing concerned that millennia of Shaligram traditions are now in danger of extinction. But there is hope yet, especially in the burgeoning platforms of digital religion and the entrance of Shaligrams into the Dark Net.

Keywords: digital religion, death, commodity, mobility, diaspora

"Death is not the opposite of life, but a part of it."
– *Haruki Murakami*

"If you trace the history of mankind, our evolution has been mediated by technology, and without technology it's not really obvious where we would be. So I think we have always been cyborgs in this sense."
– *Evgeny Morozov*

The Death of Shaligram

After leaving Mustang for the final time in the spring of 2016, I returned to Kathmandu along with a family of Shaligram pilgrims to their home in the bustling streets of the city center. I had done this many times before, but this particular trip carried a note of finality to it I had not expected.

Walters, Holly, *Shaligram Pilgrimage in the Nepal Himalayas*. Taylor & Francis Group, 2020
DOI: 10.5117/9789463721721_CH08

This would be my first experience with the death of a Shaligram brought about by the death of a friend. Like many of the pilgrims at Muktinath, the Bhandari family had lived with and cared for Shaligrams for generations, attending to their "births" during pilgrimage to the Kali Gandaki, caring for them throughout their lives in the household, and now conducting their funerals at the edge of the river. But this was more than just a symbolic funeral for an aged deity: it was an actual funeral for a beloved father.

Shrouded by thick billows of smoke, I stood on the cremation grounds of Pashupatinath temple with a mixture of discomfort and solemnity. Parul Bhandari, the elderly father of Tanuj Bhandari, had passed away two days prior and now lay at the edge of the *ghat*, wrapped in saffron blankets and surrounded by his deeply mourning family. I had known both Parul and Tanuj for over a year. As devout Hindus and as Shaligram sellers living in the Gaushala *chowk* (neighborhood), they had taken a great interest in my work from the earliest days and had invited me into their home and into Shaligram rituals on numerous occasions. It was Parul in particular who had also long lamented the lack of more accessible materials on Shaligram practices and who had continued to encourage me to draw and photograph as many *shilas* as I could with a view towards one day publishing a book (or putting it online) on Shaligram identification for devotees to read. Now, as I watched Parul's three sons carry out the rituals required to bathe him in the Bagmati River and then to send their father into the funerary pyre, I was overcome with sadness. Parul had often spoken of his anticipation of just such an event. Diagnosed with cancer six months before his death, he had taken the time to sit with me in what would be our last shared meal on the floor of his home and explain what was then to become of his cherished Shaligrams.

"First," he began. "I have set aside Vaikuntha [a Shaligram with two distinctly layered *chakras* representing the relative positions of Heaven and Earth]. Vaikuntha will go with me. We will go together, and He will carry me past the temptations of rebirth. We will go to God. Once I am there, I will send Him back. He will be reborn in the river, made new again, into a new life and find his way to a new family. I wish Him the best, of course, though I will miss Him. This Vaikuntha was my grandfather's and then my father's. He has been with us a long time."

"He will not go to Tanuj?" I asked.

"No," he replied. "Look at Him now, so old and broken. Like me. We have seen much, He and I. No, Tanuj will have all the others. They will comfort him and guide him when I am gone. They are younger, more robust, so they will go with my grandchildren when the time comes for them to leave home. But now, Vaikuntha and I will go. It is our time."

Now, just a handful of weeks later, I watched as Tanuj laid his father's body on the stone slab at the river's edge, his feet just touching the murky waters. The Vaikuntha Shaligram was then unwrapped from a cloth bag and placed in Parul's folded hands at the completion of the rituals that marked him as a man of the Chhetri (Brahmin) caste. As the family then carried him up onto the cremation stand, the attendant *pujari* offered a flame to light his mouth (to burn out any last bits of bad karma) and the fragrant sticks that would begin the cremation fires in earnest. Tanuj hesitated. Taking one final look at his father, he touched his own head to his father's forehead and then to the Shaligram in his hands.

"Be quick," he said quietly, tightening Parul's fingers over the small *shila*. "Take *buwa* home. But don't forget to hurry back. Come and visit us when you can. We miss him so much already."[1]

As Parul's wife and daughters-in-law wailed behind us, the fire was set. Later, once the family had dispersed, the *pujari* and tender-of-the-dead would gather up the ashes and the Shaligram and release them into the river. With that, no more would physically remain of Parul but the memories and memorials of his family, but this would be cause for later celebration. Parul would not be reborn into the world of suffering ever again. Rather, as a practitioner of Shaligram veneration and having passed through the material world with Shaligram at his side, he would now remain forever in the realm of *moksha* (liberation) and in a state of oneness with the divine, where all barriers of belonging and identity (of difference) would finally fall away. Unlike Clifford Geertz, who once described human beings as bestowed "with the natural equipment to live a thousand kinds of life" but who are doomed from the outset to "end in the end having only lived one" (1973: 45), Parul and his family would go on to merge the two, where a kind of life and an actual life might be repeated, ended, and begun again. The Shaligram would return to move on and begin the cycle anew. For those who venerate Shaligram, this is a cycle that cannot, and should not, be hindered.

Parul would also live on in his Shaligrams. As a beloved ancestor, his photograph and his Shaligrams would be venerated and given offerings by his family right beside all of the other household deities. They would be brought out to attend festivals and special events, such as weddings or other funeral feasts. On the first anniversary of his death and cremation, one of his Shaligrams would be chosen to return to the Kali Gandaki on pilgrimage so that his family might perform the *shraddha* ritual on the river

1 Conversations were primarily in Hindi and Nepali, with some English phrases. Transcribed from fieldnotes.

banks to honor his life and passing and to await the birth of new Shaligrams in the ever-continuing cycle of family life. Finally, he would enter into the narrative of family history and the stories of his pilgrimages to Mustang, and the Shaligrams that appeared to him there would inspire new generations of sons, daughters, grandchildren, and great-grandchildren to continue the work of extending their familial and community ties beyond the transient world of human beings and into the world of the dead and the divine.

This is how Shaligram lives and dies. Born from the Kali Gandaki River high in the Himalayas, it enters family life as a child does – bathed and wrapped in the arms of its new parents and siblings as they introduce it to the obligations and routines of daily life. Over time, Shaligram grows up, sharing meals, rest, work, milestones, and activities with everyone responsible for looking after it. It attends marriages and the births of other children, participates in festivals and celebrations, oversees the conduct of the household, and remains a careful and quiet confidante for any troubles brought before it. In many cases, they also bestow blessings and tribulations on the household in equal measure (much as human children do, my interlocutors pointed out). They are also caregivers in times of sickness, divinities for the purposes of worship and service, friends to the lonely and destitute, and teachers for the young who do not yet understand the ways in which the unseen world affects them. Then, any given Shaligram might transition between households as members of its family do – in marriage, as a gift (adoption), or in inheritance. As such, Shaligrams live multiple lives, moving down through the generations of their families and communities until the day comes that Shaligram too must pass away. Old and worn, and then cremated and returned to the river along with the body of someone recently deceased, the Shaligram carries the spirits of the dead into the presence of God and returns again to the world in their place. As a fulfillment of the debt of karma, Shaligram is then prepared to live again and be born to a new family out of the very same rivers that birthed it the first time and every time after that.

Mobility is, therefore, the unifying factor of virtually every Shaligram tradition: movement in space, movement in time, and movement through life. But as Shaligram practitioners and devotees themselves have begun to see new kinds of movement in the diaspora and in the digital age, Shaligram has not been left behind. Rather, as pilgrimage to Mustang becomes more difficult and devotees living further and further apart find obtaining Shaligrams harder and harder, Shaligrams have moved online into virtual ritual spaces, internet exchange networks, and digital *darshan*. In other words, Shaligrams and their communities have infiltrated the web and are complicating the very nature of the term "online *avatar*" in a variety of new and interesting ways.

Shaligram Online

As the diaspora moves further and further away from traditional pilgrimage networks and temples, the online world is fast becoming a new kind of proxy for the sacred rivers and temples that once organized Shaligram mobility. As one moves out from Mustang pilgrimage, Shaligram political practice transforms from a local and regional concern to a global one where concerns about the regional politics of isolation are translated into concerns about global politics of Westernization and commodification. For those unable to physically or financially undertake Shaligram pilgrimage, there is the option of buying Shaligrams online. In fact, a brief Google search will turn up any number of Shaligram sellers based in Nepal and in India as well as current listings for stones on eBay, Etsy, or Amazon. Occasionally in the hands of rock and mineral shops, Shaligrams also turn up in stores throughout South Asia as well as in Australia and the United States from time to time. For Shaligram devotees, the preponderance of Shaligram sellers in Mustang, in Pokhara, and in Kathmandu often presents something of a religious conundrum: buy a Shaligram in defiance of religious bans regarding placing monetary values on sacred stones or potentially go home without any Shaligram at all.

The abundance of Shaligrams for sale online presents an even more contentious issue: is buying a Shaligram online and having it shipped analogous to pilgrimage (specifically in terms of mobility) or does it represent a shift from the traditional kinship networks of Shaligram exchange into the more troublesome commodified networks of object exchange? In other words, as long as the Shaligram is moving and being exchanged, are the same merits acquired or are they suspect? Responses to these questions vary greatly. For some practitioners, the necessity of purchasing Shaligrams online is viewed as an inevitable consequence of the challenges and restrictions of actually going on pilgrimage to Mustang. For others, it is viewed as a result of globalization and modernity in the age of Kali Yuga: as spirituality in general is reduced and corrupted, so too must Shaligram practitioners adapt to changing times. As I sat in the living room of Priya Krishnamurthi back in Boston, she explained further:

> My family is from the area around Chennai. We moved to the US about ten years ago. My husband is a doctor here in Boston and our children now go to school here as well. Back in Chennai, though, we have strong Shaligram traditions. Many temples and many households have Shaligram, but we didn't take any with us when we moved. I regret that very much. My husband went on pilgrimage last year because of this.

> We needed to reconnect with our traditions so that our children could learn more about them and not forget them. A lot of people are afraid of even having Shaligram right now because they think the *pujas* are too difficult. This scares me because it means that our traditions won't be passed on. They'll be forgotten. Many times, parents and grandparents don't even tell their children the secret *mantras* or teach them the *pujas* because of this. My mother said we would never keep Shaligrams because there was no time to care for them properly, so best not to have them. But most of the Shaligrams he brought back we have given away now. Some to the temple here and some to families in the area that we know. It was important. There was sickness and in one family a baby died, so we gave Shaligram to the mother for her baby. A beautiful Radha-Krishna for her to look after and to ask for a new baby. But we can't afford to go on pilgrimage again and we are Americans now, so even if we went back home [to India] and started from there, we couldn't cross the border like we used to. We are separated. So, I am looking at these websites. They charge so much for Shaligram but I think of them as a blessing. Shaligram is part of our traditions, it's part of who we are as Hindus, so if this is the only way for us to get Shaligram here in America than that is how He will come.[2]

Growing markets for Shaligram stones are often explained as the result of both national tensions and economic hardship. These narratives, however, also fuel growing concerns about the continued availability of Shaligrams overall. "If sellers continue like they have been," remarked Lakshmi Muni, an Indian immigrant living in New York City, "there won't be any Shaligrams left in Kali Gandaki. I only got my Shaligram by chance. I was going down the street to get my groceries when I saw this shop filled with Indian and Nepali items. I thought maybe I should go in and see if they had any *puja* things since mine are so old and some of my things are missing. But you should imagine my surprise when I saw Shaligram in one of the display cases. The shop owner said that he had found some fossils while he was trekking in Nepal and he decided to bring them back for his shop. He had no idea what they were! Well, of course I bought it right away and I keep it with my deities for *darshan* every day. But I worry a lot now. I see so many for sale on the internet and I know sellers take them from the river to sell. Now it is almost like we have to buy them because we won't be able to get them from Gandaki anymore."

2 Conversation in English. Transcribed from audio recording.

"Do you think it is mostly people who cannot go on pilgrimage who buy them?" I asked. She leaned in conspiratorially. "I asked this of a seller in India once. He told me that he does most of his business with the Hare Krishnas, you know, the Vaishnavas from Bengal? Because so many of them are Westerners and they live here in America or in England and they can't go on pilgrimage, so they buy Shaligrams. And they have so much money, you know? I think it is also because many sacred places will not allow Westerners to come inside. You have to be born of a Hindu family to go to Pashupatinath, otherwise you have to stay outside. It is complicated. They have become Hindu, but many see them as not Hindu. So, when they go for Shaligram, some people see them as tourists even if they see themselves as pilgrims. Maybe it is better for them to buy Shaligram if that is what they want."

Among many Shaligram practitioners, concerns about the disappearance of Shaligrams from Nepal were often couched in competing arguments of cultural appropriation and Hinduism as an evangelical global religion. Stereotypes about young white Western women who visited Nepal and India looking for yoga instruction or for spiritual tourism and young white Western, men sporting deadlocks and Buddhist prayer beads featured prominently in many narratives about the blurry lines between serious religious conversion, religious commodification, and cultural appropriation. For global religions like Hinduism and Tibetan Buddhism, which have long operated on a logic of evangelical salvation, conversion rhetoric, and the equal access of all people to the means of liberation, there remains a central tension: how to reconcile universalist theology with a need to preserve their cultural and ritual practices in the face of persecution, colonialism, and dilution in the world-wide diaspora. Typically, Western converts to Hinduism are often treated as beneficiaries of Vedic theology and viewed as elevated from a culturally and spiritually impoverished homeland (meaning: the modern West) and into a new perspective of meaning and fulfillment due to the charity of teachers and *gurus*. Western converts are also sought out for their material and political privileges so that they may be leveraged as allies in advocating for specific political subjectivities (i.e., Free Tibet, preservation of Hindu India, etc.), further complicating the lines between what is appropriation and what is conversion.

For Shaligram devotees, the added concerns about the buying and selling of Shaligrams on expansive global markets echoed these issues. At what point could Western converts be considered "Hindu enough" for Shaligram practice? Did new religious movements, such as the Hare Krishnas, who claim wide networks of Western converts as well as Indian and Bangladeshi devotees, have sufficient ties to more ancient traditional contexts to warrant inclusion in Shaligram rituals and exchange? If a Shaligram appeared to a non-Hindu, was it not the same as appearing to a practicing Hindu given God's own agency in Shaligram mobility? What did it mean for Shaligram veneration when it was now all too easy just to buy a sacred stone than to chance the dangers of the Kali Gandaki in the high Himalayas of Mustang? Ultimately, concerns about the buying and selling of Shaligrams outside of South Asia underscore the challenges for Hindus, converts to Hinduism, and tourists (spiritual or otherwise) in further negotiating the links between religion and politics and in deciding how nationality and belonging can be mobilized in response. In the end, what is the responsibility of the tourist to the pilgrim and vice versa, or the convert to the native? Is it possible to import beliefs and not their accompanying cultural frameworks and objects? Where does religion or nationality end and culture begin?

Lakshmi's words cast me back to the beginning of my fieldwork in West Bengal in 2012. While traveling out from Kolkata for several months, I had visited the Hare Krishna pilgrimage town of Mayapur, and it was in a village just a short distance away that I had actually first encountered Shaligrams myself. Shaligrams were common throughout Mayapur, both in the temple of Sri Sri Radha Madhava (Radha-Krishna) and in a number of home altars. My field notes and photograph folders brimmed, in fact, with pictures of Shaligrams dressed in headwraps and shoulder scarves, sitting on pillows at the feet of Krishna in the temple or resting on brass stands for *abisheka* (ritual bathing) or artfully arranged on *puja* trays for daily home *darshan*. At the time, my questions as to the origins of the stones were usually simply met with the response of "Nepal" or "Kali Gandaki".

I hadn't thought, in those early days, to probe further into the precise methods by which each individual had acquired the Shaligrams in their collections rather than simply asking where the stone itself had come from. It would only be much later, and with much more travel of my own, that I would begin to see the extent of Shaligram mobility and accumulation in worship throughout the homes and temples of northern India – accumulation so highly prized, in fact, that it occasionally resulted in collections of Shaligrams so massive as to be cared for by virtue of entire carts of flowers, whole gardens of fruit and vegetables, and in one case, daily bathing with a garden hose. But in another sense, the narrative of my own fieldwork had come full circle: I had begun my work among the Gaudiya Vaishnavas and Hare Krishnas of West Bengal and was now ending my formal interviews with references to the very same. I had also spent time with a number of

Shaligram sellers in Pokhara and in Kathmandu, Nepal. More often than not, I was again staggered by the sheer volume of stones they possessed.

One Hindu seller near the Temple of the Sleeping God (Vishnu) in Kathmandu had eight 50-pound rice sacks filled with Shaligram stones. Another seller near Pashupatinath had an entire room dedicated to a collection of hundreds of thousands of Shaligrams accumulated over a career of nearly thirty years. And always, they were collecting more. As merchants, they were interested in the best prices for their wares, but as practitioners themselves, they also worried about the significant rise of foreign buyers, fueled by internet access, in just the last twenty years. Furthermore, many expressed growing concern about disreputable sellers flooding the market with fake Shaligrams, many of which are glued together from pieces of broken stones or are crafted from concrete or M-Seal – a type of industrial epoxy widely available in South Asia that dries to form a hard shiny black nodule that can be carved or pressed with shapes resembling the ammonite spiral. These disreputable sellers then sour potential buyers who rarely attempt to purchase Shaligrams a second time if they have had a bad experience with online exchanges.[3] Many practitioner-sellers also expressed ambiguous feelings about the expansion of Shaligram markets online, which has had the dual effect of driving up prices (sometimes into hundreds of US dollars) and further turning Indian and Nepali practitioners away, who prefer to turn to social media exchanges on Facebook and WhatsApp.

For many Buddhist sellers, the question of Puranic restrictions was not central to their concerns about buying and selling sacred stones. Rather, they viewed the selling of Shaligrams (to either Hindus or Westerners) as part and parcel of the broader commodification of religious objects in Buddhism widely. Given that sacred objects in the home were generally assumed to benefit the household whether not the people within it "believed" in the deity, tradition, or ritual, the selling of Shaligram stones was therefore often perceived to be just as meritorious to the buyer and seller as the exchange of any other object of veneration (such as prayer wheels or mantra booklets) and would not bring either to harm regardless of whether or not money was also exchanged. Hindu sellers were, on the other hand, more cautious regarding the Puranic rules and typically kept their collections separated between

3 I recorded numerous conversations in my fieldnotes to this affect. Often, new Shaligram practitioners unable to undergo pilgrimage would attempt to purchase Shaligrams online, only to find that the arriving package contained a fake stone. All, save one individual, expressed that they did not attempt to purchase another stone and took the disappointment as a karmic punishment for having attempted to buy a Shaligram in the first place.

their own household deity stones and those Shaligrams that had not yet been ritually tied and were therefore still "moveable" through donation requests. It was only a short time later, however, that the more clandestine aspects of modern Shaligram exchange took a new and even more surprising turn.

Not long before my meeting with Lakshmi and her family, I was introduced to the Shaligram Dark Net. Like all Dark Web sites, the Shaligram Dark Net is composed of a series of networked webpages and forums not indexed by internet search engines and which require a formal invitation by a current member to access. I first learned of this network of websites through a Shaligram devotee I had known since beginning my fieldwork in India. Vikram Shah, a young man in his early 20s, had come to visit me in Kolkata during a short trip near the end of my fieldwork in Nepal. While I was asking him about his experience with buying Shaligrams a few months prior, he leaned in and pursed his lips before responding. "You don't have to buy them always. There are other ways now you can reach Shaligram that don't mean going to a seller or really even to a temple if there isn't one near you."

"You mean a way to get Shaligrams without buying one or going on pilgrimage?" I asked, the skepticism likely apparent in my tone. He nodded. "We have an online community now." He pulled his laptop from his backpack. "I will show you. It's a place where we can post photos of our Shaligrams so that everyone everywhere in the world can have *darshan* and we can also write stories about Shaligram, like how they came to us and also like, how we can identify them. You can also put your contact information up here and if someone needs Shaligram or can't get Shaligram, we have forums you can go to and if someone can offer one that they already have, then you can get Shaligram by exchange. You know, like how it is supposed to be. Since so many Hindus are in America now or in England and don't have their village temples or anything like that, they can come here." The exchange Vikram was careful to mention here referred specifically to acquiring a Shaligram without the use of money. In many cases, devotees would offer other community members an "extra" Shaligram they were willing to part with or, in other cases where they had inherited more stones than they felt comfortable caring for, several from their collections. These exchanges were then finalized through private messages, where a Shaligram might either simply be free (if the recipient was willing to cover shipping costs) or was traded for other devotional items, deity icons, or ritual implements. But more importantly, even online, the relationships created via exchange were preserved such that the impersonality and potential karmic pollution of money was rejected in favor of maintaining the kinship-like mobility of each Shaligram from family to family and within the community of devotees (a la Mauss 1954).

As we moved from page to page using the haphazard links and oddly placed images and flashing icons, I took note of the confusing and rather labyrinthian design of the entire setup. "Shaligram Net was started back in 2001, I think." Vikram tapped thoughtfully at his keyboard. "It was never anything formal or official or anything like that. You just invited people to your page and then they could link their pages with your page and it just kind of grew from there." "But why make it a dark net?" I asked. "Why not have it easy to find for everyone?" Vikram shook his head sternly. "No, you can't do that. We need a space where people aren't asking questions all the time. It's just for us to talk about Shaligram with others who know Shaligram and to keep our community together. We can't always go on pilgrimage, but we can come here. Also, we don't want, you know, Westerners – not like you though, I mean regular Westerners" – I smiled at the backhanded compliment but let him continue – "coming here by accident and then wanting Shaligram."

"Do you mean that you don't want people who don't know anything about Shaligram asking for one?" "Yes, yes!" He nodded. "Like that. If Westerners know about Shaligram, then they will want one and they will probably go and buy one. If Westerners start buying Shaligram, then there won't be any more. They will buy them all and Shaligrams will become expensive like diamonds, you know?" "Expensive because they are rare?" I asked. "No, expensive because people with lots of money will buy them. Sellers [meaning Shaligram sellers who sell stones online] will raise the price because they will want Westerners to buy, not Indians like me with no money." I mentioned the parallels with similar concerns about trekkers and pilgrims in Mustang. "It's the same," Vikram hung his head sadly. "It's about money. That's why we have Shaligram Net. To hide Shaligram from money. It's also because we don't want, like I said, people who don't understand. Sometimes people just see things on the internet and then they think that is what they should do or what they should be. Shaligram requires learning and they don't want to learn. They just want to have one."[4]

Vikram's concerns regarding the buying and selling of Shaligrams echoed a common refrain among Shaligram practitioners who voiced concerns about the commodification of the stones. In many ways, their overall worries about markets for Shaligrams reflected their experiences with a variety of precious and semi-precious minerals, crystals, and other fossils typically found in the geology sections of "science" stores, in New Age gift shops, and

4 Conversation in English. Transcribed from fieldnotes.

on the shelves at specialty mall outlets. Since placing monetary values on sacred stones is expressly forbidden by Puranic scriptures (and is a widely held stricture in Shaligram practice), this anxiety about Shaligrams as "resources" (Ferry 2008: 52) remained situated within broader concerns about those who might seek to exploit religion for profit. Furthermore, where Elizabeth Ferry's discussion of "resources" is most salient here is in the view that Shaligrams would become objects whose purpose was to generate profit (through various kinds of use and exchange values) rather than to facilitate kinship and community ties and that Shaligrams would become further "objectified" in that their value might be measured and transacted rather than maintaining their status as divine persons whose value should remain ephemeral and priceless. In other discussions, there were also concerns about the view of Shaligrams as "scarce". As Vikram made note of more than once, if Shaligram pilgrimage were to become inaccessible, then the markets for selling Shaligram stones would almost undoubtably arise for no other reason than buying and selling would become the only way to acquire one.

Because mobility plays a significant role in the construction of Shaligrams both as persons and as sacred objects, practitioners' concerns over the loss of "free" access to the stones remained part of their overall fear that the movement of Shaligrams (through both physical and digital channels) would become further directed by the wealth of high-caste or diasporic pilgrims and participants and by the affluence of Western nations generally. With additional concerns about the ability of practitioners to continue passing down Shaligrams through their families, temporality and mobility became the main viewpoints that encompassed the notion that Shaligrams created spatial and temporal "paths" (Giddens 1995) of kinship and community connection through their mobility across sacred landscapes. If Shaligrams were then to take on the more abstract market relationships of producer (collector/miner) and consumer, their relationships with pilgrims and ritual practitioners as familial and community kin would be in grave danger. For Vikram and for many contributors to the forums on Shaligram Net, the metaphor was that of human trafficking: "if you wouldn't buy and sell your daughter, then you should never buy and sell Shaligram" (Forum commenter, 2016).

I returned to the Shaligram Net several times that month, reading and reviewing various Shaligram altar setups and stories of acquiring Shaligrams for the first time. In the end, there were hundreds of pages available – everything from long sections of text taken from pilgrimage literatures not widely available in print, repeated references to Rao's *Śālagrāma Kosha* and requests for copies of the book (it is often difficult to find outside of India), discussion boards dedicated to teasing out the nuances of individual Shaligram identifications, and impassioned stories about acquiring Shaligrams in strange and far-away places (including Russia and New Zealand!) to arguments over the details of various traditional ritual practices and the proper ways in which Shaligram should be cared for.

As I continued, several themes emerged: the theological and ontological contention that Westerners could not understand Shaligrams as persons (echoing Elizabeth Povinelli's ontological schism between Life and Nonlife, or geontology, which she identifies as the organizing logic of late liberalism [2016]), the concern that Shaligrams might be treated as decorations or souvenirs (opting for something similar to Anna Tsing's characterization of "non-human companion species" [2015]), and the sense that Shaligram mobility and exchange was fundamentally changing in the face of global migration, what Tim Ingold (2011) would refer to as a "line" where every being is instantiated in the world as a path of movement through a particular way of life. As Shaligrams on the web continue to expand and draw communities of practitioners together, and as practitioner communities and forums across various phone apps and online platforms grow, it is also exceedingly likely that the divine-person ammonites of Hinduism, Buddhism, and Bon will open up new avenues of inquiry in digital cultural studies.

One example of just such an avenue of inquiry is digital anthropology, which is concerned with the ways in which digital and virtual technologies change how people live their lives as well as how technology changes the methods anthropologists can use to do ethnography (Thompson 2012). This is a question I struggled with a great deal in the course of my own work. The concept of the "cyborg" also feels potentially relevant here (Haraway 2000). As the study of how humans define humanness in relationship to machines as well as the study of science and technology as activities that can shape and be shaped by culture, online Shaligrams present a fascinating new world of possibilities in demonstrating how people build networks of information exchange and community belonging. But rather than strictly through new amalgamations of organic and non-organic parts (but here stone and not a machine), these new systems might potentially include Shaligrams who are fossil, deity, and persons themselves – all at once living, dead, and divine. Ultimately, through WhatsApp, Facebook, and the Shaligram Dark Net as well as the global proliferation of Shaligrams more generally, other human distinctions like life and death, human and machine, virtual and real, may continue to disappear and in so doing redefine what it means to have or not have a body in its "natural state."[5] A short time later, I was able to contact one of the more prolific posters, Dasarath Chand Hari, and truly bring my ethnographic inquiries into the digital age.

"I have two Shaligrams in my home," he told me. Our interview was conducted over Skype: me in Nepal and he in West Bengal, India. "I have a Sri Ramkrishna Paramahansa. This is the latest incarnation of Vishnu. The Bengali guru who lived in the 1800s.[6] I also have a Kurma Shaligram given to me by my *guru*. I have never been on pilgrimage though. I wish very much that I could but there is never the money or the time." He chuckled. "That actually makes you a higher devotee than me, since you have been to Kali Gandaki!" It was an easy segue from there to ask him about Shaligram Net and the challenges of pilgrimage. "Oh yes, we have Shaligram Net now. It is good because it shows everyone that there are many different paths, many different opinions. There is no one path to Shaligram, just as there is no one Hindu path. Hinduism is a colonialist idea anyway."

Dasarath's view of Hinduism as a colonial concept was not particularly unusual. Both scholars of Hinduism and many Hindus themselves have come to view "religion" (here meaning a category that comprises a set of practices and beliefs supposedly found in every culture) as an idea distinct to the modern period. Hinduism, in this case, then emerged in the encounter between modernity's greatest colonial power, Great Britain, and the subsequent imperial control of India beginning in the 17th century. Around the turn of the 19th century, officials of the British colonial state and Christian missionaries then helped to cement the idea that the variety of regional and sectarian spiritual and ritual traditions in India were sufficiently coherent to be construed as a single systematic religion (Pennington 2005). Shaded by the articulation and development of the concept of "religion" in the West, this encounter then produced the now common idea that Hinduism is a singular unified religion – an idea that many Hindu Shaligram practitioners continue to reject.

5 Downey, Gary Lee; Dumit, Joseph; Williams, Sarah (1995). "Cyborg Anthropology". *Cultural Anthropology* 10: 264-269 – via Wiley Online Library.
Dumit, Joseph. Davis-Floyd, Robbie. (2001). *Cyborg Anthropology*. Routledge International Encyclopedia of Women.
Society, National Geographic. "Amber Case, Cyborg Anthropologist Information, Facts, News, Photos – National Geographic".
Case, Amber (2014). *An Illustrated Dictionary of Cyborg Anthropology*. p. 9.
"Robots, Robots, Everywhere – A Field Guide to Cyborg Anthropology | The World is not a desktop." *caseorganic.com*. Retrieved 31 January 2017.
"Cyborgs and Space," in Astronautics (September 1960), by Manfred E. Clynes and Nathan S. Kline.

6 Ramakrishna, 1836-1886.

I stopped to consider his words for a second. "Is Shaligram Net a way to decolonize Shaligrams?" He bounced in his chair, "This is how I see it, yes. Shaligrams weren't very important to the British, so there was never a problem of the British trying to steal them, but when they tried to say what Hindus were and what Hinduism was, Shaligrams became part of that anyway. On Shaligram Net, it is only us devotees who get to talk and who get to decide. This is our tradition: inside Hindu, inside Buddhist, inside all other religions we have Shaligram. It is the essence, I want you to understand. Shaligram physically inside *murti*, Shaligram inside religion in the same way. Our purpose is to 'see' God [finger air quotes in actual interview], you know? Shaligram is the manifestation of Mahavishnu, Param Purush. He is chosen among all gods to serve as Shaligram. I have been drawn to Shaligram since I was a small child. I'm a good student but I wanted to go beyond books. But I needed to be careful of all the businessmen selling holy *shilas*. They know how to identify them very well, but they will sometimes lie to you to get you to buy one *shila* or another. I would rather wait, though. A Shaligram comes to you because of what you have done in your past lives. If you did good deeds in previous births, Shaligram will come." In Dasarath's view, then, de-colonizing Shaligrams via internet forums meant that Shaligram practitioners did not necessarily need to identify as Hindu (or Buddhist, or Jain, etc.) but only specifically as Shaligram devotees, and as a result, outside forces (read Westerners) would not be able to further define or influence Shaligram practices or community identity in a time of relative instability.

"So, you don't think people should ever buy a Shaligram?" I pressed. "Not if you do not have to," Dasarath replied. "Shaligram is not a joke. It's not funny. It should never be part of consumerism, you know? It hurts us all that someone would buy it and just put it on a shelf or throw it away! When you just use money, you don't understand how important something is and it is easy for you to stop caring." I thanked Dasarath for his time. "Of course." He smiled. "Just remember, *Santana Dharma* [Hinduism] teaches us that you must journey inwards and outwards to reach God. Even the *Bhagavad-Gita* is about God's journey. Any way you want to reach Him, He will appear to you in that way so you can. If you must reach him on pilgrimage, He comes on pilgrimage. If you must reach him through people online because you cannot go to Him, then He comes to you online. God goes where he needs to. He will always find a way."

The movement of Shaligram pilgrimage is thus a complex and deeply integrated matter. As a performative action, the sense of movement surrounding Shaligrams affects (though not always consciously) multiple social and cultural transformations: from the life cycle of birth and death to marriage and migration and to decolonialization. In his own work, Surinder Bhardwaj explored the idea of movement as integral to creating an integrative pan-Hindu sacred space, but a less functionalist approach (e.g., Sallnow) would still see Shaligram pilgrimage and practice as involving the mapping and embodiment of certain kinds of spaces and the claiming of specific places as belonging to one identity or another.

Shaligram pilgrimage also resonates with De Certeau's notion of walking as constitutive of social spaces in much the same way as speech acts are said to constitute language. But for those constrained to view (to take *darshan*) and purchase Shaligrams online, the movement of Shaligram from seller to devotee becomes a kind of pilgrimage by proxy. Shaligram pilgrimage from this vantage point entails physical immobility with movement, cultivated through the religious imagination, realized through the transportation of sacred objects, and sacralized through secular networks of economic trade. In other words, just as ammonites are "fossilized" through various discourses and networks involving geological time, scientific discovery, and the Anthropocene's preoccupation with human destruction of nature, Shaligram sellers (as opposed to devotees on the Shaligram Dark Net and elsewhere) make Shaligrams vendible by using economic mobility as a stand-in for pilgrimage and the seller's home collection as a surrogate for temple exchange (Povinelli 2016).

Shaligram devotees, however, tend to view Shaligram pilgrimage (in whatever form it might take) as an embodied action. For this reason, I am inclined to leverage the phenomenological approach of Karve and Frey (and to some degree Sallnow and DeJarlais' ethnographic work) to reorient this analysis towards seeing Shaligram pilgrimage as the catalyst for certain kinds of bodily experiences – namely a kind of recapitulation of birth, bodily exchange, and physical human interaction. In other words, though Shaligram sellers might legitimate their exchanges using a symbolic doubling of a pilgrim's physical mobility with virtual or spiritual mobility, many Shaligram practitioners tend to reject this linkage in favor of reiterating the agency of Shaligrams themselves as divine persons who come and go in the lives of people of their own accord and for their own reasons. Recalling Urry (2002), such corporeal co-presences like stone and body (and with whatever correspondences between the two are perceived) result in shared experience between the living and the divine, between material bodies and immaterial consciousness. Such an embodied experience cannot be manufactured through commodified exchanges no matter how much the kinds of mobility involved in Shaligram practices might correspond.

Taking Shaligram pilgrimage as movement within layered fields of meaning ultimately helps to further contextualize the meaning of "pilgrimage" within Shaligram practitioners' own understandings of mobility (see also Eickelman and Piscatori 1990 as well as Tapper in the same volume), where such notions of space, place, and landscape are blurred between actual physical and geographic spaces, sacred sites (*dhams*), and mythological locations and ideological states of being. The various methods of Shaligram mobility may then take on especially charged meanings as markers of difference (pilgrim versus tourist), just as they may become markers of sameness (Hindu is Buddhist, Nepal is India). And finally, it may yet prove continually troublesome for devotees as they move into the digital realm, negotiating their place and the place of Shaligrams in new and untested virtual worlds. Indeed, it is a delicate time for post-colonial Shaligram revival in an increasingly digital age.

In the end, Shaligram pilgrimage refutes the framework of pilgrimage as a nostalgic heritage journey where tourists from a broader diaspora return "home" to get in touch with their historical past. This is because, for the most part, virtually no Shaligram pilgrim comes from a family, tradition, or historical community originating in Mustang (and local Mustangis view Shaligrams as part of broader networks of spiritual versus political mobility that are marked as indigenous to the landscape but not necessarily to the people). However, they do experience the same intense emotions and strong sense of belonging associated with expulsion and exile (exile from Shaligram pilgrimage itself or historical and national exile from elsewhere). Therefore, Shaligram pilgrimage isn't so much a symbolic homecoming as it is a reconnection with the subjective reality of myths and a new narrative past that provides a degree of ontological security and community connection within the whirl of contemporary political change.

Shaligram pilgrims are not attempting to reassert ties of blood and territory in the way that other pilgrimage contexts might (such as "roots tourism") but rather seek to celebrate the freedom of their long-held traditions from fixity. This is why Shaligram pilgrims shift between contexts just as much as they walk from place to place, where mobility invokes, plays on, appropriates, and leverages other understandings of journeying, from karmic rebirth to international migration. This is also why the movement of Shaligram communities into online worlds might be so initially appealing – where the elusive and fluid nature of digital interactions can continue to upend authoritative (or colonialist) pronouncements about what constitutes proper Shaligram practice, proper Hinduism, or a proper devotee. Movement, therefore, cannot be taken in this case as any sort of essentialized category (see also Rapport and Dawson 1998: 23) but rather as a continuous layering of cultural contexts that attend to change over time.

Lastly, as a kind of movement as metaphor, Shaligram pilgrimage invokes a variety of other pilgrimage and mobility-based discourses aside from purely physical instantiations – with Shaligrams as the core link between diaspora and homeland. Just as Morinis referred to concepts of "inner pilgrimage" or "life as a journey" metaphors common to Hinduism and Buddhism, the Shaligram pilgrimage example demonstrates the ways in which daily life is distilled in pilgrimage (and pilgrimage distilled in daily life) as a way of socially commenting on the state of being in transit or continuously becoming as one lives life. Given the oft-mentioned historical contexts of Indian and Pakistan's Partition, the Partition of Bangladesh, the invasion of Tibet, and the closing of Nepal's borders, these concerns over the sovereignty of mobility, such as they are expressed in transnational Shaligram pilgrimage, thus perpetuate an ideology of person, place, and object as belonging to identities that lie outside of national contexts, as being *in* Nepal, but *not* Nepali.

Paths in Stone

The connecting thread that links Shaligram dimensions of mobility in pilgrimage, in death, and in the digital world is therefore movement as power, movement as transformation, and of accepting or rejecting previous structures of belonging. From the perspective of anthropology, this work then contributes an analysis that links networks of place-making and movement across landscapes with networks of kinship, identity, and exchange throughout the life-course of families and communities. This does not mean, however, that Shaligram mobility can be reduced to a physiological act and a cultural performance. Rather, Shaligrams intersect with multiple different contexts and processes of meaning, movement, and identity-making, from the macro-national level in Nepal to the micro-localized level of individual households and into the global reach of the South Asian diaspora.

Within the macro-context of Nepal's political economy of travel and the globalization of religious cultures, Shaligram pilgrimage dynamically interplays with issues of outward transnational migration and inward tourism, where the religious practitioner seeks to reconnect with cultural histories and traditions by returning to sacred sites but views the tourist and the merchant as potentially threatening to those traditions by replacing sacred travel with commodified travel. At the micro-localized level, the embodiment of and continuous replaying of idealized (karmic) life cycles within the practice of pilgrimage and outward mobility locates it within broader fields of action where multiple potential lives, both divine and human, are played out across sacred geographies and architectures that provide the material and symbolic background for the motion itself. This is why so many Shaligram practitioners express concern about the growing inaccessibility, militarization, and national homogenization of sacred landscapes. On the global level, Shaligram mobility speaks to concerns about the loss of community within the diaspora and about the rise of neo-liberal capitalism as a way of converting the West using a watered-down religious practice indicative of cultural decline. Shaligram life cycles as ritual and performance involves a number of unpredictable encounters between entities, cultural and theological forms, personal experiences, and memory, which are all translated through physical acts of the body (human and stone).

Over the course of this research, I often found the rather static terms of 'frame' and "landscape" difficult to manage in an attempt to describe a practice – and to some degree a community – that is and remains incredibly fluid. This is especially true because Shaligram practitioners do not form any kind of easily discernable bounded community. Rather, they are linked through shared practice and shared experience even though they may ascribe to various religious traditions, nationalities, ethnicities, and backgrounds otherwise incommensurate with one another. My aim, therefore, has not been to force Shaligram mobility into any particular theoretical framework or category of my own devising but to demonstrate how the phenomena of Shaligram veneration is both transient and permanent depending on the contexts within which the mobility is viewed and how this very mobility is often translated into political practice.

Further examinations of Shaligrams will then likely be productive outside of the "movement frame": for instance, examining Shaligram mobility and practice in relation to other types of social theory involving consumption and the commodification of religion, the post-modern contestation of symbols, modernity and popular culture, economics and exchange, and the gendering of objects and space. These are all ideas that, for lack of time and page space, cannot be extensively examined in the present work. Needless to say, the river metaphor of Shaligram mobility continues to hold water and there are still many streams for us to trace.

Shaligram practices involve representations of mobility within multiple diverse cultural, social, and religious contexts. One obvious connecting theoretical theme is the continuous sacralization and re-sacralization of movement and space. This active sacralization, as opposed to the label of sacred, emphasizes the often partial, performative, and contested character of Mustang's appropriated landscapes and people as distinctly "holy". In that way, the "meta-movement" of Shaligram pilgrimage (Coleman and Eade 2004) – the combination of mobility and some degree of reflexivity as to its meaning or function – shows how Shaligram pilgrims often reflect upon, embody, and sometimes even retroactively reform past journeys and experiences through the context of Shaligram veneration. The Shaligram stones themselves also provide an additional set of ambiguities – themselves pilgrims on a geologic and mythological scale.

Constituted as divine persons through their relationships with other persons, places, and things, Shaligrams are often metonymically associated with the self, the family, and the community at large and help to structure social relationships concerning "proper" or "meritorious" interactions among all involved. This reflects, in many ways, the overall constitution of persons in South Asia, not as Western-type individuals but as DeLeuzian dividuals – where it is one's interactions with and relative ties (i.e., *maya*) to family, birthplaces, objects, and identities that constitute personhood over any sense of internal emotional and cognitive awareness within the confines of a physical body (Lamb 2004). As the next chapter will detail, this is also the very process by which Shaligrams themselves "return home" as kin and become persons – bodies who happen to be stones. Shaligram pilgrimage is then a journey within many and about many other journeys, where history is transformed into myth and ritual and back again, and theology is a matter of experience and practice and not simply textual exegesis. Shaligram is the journey without end.

The Lee of the Stone

When a Shaligram dies, it sets out on a new journey. Cremated in the hands of the beloved dead, it first carries the soul of the deceased person into the presence of the divine and then returns, in their place, to be born again out of the river that received the ashes of the funeral pyre and into a new life as part of the never-ending karmic cycle. Conversely, a Shaligram may be inherited: handed down to the next generation to continue the life cycle similarly until such a day as it is simply too used and deteriorated to be ritually recognizable any longer. The Shaligram then meets its end in cremation or "retirement" at a temple where it is summarily returned to the river afterwards either way.

But in either case, the ever-present issue of mobility continues to foreground both the life of a Shaligram and the lives of the practitioners who venerate them. For a Shaligram to truly be Shaligram, it must not only continuously move through time and space but also through the progressions of life in the hands and homes of the devoted. In an ever-increasing age of diaspora and digitalization, keeping up with changes in communities and families living abroad has been a significant challenge for those maintaining Shaligram practices. For many people in the South Asian diaspora, pilgrimage to Mustang, Nepal is an insurmountable economic hardship, not to mention the political barriers to getting permits and passports to traverse the heavily militarized borders. For others, limited access to their home communities, temples, and gurus has meant that their local and regional Shaligram traditions can no longer be passed down effectively and children are now commonly ignorant of their parent's and grandparent's practices. As a result, many Shaligram practitioners are growing more and more concerned that millennia of Shaligram traditions are now in danger of extinction within just a few generations and that what is still mostly comprised of oral traditions will one day vanish.

The answer, however, for many Shaligram practitioners and communities has come in the form of digital religious revival: where online forums, listservs, and websites can become proxies for pilgrimage and for temple Shaligram exchange. Most importantly, many Hindu and Buddhist Shaligram devotees view these online communal spaces as a method for protecting Shaligram traditions both by being able to record and share them across the world but also shielding Shaligram exchanges from the potentially corrupting influence of money. Given that proper Shaligram movement is viewed as taking place through structures of kinship and gift-giving, the shift to commercialized and commodified forms of exchange are seen as a threat to the sacrality of Shaligram practice. For this reason, many Shaligram websites online are part of the Dark Web: a framework of online portals that are kept intentionally hidden from search engines and from advertising. Therefore, as a space only for invited members who must demonstrate their knowledge and understanding of Shaligrams before they are allowed access, these online spaces are also seen as a way to decolonize Shaligram traditions and to keep them out of the purview of souvenir hunters, fossil prospectors, and insincere marketers looking to profit off of growing Shaligram scarcity.

Bibliography

Bhardwaj, Surinder Mohan. 1973. *Hindu Places of Pilgrimage in India: A Study in Cultural Geography*. Berkeley: University of California Press.

Coleman, Simon, and John Eade, eds. 2004. *Reframing Pilgrimage: Cultures in Motion*. European Association of Social Anthropologists. London: Routledge.

Dejarlais, Robert R. 1992a. *Body and Emotion: The Aesthetics of Illness and Healing in the Nepal Himalayas*. Philadelphia: University of Pennsylvania Press.

—. 1992b. "Yolmo Aesthetics of Body, Health and 'Soul Loss'," *Social Science and Medicine* 34 (10): 1105-1117.

Eade, John, and Michael J. Sallnow. [1991] 2000. *Contesting the Sacred: The Anthropology of Christian Pilgrimage*. Chicago: University of Illinois Press.

Eickelman, Dale, and James Piscatori (eds.). 1990. *Muslim Travellers: Pilgrimage, Migration, and the Religious Imagination*. Berkeley and Los Angeles: University of California Press.

Ferry, Elizabeth and Mandana E. Limbert. 2008. *Timely Assets: The Politics of Resources and Their Temporalities*. Santa Fe: School for Advanced Research Press.

Frey, Nancy. 1998. *Pilgrim Stories: On and Off the Road to Santiago*. Berkeley: University of California Press.

Geertz, Clifford. 1973. *The Interpretation of Cultures*. New York: Basic Books, Basic Books Classics.

—. 1973 [1966]. "Person, time and conduct in Bali." In *The Interpretation of Cultures*, 360-411. New York: Basic Books.

Giddens, Anthony. 1995. *Politics, Sociology and Social Theory: Encounters with Classical and Contemporary Social Thought.* Stanford: Stanford University Press.

Haraway, Donna. 2000. "A Cyborg Manifesto." In *The Cybercultures Reader*, edited by David Bell and Barbara Kennedy, pp. 291-324. London: Routledge (via Georgetown University Online).

Ingold, Tim. 1994. *What is an animal.* London: Routledge.

—. 2000. *The Perception of the Environment: Essays in Livelihood, Dwelling, and Skill*. London: Routledge.

—. 2011. *Redrawing Anthropology: Materials, Movements, Lines*. Ashgate: Aldershot.

Karve, Irawati. 1962. "On the Road: A Maharashtrian Pilgrimage," *The Journal of Asian Studies* XXII (1): 13-31.

Lamb, Sarah. 1997. "The Making and Unmaking of Persons: Notes on Aging and Gender in North India," *Ethos* 25 (3): 279-302.

—. 2000. *White Saris and Sweet Mangoes: Aging, Gender, and Body in North India*. Berkeley: University of California Press.

—. 2004. "Making and Unmaking of Persons: Gender and Body in Northeast India." In *Gender in Cross-Cultural Perspective*, 4th ed., edited by Caroline Brettell and Carolyn Sargent, pp. 230-240. Upper Saddle River, NJ: Prentice Hall.

Mauss, Marcel. 1954. [1925] *The Gift; forms and functions of exchange in archaic societies*. Expanded edition. Translated by Jane I. Guyer. Chicago: Hau Books.

Pennington, Brian K. 2005. *Was Hinduism Invented? Britons, Indians, and the Colonial Construction of Religion*. Oxford: Oxford University Press.

Povinelli, Elizabeth. 2006. *The Empire of Love: Towards a Theory of Intimacy, Genealogy, and Carnality*. Durham: Duke University Press.

—. 2016. *Geontologies: A Requiem to Late Liberalism*. Durham: Duke University Press.

Rao, S.K. Ramachandra. 1996. *Shaligram Kosh* [*Śālagrāma – Kosha*]. Sri Satguru Publications. Delhi: Indian Books Center.

Rapport, Nigel. 1998. "Coming Home to a Dream: A Study of the Immigrant Discourse of 'Anglo-Saxons' in Israel." In *Migrants of Identity: Perceptions of Home in a World of Movement*, edited by N. Rapport and A. Dawson. Oxford: Berg.

Thompson, Matt. 2012. "Digital Anthropology Group Is Happening Now," *Savage Minds*. Retrieved 31 January 2017.

Tsing, Anna Lowenhaupt. 1993. *In the Realm of the Diamond Queen*. Princeton: Princeton University Press.

—. 2005. *Friction: An Ethnography of Global Connection*. Princeton and Oxford: Princeton University Press.

—. 2016. *The Mushroom at the End of the World: On the Possibility of Life in Capitalist Ruins*. Princeton: Princeton University Press.

Urry, John. 2000. *Sociology Beyond Societies: Mobilities for the Twenty-First Century*. London: Psychology Press.

—. 2002. "Mobility and Proximity," *Sociology* 36 (2): 255-273.

Conclusion

Touch Stones

Abstract
The ever-changing conditions, constraints, and challenges under which a widely disparate community of ritual practitioners have forged ties of relations, embodied experiences, and histories have demonstrated the ways in which Shaligram veneration crosses the boundaries of nation, religion, ethnicity, caste, and generation. As an exploration in political conflict and resistance, religious innovation in the digital age, and symbolic accommodation for decreasing physical mobility within and outside of South Asia, this work on Shaligrams reveals how contemporary understandings of personhood, place, and family remain fluid and how they are continuously constructed as a result of the interactions between people, landscapes, material life, and the immaterial divine.

Keywords: Shaligram, cultural preservation, Hinduism, Buddhism, diaspora

"Did you see Aarav's new Shaligram?" Vihaan Bansal discreetly whispered one morning. "It's just come from a temple in India. He says it is Narasimha, but I don't think so. It doesn't feel like Narasimha and I don't see any teeth. It's probably Lakshmi. Which would be good, actually. Since he can't afford to go on pilgrimage [to Mustang], I'm thinking that maybe she will bring fortune enough to see Kali Gandaki one day. It's what we all hope for."

The identity of Aarav's Shaligram remained a consistent topic of discussion among the local community of Hindu devotees in Boston for the next several weeks. Reading a Shaligram is generally something that takes a fair amount of time and training as well as access to a plethora of religious texts in order to properly carry out. But without access to a knowledgeable *guru* or proficiency in some of the more esoteric complexities of Sanskrit descriptions, many practitioners noted that the ultimate identity of the

Walters, Holly, *Shaligram Pilgrimage in the Nepal Himalayas.* Taylor & Francis Group, 2020
DOI: 10.5117/9789463721721_CONCL

murti would likely remain obscure despite everyone's best efforts. This wasn't to say that the Shaligram could not be worshipped or could not enter into family life as a household deity but that the precise nature of its interactions with the community over time would have to be carefully monitored so as to ensure that the Shaligram was being cared for in the manner in which it preferred. Often, this involved a measure of trial and error in weighing the family's overall fortune or misfortune against the precision of their ritual observances or in reading signs of acceptance or rejection in the state of the daily offerings. But beyond the immediate, discussions regarding the nature of this particular *shila* continued to reflect many of the continued concerns practitioners have about the sustainability of pilgrimage to Mustang and about the ultimate longevity of Shaligram veneration in light of many current political and economic challenges. But as a way of temporarily circumventing some of these issues, receiving a new Shaligram from a temple collection (especially one in either India or Nepal) often became tantamount to having experienced pilgrimage by proxy. It was also a promise that pilgrimage may yet be to come.

In the years following my last trip to Mustang, I have continued to work with Shaligram practitioners the world over, forging connections in India, Nepal, the United States, Australia, and the UK. And the more these networks grow, the more I realize how critical mobility is and always has been to the making of community, kin, and belonging – not just among Shaligram devotees but everywhere. In Shaligrams, practitioners find the tangible verification of the stories they tell, from national or family histories to tales of the mythic landscape to chronologies accounting for the creation of the world and everything in it. In Shaligram pilgrimage, those people, places, and deities then enter into continuous cycles of identity formation, cultural tradition-making, and kinship. Through Shaligram veneration, each of these things then carries materially forward into successive generations who retain the literal touchstone that binds disparate threads of time, space, and relatedness together. In the end, I found that Shaligrams are a process wherein belonging and identity are both left behind and taken up as well as a place, a person, and an object of divine nature.

Throughout this book I have examined the ever-changing conditions, constraints, and challenges under which a widely variable community of ritual practitioners, primarily united by a shared reverence for a particular type of sacred stone, have forged ties of relations, embodied experiences, and histories across the boundaries of nation, religion, ethnicity, caste, and generation. This volume has explored conflicts and resistance just as much as it has attended to innovation in the digital age and accommodation for decreasing opportunities to move as people once did. In part, the case of Shaligrams reveals how contemporary understandings of personhood, place, and family remain fluid and how they are continuously constructed as a result of the interactions between people, landscapes, material life, and the immaterial divine. What then might appear as a kind of ritualized daily life rendered in miniature during Shaligram *puja* actually reveals a complicated set of political and religious alignments, familial entanglements, and diasporic identity choices influenced by the outward limitations placed on both people and stones.

As an examination of the myriad conditions under which people and their divine kin throughout South Asia and the world foment their own understandings of homelands, community, and history, this study complicates our broader understandings of how nationality, social change, culture, and identity interact. Rather than as an ethnographic intervention into a group of people located in a specific place or time, research on Shaligrams challenges us to expand our methodologies to include not just object-persons or divine persons in the process of community and kin-making but also the extensive and ever-changing relationships between places, ideas of places, continuous movement, and the political acts of labeling any of these things as belonging to one category or another. In other words, studying Shaligrams means examining how alternative communities might come to exist and be revealed when we dispense with typical notions of Hindu-ness, Indian-ness, or Nepali-ness. The discussion of Shaligram lifetimes presents one approach in understanding the ongoing ways by which ritual practitioners create a unique kind of society that shapes individuals as they attempt to confront and adapt to rapidly changing social realities, all the while working to preserve the practices that have brought them this far.

Shaligrams are particularly notable in terms of their ambiguous ontologies. The otherwise blanket assumption of Shaligram veneration as a specifically Hindu practice, which tends to play into a Brahminically elite version of Hinduism as a global religion inevitably moving towards homogenization, obscures both the complex histories of Nepal, India, and Tibet from prehistory and onward as well as the fact that active Shaligram traditions still exist among Buddhists, Bonpos, and even Jains. But a growing revival of Shaligram practices among second and third-generation Hindus (as well as Hindu converts) in the South Asian diaspora cannot be ignored, and it is their capacity to introduce digital platforms and new forms of non-monetary exchange into the traditional kinship-based networks of Shaligram movement that may yet save the practice from generational obscurity. Especially considering continued political instability in Mustang,

Nepal and the rising prices of travel to the Himalayas, Shaligrams as both divine persons and divine places help to layer the kinds of semiotic associations necessary for other temples and sites to become the land of Śālagrāma, for any divinely coded mobility to become pilgrimage, and for flows of personal exchange to become the Kali Gandaki. As a result, Shaligrams may continue to move without engendering any kind of cultural or religious inconstancy. Rather, they accomplish quite the opposite by destabilizing essential categories and offering a view into a dialectic that conjoins authenticity with transformation.

As noted earlier, any analysis of Shaligram practices must be cognizant of the fact that while each instance of veneration might reveal something specific about the times, places, and people within which and for whom it takes place, it is only by attending to the larger systems of mobility and exchange that one can glimpse the far-reaching connections that make Shaligrams what they are. It is also through this method that we begin to understand how the layering of local onto national, personal onto cosmological, and religious onto scientific expands the relevant contexts through which sacred stones are interpreted. As an intervention in the study of South Asia, Shaligrams help to normalize the extent to which devotees engage in cross-cultural and cross-traditional borrowing and hybridity, where any number of symbols or categorical distinctions may be leveraged for interpretation, re-interpretation, synthesis, parody, or play. This holds equally true for weighing Shaligram practices as they are described in texts and for how people actually use them.

Modern-day Shaligram practice is still largely passed down generationally through oral traditions, and so many of the scriptural restrictions that exist in the Shastras and Puranas are almost never referenced in conversation and even more infrequently practiced when it comes to Shaligrams already established within community and kinship networks. Though texts comprise an important aspect of Shaligram veneration and are still referred to as authoritative in terms of spiritual ideals, actual ritual practice, the incorporation of Shaligrams into daily life, the respectful care for and interaction with the manifest deity, and the intentions of the devotee weigh far more heavily on the minds of practitioners than strict adherence to multiple and often conflicting scriptural authorities. As one Nepali pilgrim explained, "I have read the texts and I understand them, but they can only tell you so much. They are good guides, but they cannot tell you everything you need to know. You must 'see' (*darshan*) Shaligram, you must hear it and touch it and experience it. Shaligram must speak to you, and when it does, you must listen. Without this, there is nothing."

This focus on praxis and on individual ties with specific Shaligrams rather than on authoritative doctrine is one of the hallmarks of Shaligram veneration worldwide. With great care given to the preservation of traditions, *shilas* are routinely passed down from parents to children as re-instantiations of family relationships. With an eye towards maintaining community and familial identity, thousands of pilgrims make their way to the Kali Gandaki River Valley each year to find new Shaligrams for homes and temples, and with a desire to maintain these connections across great distances of time and space, devotees send their Shaligrams far and wide – to children, to grandchildren, and across the diaspora to remind themselves and one another that wherever the Shaligram goes, their history and their culture, their vital connection to the intangibility of life itself, go with them. What makes a Shaligram a Shaligram, then, is not only its narratives and iconography as written in historical and religious texts or its place in the paleontological exploration of earth's ancient past alone but also its movements through geological and mythological time as they are carried contiguously forward into its movements through lived spaces and human experiences. With these profound "ties of *maya*", the Shaligram is then ready to begin its karmic life anew: to be born, to live, and to die with the people as one of the people.

A Shaligram is thus a living fossil in a very different sense of the term. As ammonites, they hold clues to the existence of an ancient world filled with living creatures who once swam the waters of a primordial ocean back in an era when the earth was young. As deities, they live on, born out of the geological processes that once transformed animal into mineral, they transform once more from stone to person through a journey across a vast mythic landscape. Shaligrams challenge the notion of life's progress as a series of linear assumptions from birth to death that embody a belief in the progressive march of history. Rather, they lived once in another form, in another kind of history, and then pass on, only to be reborn as stone and then as deity through four thousand years of human movement, heritage, practice, and spiritual imagination. They live as gods and family members in temples and homes. They are born from the river, are given names, travel the world, and share in home-cooked meals and daily bathing. They become persons, participate in relationships, and eventually even die, returning to the river with the cremated remains of their loved ones or retired to temples, too old and worn to carry on. And then they are reborn again, reappearing in the river to a new devotee or passing down into the hands of a new generation, continuing along in the karmic life cycle of the world.

This is also a world where landscapes themselves are bodies, constructed from the bits and pieces of the gods and men who traverse them. In this land as body, the endless cycle of erosion and reconstruction recapitulates the karmic life cycle of all living things, where geological space has become mythological time. Given enough time, both the works of Nature and the works of Man are then reduced to the same dust. Rocks are eroded to sediment, sediment is hardened into rocks, rocks are elevated above sea level through the movement of tectonic plates and transformed into mountains. Mountains weather away and become sediment once again. So too do homes and temples and bodies live and die and live again. This is the great wheel of the ages, and Shaligram has endured them all.

Shaligram Stones in an Ammonite World

The article is called "Preserving Nepal's Soul" and it appears in the March 19th, 2015 edition of the Nepali Times. "For many of Nepal's development partners, the priority is poverty-reduction, health and education", Stéphane Huët begins, "But as Nepal makes progress in literacy and mother-child survival, some have turned to preserving Nepal's unique and rich cultural heritage."

While it is interesting to note that literacy and mother-child survival are excluded from Nepal's otherwise rich cultural heritage in this setup, the gist of the article is actually about the particular interest the United States has taken in Nepali cultural preservation projects over the past decade or so. For the most part, this interest involves funding for the architectural and artistic restoration of ancient shrines and temples as well as programs for eco-tourism that promote the reinvigoration of cultural performances. For example, the article continues: "In 2012, the program supported Alliance for Ecotourism for the preservation of intangible heritage, the Kartik Nach dance which had not been performed in its full form since 1949. Ambassador Bodde says he was touched when he sat through the performance and watched hundreds of young Nepalis proud of a revival of a nearly lost part of their heritage. 'If we can help do that, we've done something special,' he said."

It might be confusing as to why my tone here appears circumspect. After all, doesn't this sound exactly like the kinds of helpful, culturally sensitive partnerships we should be engaging our post-colonial globalized dollars in? In light of recent tragedies showing the destruction of museums and artifacts by extremist groups elsewhere in the world, is this not the noble way to ensure the cultural continuity of marginalized peoples? In principle, that would be the hope, but unfortunately, such is not always the reality.

In Sienna Craig's "A Tale of Two Temples: Culture, Capital, and Community in Mustang, Nepal", we're introduced to a more sobering account of preservation efforts through the case of Thubchen Lhakhang, a temple built in 1472 and located in the heart of Monthang in Mustang. Bringing together a veritable Greek chorus of characters – a US-based preservation foundation, a couple of Nepalese conservation and development organizations, a Kathmandu-based international architecture and restoration firm, a host of foreign and Nepali subcontractors, and the "rather nebulous category of 'community support'" (2001: 15) – the project to restore and conserve Thubchen Lhakhang had something of an unanticipated result: the Loba people who once used it now cannot.

While it is true that the physical monument has been restored (for the time being, at least) and a gripping NOVA documentary has been produced and shown worldwide about the project, that "intangible heritage" so lauded by the Nepali Times so recently seems to have gone missing. The Loba people of Mustang, the "cultural owners" of the monastery, somehow became antithetical to their own culture in the process of saving it. What I mean by this more plainly is that the crumbling edifice of the building itself had been taken to mean neglect by the people, now culturally bankrupted by the forces of modern society. In other words, the peoples of Mustang, in their efforts to reconcile their lived realities with pressures to "modernize", became agents in their own cultural demise. For the agencies involved, Loba culture needed to be saved from the Loba people. What was important now was preserving culture – a culture that now only existed within the construct of a fifteenth-century religious complex. As Craig describes it, "the socio-economic, political, and even aesthetic underpinnings of Thubchen's neglect – as well as the place of Mustang's people as agents of this change – is rendered superfluous to the larger mission: to preserve, protect, and restore cultural heritage as a catalyst for what the arbiters of this perspective see as 'positive' local practice." (16)

While this ideological split is frustrating in and of itself, the final result was that, in the end, new centers of Loba cultural identity and activity now no longer appear in Mustang but in the urban center of Kathmandu, where many groups located in outlying districts have begun to relocate. As for the preserved centers of "culture" still located in ancestral lands, they are now visited only by tourists. A new kind of "cultural zoo" has come about: traditions frozen in time and space have been set aside for the paying interests of foreign visitors, donors, and trekking agencies.

By designating specific sites as places of "cultural heritage", local peoples are often erased or removed from notions of universal heritage, especially in cases where money, government agencies, and international collaboration are centered around an idea that culture is, somehow, disappearing (even though, ironically, it is because more and more people leave Mustang behind for the less destitute opportunities of urban life). And what is more, that which is preserved is no longer produced by the practices and traditions of the peoples located there. Instead, it is produced through an Italian conservationist's paintbrush, an American development specialist's spreadsheet, and an Indian architect's blueprint. It may not be their intent, but this culture is quite different than the one that came before it.

One might not imagine that fossils would enter into a discussion about "cultural preservation", but the problem of buying and selling Shaligram stones in fact highlights exactly the same problem as laid out above. Selling Shaligrams as religious objects or as souvenir fossils is often narrated as the result of both national tensions and economic hardship. In this case, impoverished local people are selling the stones to wealthier foreign practitioners who cannot access Mustang due to its political restrictions or to tourists. These narratives, however, also bolster growing concerns about the continued availability of Shaligrams overall.

When practitioners talk about their concerns with the potential loss of Shaligram practices, it tends to take two forms. The first is the fact that, outside of South Asia, few people think of fossils as distinctly cultural objects and, therefore, the global trade in Shaligram stones has much the same character as the usual consumerist trade in souvenir minerals. The second is the problem of passing on Shaligram ritual practices to younger generations who either have little connection to the religious practices of their natal communities or who will never have the opportunity to undertake pilgrimage to the Kali Gandaki. In this way, concerns about cultural preservation as it relates to the community of Shaligram practitioners mirrors many of the Loba concerns about the preservation of religious buildings: once the people are absent, the material might remain but the "culture" ends. In other words, Shaligrams themselves will always be there, but the *seva* could potentially disappear forever. And should that happen, Shaligrams will effectively revert to what the uninitiated have often presumed them to be all along: inanimate rocks.

When I interviewed Shaligram devotees, however, they often spoke hopefully of the ways in which they believed that these wily little deities could ultimately change the world. They mentioned the links between scientific exploration of the past and the modern need for spiritual connections as something Shaligrams were uniquely suited to address. As global in both geologic scale and portability, Shaligrams could also be the answer to breakdowns in national affiliation, caste or class identification, gender fluidity, and religious orthodoxy because they inhabit a variety of such categories themselves as well as transcend them. Such mobility and instability could then inspire more devotees to take up the practice of Shaligram veneration, including those who might never have otherwise encountered Hindu or Buddhist rituals, pilgrimage, or the Himalayas. But this is also a double-edged sword. Where the boundaries of belonging blur and twist, it can sometimes be difficult to say what is genuine reverence and what amounts to rock collecting.

In the early spring of 2017, I found myself having just such a conversation with a Hindu community center located just outside of London. They had gathered together specifically to discuss whether or not the attendants should begin Shaligram practices in their homes. The community had only recently begun venerating a large Lakshmi-Narayan Shaligram in their newly renovated temple and now had the opportunity, thanks to a recent donation, to disseminate a group of Shaligrams among the congregation.

"They're not monsters," their Brahmin priest explained when asked if the Shaligrams so presented would bring about misfortune to those untrained in the ritual practice. "They're not here to cause you problems. They're here to help. They'll help us all learn." In repositioning the Shaligrams from inaccessibly sacred objects to teachers in their own right, the priest was then able to counteract much of the distress the community felt in their alienation from their home cultural contexts. This wasn't to say that he openly eschewed formal theological orthodoxies but that the focus was redirected towards the embodied orthopraxy he sought to return to. In the end, it was more important that the community practiced, even if they did so incorrectly, and then introduced the practice to their children and grandchildren as a way of helping them connect with a culture and homeland most of them had never seen. As far as he was concerned, the Shaligrams wouldn't mind a few mistakes here and there, and it would further serve as potent ritual grounds for re-instantiating community ties, given that so many members of the cultural center were from disparate nations, castes, and backgrounds. In short, it was felt that the reintroduction of Shaligram *seva* would create a community of kin within this particular diasporic group in a way that their shared religion had not.

In bringing this book to a close, then, I look towards the ways in which Shaligrams have engendered a cultural reach quite unlike anything else before them: in particular, how devotees have been able to ritualize and reinvigorate their lives in ways that reject the discourses of disillusionment typical of capitalist modernity as well as how they have come to resist growing consumerism in religion by re-making networks of exchange in the digital world. Through their engagements with Shaligram *darshan* and *puja* and the endless semiotic diversity of Shaligram narratives, practitioners redefine the world around them as more than a product of droll materialism but as something alive, interactive, and curious. It may well be that, for some, the act of perceiving agency in a fossil ammonite is a way to reject Euro-American spiritual ontologies that posit a distant God dwelling elsewhere from a fundamentally corrupt physical world doomed to apocalypse. But it is more likely the case that it is a continuation of broader social and cultural mores that view the natural world as something fundamentally active and conscious, and in a constant reciprocal relationship with humanity on a level that is equally agentive with us. For my part, I have been participating in Shaligram *seva* and with the people who practice it for close to a decade now, and these experiences – along with the years of work that has gone into writing this book – have changed the course of my life in more ways than I can count. So in that way, Shaligrams have not only had a profound effect on their own world of devotees but, through this work, on mine and on the broader world beyond us.

In the end, Shaligrams are a complicated mess of the fascinating, the provocative, and the strange, and there are undoubtedly as many ways to understand their meanings as there are people who practice their veneration. Given my own trajectory, I have chosen to begin by looking at their extensive reach in and around South Asia and the contexts of political and historical mobility that frame their ritual worship. Shaligrams are objects of wonder and amazement just as much as they are divine persons with needs, desires, and intents. They are also places of divine dwelling and repositories of history. They bring out the best and the worst – from admirable gifts to beloved family members to high-priced commodities for those looking to prey on religious sentiments to get rich. For me to attempt to reduce any of these moments to some singular meaning would obscure the far grander diversity that gives Shaligrams their heart and soul. This, then, has been a book that seeks to uncover just a few of the partial truths of Shaligram spirituality. By providing a unique ethnographic point of access with which to approach the utterly ambiguous boundaries of identity and belonging in South Asian religion, these stones demonstrate how conceptual dichotomies are so often more the province of analytical or political alignments than they are meaningful statements of origin. Yet, in providing such vibrant frameworks for ritual, narrative, or history, Shaligrams become a truly unique space for the transformation of anyone willing to reach out and be touched by them.

Bibliography

Primary Sources

Chand, Devi. 1997 (2002). *Atharvaveda, Sanskrit Text with English Translation.* Delhi: Munshiram.

Prana-toshini tantra. 1983. Ramatosana Bhaṭṭa and Ramadatta Shukla, translated by Prayaga: Shakta Sadhana Piṭha Publishers.

Puranas. 2007. Translated into English by a board of scholars. Motilal Banarsidas Publishers Pvt. Ltd. 74 Volumes.

Rao, S.K. Ramachandra. 1996. *Shaligram Kosh* [*Śālagrāma – Kosha*]. Sri Satguru Publications. Delhi: Indian Books Center.

Sharma, Ram Charan. 2000. *Shaligram Purana.* S.R.C. Museum of Indology & Universal Institute of Orientology Trust, 199. Digitized 2009. Ann Arbor: University of Michigan.

Singh, Chandra Shekar et al. 2017. *The Puranas.* Amazon Digital Services.

Secundary Sources

Abu-Lughod, Lila. 1986. *Veiled Sentiments: Honor and Poetry in a Bedouin Society.* Berkeley: University of California Press.

—. 2008. *Writing Women's Worlds: Bedouin Stories.* Berkeley: University of California Press.

Agamben, Giorgio. 1998. *Homo Sacer: Sovereign Power and Bare Life.* Translated by Daniel Heller-Roazen. Stanford: Stanford University Press.

Ahearn, Laura. 2006. *Invitations to Love: Literacy, Love Letters, & Social Change in Nepal.* Ann Arbor: University of Michigan Press.

Allen, Michael, ed. 1994. *Anthropology of Nepal; Peoples, Problems and Processes.* Kathmandu: Mandala Book Point.

Anderson, Benedict. 1991. *Imagined Communities: Reflections on the Origin and Spread of Nationalism.* London: Verso.

Appadurai, Arjun. 1986. *The Social Life of Things, Commodities in Cultural Perspectives.* Cambridge: Cambridge University Press.

Arkell, W.J., B. Kummel, and C.W. Wright. 1957. "Mesozoic Ammonoidea. Treatise on Invertebrate Paleontology, Part L, Mollusca 4." Lawrence, Kansas: Geological Society of America and University of Kansas Press.

Aryal, M. 1999. "Motar pugyo Mustanma. Himal" [Nepali language version] – (Reaching Mustang by Motor). 1-15 *Mansir* 2056 (Nov 17-Dec 1, 1999).

Asad, Talal. 2002. "The Construction of Religion as an Anthropological Category." In *A Reader in the Anthropology of Religion*, edited by Michael Lambek. Oxford: Blackwell Publishing Ltd.

Baan, Stephen. 1995. "Shrines, Curiosities, and the Rhetoric of Display." In *Visual Display: Culture Beyond Appearances*, edited by Lynne Cooke and Peter Wollen. Seattle: Bay Press.

Badone, Ellen, and Sharon R. Roseman. 2004. *Intersecting Journeys: The Anthropology of Pilgrimage and Tourism*. Chicago: University of Illinois Press.

Ball, Christopher, and Nicholas Harkness, eds. 2015. "Kinship Chronotopes: A special issue," *Anthropological Quarterly* 88 (2).

Basset, M.G. 1982. "Formed stones, folklore and fossils." National Museum of Wales. Geological Series No. 1.

Basso, Keith H. 1979. *Portraits of 'The Whiteman': Linguistic Play and Cultural Symbols Among the Western Apache*. Cambridge: Cambridge University Press.

—. 1996. "Wisdom Sits in Places." In *Wisdom Sits in Places: Landscape and Language among the Western Apache*, pp. 105-149. Albuquerque: University of New Mexico Press.

Bastide, Roger. 2007. "The Problem of Religious Syncretism." In *The African Religions of Brazil: Toward a Sociology of the Interpenetration of Civilizations* (Johns Hopkins Studies in Atlantic History and Culture). Baltimore: Johns Hopkins University Press.

Bennett, Lynn. 1983. *Dangerous Wives and Sacred Sisters: Social and Symbolic Roles of High Caste Women in Nepal*. New York: Columbia University Press.

Bhardwaj, Surinder Mohan. 1973. *Hindu Places of Pilgrimage in India: A Study in Cultural Geography*. Berkeley: University of California Press.

Bhargava, Rajeev. 2001. "Are There Alternative Modernities." In *Culture, Democracy, and Development in South Asia*, edited by N.N. Vohra. New Delhi: Shipra Publications.

Bishara, Amahl. 2015. "Driving while Palestinian in Israel and the West Bank: The politics of disorientation and the routes of a subaltern knowledge," *American Ethnologist* 42: 33-54.

Blanes, Ruy, and Diana Espírito Santo, eds. 2014. *The Social Life of Spirits*. Chicago: University of Chicago Press.

Bloch, Maurice, and Parry, Jonathan. 1982. "Introduction: Death and the Regeneration of Life." and "Death, Women and Power." In *Death and the Regeneration of Life*. Cambridge: Cambridge University Press.

Bourdieu, Pierre. 1977. *Outline of a Theory of Practice*. Cambridge: Cambridge University Press.

Brooks, Charles R. 1990. "Hare Krishna, Radhe Shyam: The Cross-Cultural Dynamics of Mystical Emotions in Brindaban". In *Divine Passions: The Social Construction of Emotion in India*, edited by Owen M. Lynch. Berkeley and Los Angeles and Oxford: University of California Press.

Brown, Bill. 2004. *A Sense of Things*. Chicago: University of Chicago Press.

Bryant, Edwin F. 2007. *Krishna: A Sourcebook*. Oxford and New York: Oxford University Press.

Bubandt, Nils, and Martijn van Beek. 2012. *Varieties of Secularism in Asia: Anthropological Explorations of Religion, Politics and the Spiritual*. New York: Routledge.

Burghart, Richard. 1984. "The Formation of the Concept of Nation-State in Nepal," *Journal of Asian Studies* 44 (1): 101-26.

Candea, Matei. 2010. "Debate: Ontology Is Just Another Word for Culture," *Critique of Anthropology* 30 (2): 172-179.

Carsten, Janet. 2000. *Cultures of Relatedness: New Approaches to the Study of Kinship*. Cambridge: Cambridge University Press.

Cassaniti, Julia, and Luhrmann, Tanya. 2011. "*Encountering the Supernatural*: A Phenomenological Account of Mind," *Religion and Society: Advances in Research* 2: 37-53.

Castells, Manuel, ed. 2004. *Network Society: A Cross Cultural Perspective*. Cheltenham: Edward Elgar Publishing Ltd.

Chakrabarty, Dipesh. 1992. "Postcoloniality and the Artifice of History: Who Speaks for 'Indian' Pasts," *Representations* 37: 1-26.

Chetri, M., L.K. Maskey, N.R. Chapagain, and B.D. Sharma, eds. 2004. *Mustang – The Land of Fascination*. King Mahendra Trust for Nature Conservation, Nepal.

Clark, Herbert H. 1996. *Using Language*. Cambridge: Cambridge University Press.

—, and Thomas Carlson. 1982. "Hearers and Speech Acts," *Language* 58: 332-73.

Coleman, Simon. 2002. "Do you believe in pilgrimage? Communitas, contestation, and beyond," *Anthropological Theory* 2 (3): 355-68.

—, and Peter Collins, eds. 2006. *Locating the field: Space, place and context in anthropology*. Oxford: Berg.

—, and John Eade, eds. 2004. *Reframing Pilgrimage: Cultures in Motion*. European Association of Social Anthropologists. London: Routledge.

Comaroff, Jean, and John L. Comaroff. 1993. Introduction to *Modernity and its Malcontents: Ritual and Power in Post-Colonial Africa*, edited by John L. Comaroff and Jean Comaroff. Chicago: University of Chicago Press.

Craig, Sienna. 2001. "A Tale of Two Temples: Culture, Capital, and Community in Mustang, Nepal." Paper presented at the New York Conference on Asian Studies, Ithaca, NY.

—. 2008. *Horses Like Lightening: A Story of Passage Through the Himalayas*. Somerville: Wisdom Publications.

Csordas, Thomas J. 1999. "Embodiment and Cultural Phenomenology." In *Perspectives on Embodiment: The Intersections of Nature and Culture*, edited by Gail Weiss and Honi Fern Haber. New York and London: Routledge.

Dahal, Sharada Prasad. 1988. *Muktikshetra: A Historical, Religious and Cultural Explorations* (Original text in Nepali); 'Muktikshetra' Publication Committee. Translated under the direct supervision of Dr. Ramesh K. Dhungel by Rana, Binaya SJB: Post Graduate Diploma in Buddhist Studies, Buddhist Studies Department, Tribhuvan University, Kathmandu, Nepal. Translation sponsored by André Kalden, President of Muktinath Foundation International, 2002.

Dana, Jessamine. 2011. "The Politics of Encounter: Hindu Belonging in a Multi-Faith Pilgrimage Site in Nepal." In *The Politics of Belonging in the Himalayas: Local Attachments and Boundary Dynamics* (*Governance, Conflict and Civic Action*), 1st edition, edited by Joanna Pfaff-Czarnecka and Gerard Toffin. Los Angeles: Sage Publications.

Dancygier, Barbara, and Sweetser, Eve. 2012. *Viewpoint in Language: A Multimodal Perspective*. Cambridge: Cambridge University Press.

Daniel, E. Valentine. 1984. *Fluid signs: Being a person the Tamil way*. Berkeley: University of California Press.

Dawson, Allan Charles, Laura Zanotti, and Ismael Vaccaro, eds. 2014. *Negotiating Territoriality: Spatial Dialogues Between State and Tradition*. New York: Routledge.

Dejarlais, Robert R. 1992a. *Body and Emotion: The Aesthetics of Illness and Healing in the Nepal Himalayas*. Philadelphia: University of Pennsylvania Press.

—. 1992b. "Yolmo Aesthetics of Body, Health and 'Soul Loss'," *Social Science and Medicine* 34 (10): 1105-1117.

de Bruijn, Mirjam, and Rijk van Dijk. 2012. "Connecting and Change in African Societies: Examples of 'Ethnographies of Linking' in Anthropology," *Anthropologica* vol. 54 (1): 45-59.

De la Cadena, Marisol. 2010. "Indigenous Cosmopolitics In the Andes: Conceptual Reflections beyond 'Politics'," *Cultural Anthropology* 25: 334-370.

De León, Lourdes. 2012. "Language Socialization and Multiparty Participation Frameworks". In *The Handbook of Language Socialization*, edited by Alessandro Duranti, Elinor Ochs, and Bambi B. Schieffelin. Oxford: Wiley-Blackwell, Blackwell Publishing Limited.

Deleuze, Gilles. 1992. "Postscript on the societies of Control." Article available at *libcom.org*.

—. 1994. *Difference and Repetition*. London: Athlone.

Descola, Philippe. 2005. *Par-delà nature et culture*. Paris: Gallimard.

Dey, Nundo Lal. 1971. *Geographical Dictionary of Ancient and Mediaeval India*. Calcutta: W. Newman & Co.

Dhital, Megh Raj. 2015. *Geology of the Nepal Himalaya: Regional Perspective of the Classic Collided Orogen* (Regional Geology Reviews). New York: Springer.

Dhungel, Ramesh K. 2002. *The Kingdom of Lo (Mustang): A Historical Study*. Kathmandu: Tashi Gephel Foundation.

Dhruba, Kumar. 2013. *Social Inclusion: Human Development and Nation Building in Nepal*. Imprint Unknown. 1st edition.

Dilger, Hansjörg, Abdoulaye Kane, and Stacey A. Langwick, eds. 2012. *Medicine, Mobility, and Power in Global Africa: Transnational Health and Healing*. Bloomington: Indiana University Press.

Droysen, J.G. 1952-53 [1836-78]. *Geschichte des Hellenismus* (The History of Hellenism). vol 1, 2, 3, edited by E. Bayer. München: Deutscher Tashenbuch.

Dumont, Louis. 1980. *Homo hierarchicus: The caste system and its implications*, revised, edited, and translated by Mark Sainsbury, Louis Dumont, and Basia Gulati. Chicago: Chicago University Press.

—. 1983. *Affinity as a Value: Marriage Alliances in South India, with Comparative Essays on Australia*. Chicago: University of Chicago Press.

Durkheim, Emile. (1912) 2001. *The Elementary Forms of Religious Life*. Oxford: Oxford University Press.

Eade, John, and Michael J. Sallnow. [1991] 2000. *Contesting the Sacred: The Anthropology of Christian Pilgrimage*. Chicago: University of Illinois Press.

Eck, Diana. 1986. "Darshan of the Image," *India International Centre Quarterly* vol. 13, no. 1, IMAGES (March): 43-53.

—. 1998. *Darśan: Seeing the Divine Image in India*. New York: Columbia University Press.

—. 2012. *India: A Sacred Geography*. New York: Three Rivers Press.

Ehrhard, F.K. 1999 (1993). "Tibetan Sources on Muktinath: Individual Reports and Normative Guides." In *Archaeological, Historical, and Geographical Reports on Research Activities in the Nepal-Tibetan border area of Mustang during the years 1992-1998, Part IV*, H.14, Dieter Schuh (ed.), pp. 23-29. Bonn: Archiv für Zentralasiatische Geschichtsforschung (Archive for Central Asian History).

Eickelman, Dale, and James Piscatori, eds. 1990. *Muslim Travellers: Pilgrimage, Migration, and the Religious Imagination*. Berkeley and Los Angeles: University of California Press.

Eidsiedel, Sebastian von, David M. Malone, and Suman Pradhan. 2012. *Nepal in Transition: From People's War to Fragile Peace*. Cambridge: Cambridge University Press.

Eliade, Mircea, and William R. Trask. 1959. *The sacred and the profane: The nature of religion*. New York: Harcourt; Chicago: Brace & World.

Enay, Raymond, and Elie Cariou. 1997. "Ammonite faunas and palaeobiogeography of the Himalayan belt during the Jurassic: Initiation of a Late Jurassic austral ammonite fauna," Palaeogeography, *Palaeoclimatology, Palaeoecology* 134 (1997): 1-38.

—. 1999. "Jurassic ammonite faunas from Nepal and their bearing on the palaeobiogeography of the Himalayan belt," *Journal of Asian Earth Sciences* 17: 829-848.

Engelke, Matthew. 2007. *A Problem of Presence: Beyond Scripture in an African Church* (*The Anthropology of Christianity*). Berkeley: University of California Press.

Evans-Pritchard, E. 1940. *The Nuer*. Oxford: Clarendon

Fabian, Johannes. 1983. *Time and the other: How anthropology makes its object*. New York: Columbia University Press.

Ferry, Elizabeth, and Mandana E. Limbert. 2008. *Timely Assets: The Politics of Resources and Their Temporalities*. Santa Fe: School for Advanced Research Press.

Fisher, William F. 1987. *The Re-Creation of Tradition: Ethnicity, Migration, and Social Change Among the Thakali of Central Nepal*. PhD diss., Columbia University.

—. 2001. *Fluid Boundaries: Forming and Transforming Identity in Nepal*. New York: Columbia University Press.

Flood, Gavin. 1998. *An Introduction to Hinduism*. Cambridge: Cambridge University Press.

Fowler, Chris. 2010. "From Identity and Material Culture to Personhood and Materiality." In *The Oxford Handbook of Material Culture Studies*, edited by H.D. and M.C. Beaudry, pp. 352-85. Oxford: Oxford University Press.

Fowles, Severin. 2010. "People Without Things." Springer Science and Business Media. http://www.campusincamps.ps/wp-content/uploads/2014/04/People-Without-Things.pdf. Accessed 19 September 2019.

Franklin, Sarah, and Susan McKinnon. 2001. *Relative Values: Reconfiguring Kinship Studies*. Durham: Duke University Press.

Freedberg, David. 1989. *The Power of Images: Studies in the History and Theory of Response*. Chicago and London: University of Chicago Press.

Frey, Nancy. 1998. *Pilgrim Stories: On and Off the Road to Santiago*. Berkeley: University of California Press.

Fuller, Christopher John. 2004. *The Camphor Flame: Popular Hinduism and Society in India*, Princeton: Princeton University Press.

Fürer-Haimendorf, Christoph von. 1957. "The Interrelations of Caste and Ethnic Groups in Nepal," *Bulletin of the School of Oriental and African Studies* 20: 243-60.

—. 1966. "Caste Concepts and Status Distinctions in Buddhist Communities of Western Nepal." In *Caste and Kin in Nepal, India, and Ceylon: Anthropological Studies in Hindu-Buddhist Context*, edited by Chrisoph von Fürer-Haimendorf, pp. 140-60. London: Asia Publishing House.

—. 1975. *Himalayan Traders*. New York: St. Martin's.

—. 1978. "Trans-Himalayan Traders in Transition." In *Himalayan Anthropology*, edited by James Fisher, pp. 339-57. The Hague: Mouton.
—. 1981. "Social Change in a Himalayan Region." In *The Himalaya*, edited by J.S. Lall, pp. 175-203. Delhi: Oxford University Press.

Garzanti, Eduardo, and Marco Pagni Frette. 1991. "Stratigraphic succession of the Thakkhola region (Central Nepal). Comparison with the north-western Tethys Himalaya," *Rivista Italiana Paleontologia Stratigrafa* 97 (1): 3-26.
Geertz, Clifford. 1973. *The Interpretation of Cultures*. New York: Basic Books.
—. 1973 [1966]. "Person, time and conduct in Bali." In *The Interpretation of Cultures*, pp. 360-411. New York: Basic Books.
—. 1975a. "Common Sense as a Cultural System," *The Antioch Review* 33 (1): 5-26.
—. 1975b. *Kinship in Bali*. Chicago: University of Chicago Press.
Geismar, Haidy. 2011. "Material Culture Studies and other Ways to Theorize Objects: A Primer to a Regional Debate," *Comparative Studies in Society and History* 53 (1): 210-218.
Gell, Alfred. 1992. *The Anthropology of Time*. London: Berg.
—. 1998. *Art and Agency: An Anthropological Theory*. Oxford and New York: Oxford University Press.
Gellner, David N. 1994. "Priests, Healers, Mediums and Witches: The Context of Possession in the Kathmandu Valley, Nepal," *Man* 29 (1): 27-48.
Gentry, James Duncan. 2016. *Power Objects in Tibetan Buddhism*. (Brill's Tibetan Studies Library). Boston: Brill Publishers.
Geological Survey of India. 1904. *Geological Survey of India, Vol. XXI, Part 1*. Government of India. London and Calcutta.
George, David E.R. 1987. "Ritual Drama: Between Mysticism and Magic," *Asian Theater Journal* vol. 4, no. 2. (Autumn): 127-165.
Ghassem-Fachandi, Parvis. 2011. "Religious Synthesis at a Muslim Shrine." In *A Companion to the Anthropology of India*, edited by I. Clark-Decès. Oxford: Wiley-Blackwell.
Gibling, Martin R., Felix M. Gradstein, Inga L. Kristiansen, James Nagy, Maurizio Sarti, and Jörg Wiedmann. 1994. "Early cretaceous strata of the Nepal Himalayas: conjugate margins and rift volcanism during Gondwanan breakup," *Journal of the Geological Society, London* 151: 269-290.
Giddens, Anthony. 1995. *Politics, Sociology and Social Theory: Encounters with Classical and Contemporary Social Thought*. Stanford: Stanford University Press.
Ginzburg, Carlo, and Davin, Anna. 1980. "Morelli, Freud, and Sherlock Holmes: Clues and Scientific Method," *History Workshop Journal* vol. 9, no. 1 (Spring): 5-36.
Goffman, Erving. 1974. *Frame Analysis: An Essay on the Organization of Experience*. Cambridge, MA: Harvard University Press.

—. 1976. "Performances." In *Ritual, Play, and Performance*, edited by R. Schechner, pp. 89-96. New York: Seabury Press.

—. 1981. *Forms of Talk*. Philadelphia: University of Pennsylvania Press.

Goodwin, Charles. 1986. "Audience Diversity, Participation and Interpretation". *Text* 6 (3): 283-316. Web. http://www.sscnet.ucla.edu/clic/cgoodwin/86aud_diver.pdf. Accessed 21 November 2012.

Gosden, Christopher. 1994. *Social Being and Time*. Oxford: Blackwell Publishers.

Goswami, Vaisnavacharya Chandan. 2017. *The Sacred Shilas*. Delhi: Anupam Goswami.

Graburn, Nelson. [1977] 1989. "Tourism: The Sacred Journey." In *Hosts and Guests: The Anthropology of Tourism*, edited by Valene L. Smith. Philadelphia: University of Pennsylvania Press.

Gramsci, Antonio. 1971. *Selections from the Prison Notebooks*. New York: International.

Griffith-Jones, Susan M. 2009. *Muktinath, Secret Treasure of Annapurna... The Rainbow Bridge*. Shanti Bhawan: Vibgyor Publishing House.

Guneratne, Arjun. 2002. *Many Tongues, One People: The Making of Tharu Identity in Nepal*. Ithaca, NY: Cornell University Press.

Gunkel, Herman. 1903. "The Religio-Historical Interpretation of the New Testament," Translated by W.H. Carruth. *The Monist* vol. 13, issue 3 (April): 398-455.

Gupta, Akhil, and James Ferguson. 1992. "Beyond 'Culture': Space, Identity, and the Politics of Difference," *Cultural Anthropology* 7 (1): 6-23; reprinted in *Culture, Power, Place* (edited by Gupta and Ferguson, 1997: 33-51) and in *Anthropology of Globalization* (edited by Inda and Rosaldo, 2002: 65-80).

—. 1997. "Discipline and Practice: 'The Field' as Site, Method, and Location in Anthropology." In *Anthropological Locations: Boundaries and Grounds of a Field Science*, edited by A. Gupta and J. Ferguson. Berkeley: University of California Press.

Gutschow, Kim. 2004. *Being a Buddhist Nun: The Struggle for Enlightenment in the Himalayas*. Cambridge, MA: Harvard University Press.

Haenn, Nora, and Richard Wilk, eds. 2016. *The environment in anthropology: A reader in ecology, culture and sustainable living*. 2nd ed. New York: New York University Press.

Hagn, H. 1977. "Shaligram – Cobbles of Malm Limestones with Ammonites as Cult Objects of India," *Communications of the Bavarian State Collection of Paleontology and Historical Geology* 17: 71-102. ("Saligrame – Gerölle von Malm-Kalken mit Ammoniten als Kultgegenstände Indiens." *Mitteilungen der Bayerischen Staatssammlung für Paläontologie und Historische Geologie*).

—. 1988. "On Shaligrams – The Holy Stones of India," *Contributions to Himalayan Geology* 4: 53-62.

Hanks, William F. 1996. "Exorcism and the Description of Participant Roles." In *Natural Histories of Discourse*, edited by Michael Silverstein and Greg Urban, pp. 160-202. Chicago: University of Chicago Press.

—. 1990. *Referential Practice: Language and Lived Space Among the Maya*. Chicago: University of Chicago Press.

Haraway, Donna. 2000. "A Cyborg Manifesto." In *The Cybercultures Reader*, edited by David Bell and Barbara Kennedy, pp. 291-324. London: Routledge (via Georgetown University Online).

Hausner, Sondra L. 2007. *Wandering with Sadhus: Ascetics in the Hindu Himalayas*. Bloomington: Indiana University Press.

Hawley, John Stratton. 1992. *At Play with Krishna: Pilgrimage Dramas from Brindavan*. Delhi: Motilal Banarsidass Publishers Private Limited.

—. 2000. "Who Speaks for Hinduism: And Who Against?" *Journal of the American Academy of Religion* vol. 68, no. 4 (Dec): 711-720.

Hayden, Corinne P. 1995. "Gender, Genetics, and Generation: Reformulating Biology in Lesbian Kinship," *Cultural Anthropology* 10 (1): 41-63.

Heimsath, Kabir. 2011. *Urban Space of Lhasa*. PhD diss., University of Oxford.

Helmreich. Stefan. 2009. *Alien Ocean: Anthropological Voyages in Microbial Seas*. Berkeley and Los Angeles: University of California Press.

Henare, Amiria, Martin Holbraad, and Sari Wastell, eds. 2006. *Thinking Through Things: Theorising Artefacts Ethnographically*. London: Routledge.

Herskovits, Melville. 1958 [1941]. *The Myth of the Negro Past*. 2nd ed. Boston: Beacon Press.

Heywood, Paolo. 2012. "Anthropology and What There Is: Reflections on 'Ontology'," *Cambridge Anthropology* 30 (1): 143-151.

Hirsch, Eric, and Michael O'Hanlon. 1995. *Anthropology of Landscape: Perspectives on Place and Space*. Oxford: Clarendon Press.

Hitchcock, James T., and Rex L. Jones. 1976. *Spirit possession in the Nepal Himalayas*. New Delhi: Vikas Publishing House.

Hodder, Ian. 2003. "The 'social' in Archaeological Theory: An Historical and Contemporary Perspective." In *A Companion to Social Archaeology*, edited by Lynn Meskell and Robert Pruecel, pp. 23-42. Malden: Blackwell.

Hodges, Matt. 2008. "Rethinking Time's Arrow: Bergson, Deleuze and the Anthropology of Time," *Anthropological Theory* 8 (4): 399-429.

Höfer, Andras. 1979. *The Caste Hierarchy and State in Nepal: A Study of the Muluki Ain of 1854*. Innsbruck: Universitätsverlag Wagner.

Holbraad, Martin. 2012. *Truth in Motion. The Recursive Anthropology of Cuban Divination*. Chicago: Chicago University Press.

—, and Morten Axel Pedersen. 2009. "Planet M: The intense abstraction of Marilyn Strathern," *Anthropological Theory* 9 (4): 371-394.

Hooker, J.D. 1854. *Himalayan Journals*. London (J. Murray), 2 vols. Johnson, M.R.W., and Rogers, G., 1997, Rb-Sr ages of micas from the Kathmandu complex, Central Nepalese Himalaya. implications for the evolution of the Main Central Thrust: *Journal of the Geological Society, London* vol. 154, pp. 863-869.

Horst, Heather A., and Daniel Miller. 2005. "From Kinship to Link-up. Cell Phones and Social Networking in Jamaica," *Current Anthropology* 46 (5): 755-778.

Hoskins, Janet. 2006. "Agency, biography. And objects." In *Handbook of Material Culture*, edited by C. Tiller et al., pp. 74-84. London: Sage Publications.

Huët, Stéphane. 2015. "Preserving Nepal's Soul," *Nepali Times*. 19 March 2015 http://nepalitimes.com/article/Nepali-Times-Buzz/ambassador-fund-cultural-preservation-nepal,2092. Accessed 10 January 2016.

Huizinga, Johan. 1955. *Homo Ludens: A Study of the Play-Element in Culture*. Boston: Beacon Press.

Hymes, Dell. 1972. "Models of the Interaction of Language and Social Life." In *Directions in Socio-linguistics: The Ethnography of Speaking*, edited by J. Gumperz and D. Hymes, pp. 35-71. New York: Holt, Rinehart, & Winston.

Iijima, Shigeru. 1960. "The Thakali: A Central Himalayan Tribe," *Japanese Journal of Ethnology* 24 (3): 1-22 (Japanese text, English summary).

—. 1963. "The Hinduization of a Himalayan Tribe in Nepal," *Krober Anthropological Society Papers* 29: 43-52.

Inden, Ronald B., and Ralph Nicholas. 1977. *Kinship in Bengali culture*. Chicago: Chicago University Press.

Ingold, Tim. 1994. *What is an animal*. London: Routledge.

—. 2000. *The Perception of the Environment: Essays in Livelihood, Dwelling, and Skill*. London: Routledge.

—. 2011. *Redrawing Anthropology: Materials, Movements, Lines*. Ashgate: Aldershot.

Irvine, Judith. 1996. "Shadow Conversations: The Indeterminacy of Participant Roles." In *Natural Histories of Discourse*, edited by Michael Silverstein and Greg Urban, pp. 131-159. Chicago: University of Chicago Press.

Irvine, Leslie. 2004. *If You Tame Me: Understanding Our Connection with Animals (Animals Culture and Society)*. Philadelphia: Temple University Press.

Jackson, David. 1976. "The Early History of Lo (Mustang) and Ngari," *Contributions to Nepalese Studies* 4: 39-56.

—. 1978. "Notes on the History of Se-Rib and Nearby Places in the Upper Kali Gandaki Valley," *Kailash* 6 (3): 195-227.

—. 1980. "A Genealogy of the Kings of Lo (Mustang)." In *Tibetan Studies in Honour of Hugh Richardson: Proceedings of the International Seminar on Tibetan Studies,*

Oxford 1979, edited by Michael Aris and Aung San Suu Kyi, pp. 133-37. New Delhi: Vikas.

—. 1984. *The Mollas of Mustang: Historical, Religious, and Oratorial Traditions of the Nepalese-Tibetan Borderland*. Dharamsala: Library of Tibetan Works and Archives.

James, William. 1982 [1902]. *The Varieties of Religious Experiences: A Study in Human Nature*. New York: Penguin Classics.

Karve, Irawati. 1962. "On the Road: A Maharashtrian Pilgrimage," *The Journal of Asian Studies* XXII (1): 13-31.

Kawakita, Jiro. 1955. "Some Ethnographic Observations in the Nepal Himalaya," *The Japanese Journal of Ethnology* 19 (1): 1-57.

—. 1957. "Ethno-geographical Observations on the Nepal Himalaya." In *Peoples of the Nepal Himalaya*, edited by H. Kihara, pp. 1-362. Kyoto: Flora and Fauna Society.

Keane, Webb. 2009. "On Multiple Ontologies and the Temporality of Things." Material World blog, 7 July 2009. URL: http://www.materialworldblog.com/2009/07/on-multiple-ontologies-and-the-temporality-of-things/. Accessed 26 October 2017.

—. 2014. "On Spirit Writing: Materialities of Language and the Religious Work of Transduction," *Journal of the Royal Anthropological Institute* vol. 19 (1): 1-17.

Kirksey, S. Eben, and Stefan Helmreich. 2010. "The Emergence of Multispecies Ethnography," *Cultural Anthropology* 25 (4): 545-576.

Klassen, Pamela E. 2011. *Spirits of Protestantism: Medicine, Healing, and Liberal Christianity*. Berkeley: University of California Press.

Knapp, Stephen. 2011. *Krishna Deities and Their Miracles: How the Images of Lord Krishna Interact with Their Devotees*. Detroit: The World Relief Network.

Knappett, Carl. 2005. "Animacy, Agency, and Personhood." In *Thinking through material culture: An interdisciplinary perspective*, pp. 11-34. Philadelphia: University of Pennsylvania Press.

Kohl-Bonn, Joseph. 1936. "To the Indian Stone Cult," *Journal of the German Oriental Society* vol. 90, No. 2 (1936): 432-440. ("Zum indischen Steinkult." *Zeitschrift der Deutschen Morgenländischen Gesellschaft*).

Kohn, Eduardo. 2007. "How Dogs Dream: Amazonian Natures and the Politics of Transspecies Engagement," *American Ethnologist* 34 (1): 3-24.

—. 2013. *How Forests Think: Toward an Anthropology Beyond the Human*. Berkeley: University of California Press.

—. 2014. "What an Ontological Anthropology Might Mean." Theorizing the Contemporary, Cultural Anthropology website. https://culanth.org/fieldsights/463-what-an-ontological-anthropology-might-mean. Accessed 13 January 2014.

Korom, Frank. 2001. "Introduction: Fieldwork, Ethnography, and the History of Religions," *Method & Theory in the Study of Religion* vol. 13 (1): 3-11.

—. 2006. *South Asian Folklore: A Handbook*. Westport: Greenwood Press.

Koselleck, Reinhart, and Keith Tribe. 1985. *Futures Past: On the Semantics of Historical Time*. New York: Columbia University Press.

Kraemer, Hendrik. 1962. "Syncretism." In *Religion in History and the Present. Hand-dictionary for Theology and Religious Studies*. ("Synkretismus." In *Die Religion in Geschichte und Gegenwart. Handwörterbuch für Theologie und Religionwissenschaft*) 3rd ed., vol. 6, pp. 563-68. Tübingen: Gregorian Biblical Press.

Laidlaw, James. 2012. "Ontologically Challenged." *Anthropology of This Century*, vol. 4, London. http://aotcpress.com/articles/ontologically-challenged/. Accessed 10 October 2017.

Lamb, Sarah. 1997. "The Making and Unmaking of Persons: Notes on Aging and Gender in North India," *Ethos* 25 (3): 279-302.

—. 2000. *White Saris and Sweet Mangoes: Aging, Gender, and Body in North India*. Berkeley: University of California Press.

—. 2004. "Making and Unmaking of Persons: Gender and Body in Northeast India." In *Gender in Cross-Cultural Perspective*. 4th ed., edited by Caroline Brettell and Carolyn Sargent, pp. 230-240. Upper Saddle River, NJ: Prentice Hall.

Lambek, Michael. 2008. "Provincializing God? Provocations from an Anthropology of Religion." In *Religion: Beyond a Concept*, edited by Hent de Vries, pp. 120-138. New York: Fordham University Press.

Latour, Bruno. 1988. *The Pasteurization of France*. Translated by Alan Sheridan and John Law. Cambridge, MA: Harvard University Press.

—. 1993. *We have never been modern*. Cambridge, MA: Harvard University Press.

—. 1999. "On Recalling ANT." In *Actor Network Theory and After*, edited by Law and Hassard, pp. 15-26. Oxford: Blackwell Publishers.

Lavecchia, Giusy, and Giancarlo Scalera, eds. 2003, "Frontiers in Earth Sciences: New ideas and interpretations," *Annals of Geophysics*, Supplement to v. 49.

Le Fort, Patrick. 1975. "Himalayas: The collided range. Present knowledge of the continental arc," *American Journal of Science*, 275-A, pp.1-44. (1981, Manaslu leucogranite: A collision signature of the Himalaya, a model for its genesis and emplacement. *Journal of Geophysical Research*, 86: 545-568.)

Leeuw, Gerardus van der. 1938 [1933]. *Religion in Essence and Manifestation: A Study in Phenomenology*. Translated by J.E. Turner. London: G. Allen & Unwin.

Leopold, Anita Marie, and Jeppe Sinding Jensen, eds. 2005. *Syncretism in Religion: A Reader*. New York: Routledge.

Levi-Strauss, Claude. 1955. "The Structural Study of Myth." *Structural Anthropology. The Journal of American Folklore* vol. 68, no. 270, Myth: A Symposium (Oct-Dec 1955), pp. 428-444. Published as a separate pamphlet by American Folklore Society (pp. 206-131).

—. 1969. *The Elementary Structures of Kinship.* Translated by James Harle Bell and John Richard von Sturmer. Boston: Beacon.

Levine, Nancy. 1987. "Caste, State, and Ethnic Boundaries in Nepal," *Journal of Asian Studies* 46: 71-88.

Levinson, Stephen C. 1983. *Pragmatics*. Cambridge: Cambridge University Press.

—. 1988. "Putting Linguistics on a Proper Footing: Explorations in Goffman's Participation Framework." In *Goffman: Exploring the interaction order*, edited by P. Drew and A. Wootton, pp. 161-227. Oxford: Polity Press.

Low, Setha, ed. 1999. *Theorizing the city: The new urban anthropology reader.* New Brunswick: Rutgers University Press.

Mahmood, Saba. 2004. *Politics of Piety: The Islamic Revival and the Feminist Subject.* Princeton: Princeton University Press.

Marcus, George E. 1995. "Ethnography in/of the World System: The Emergence of Multi-Sited Ethnography," *Annual Review of Anthropology* vol. 24 (October): 95-117.

Marriot, McKim. 1976. "Hindu transactions: Diversity without dualism." In *India through Hindu categories*, edited by Bruce Kapferer, pp. 109-42. New Delhi: Sage Publications.

—, and Ronald Inden. 1977. "Toward an ethnosociology of South Asian caste systems." In *The new wind: Changing identities in South Asia*, edited by Kenneth David, pp. 227-38. The Hague: Moulton Publishers.

Maskarinec, Gregory G. 1992. "A Shamanic Etiology of Affliction from Western Nepal," *Social Science and Medicine* 35 (5): 723-734.

Mason, David. 2005. *Theatre and Religion on Krishna's Stage: Performing in Vrindavan*. New York: Palgrave Macmillan Publishers.

Masquelier, Adeline. 2005. *Dirt, Undress, and Difference: Critical Perspectives on the Body's Surface*. Bloomington and Indianapolis: Indiana University Press.

Mauss, Marcel. 1954. [1925] *The Gift; forms and functions of exchange in archaic societies*. Expanded edition. Translated by Jane I. Guyer. Chicago: Hau Books.

Mayor, Adrienne. 2005. *Fossil legends of the first Americans*. Princeton: Princeton University Press.

McIntosh, Janet. 2005. "Liminal Meanings: Sexually Charged Giriama Funerary Ritual and Unsettled Participant Frameworks," *Language and Communication* 25: 39-60.

—. 2009. *The Edge of Islam: Power, Personhood, and Ethnoreligious Boundaries on the Kenya Coast*. Durham: Duke University Press.

McManners, John. 1992. "Enlightenment: Secular and Christian (1600-1800)." In *The Oxford Illustrated History of Christianity*, edited by John McManners, pp. 267-300. Oxford: Oxford University Press.

Menon, Ramesh. 2000. *Blue God: A Life of Krishna*. Bloomington: Writers Club Press.

Merleau-Ponty, Maurice. 1981. *The Phenomenology of Perception*. Evanston and Chicago: University of Chicago Press.

Messerschmidt, Donald A. 1984. "The Thakali of Nepal: Historical Continuity and Socio-Cultural Change," *Ethnohistory* 29: 265-80.

—. 1989. "The Hindu Pilgrimage to Muktinath, Nepal. Part 1 and Part 2: Natural and Supernatural Attributes of the Sacred Field," *Mountain Research and Development* vol. 9, no. 2 (May): 89-104.

—, and Nareshwar Jang Gurung. 1974. "Parallel Trade and Innovation in Central Nepal: The Cases of the Gurung and Thakali Subbas Compared." In *Contributions to the Anthropology of Nepal*, edited by C. von Fürer-Haimendorf, pp. 197-221. Warminster: Aris and Phillips.

Mieu, George Paul. 2015. "'Beach-Boy Elders' and 'Young Big-Men': Subverting the Temporalities of Ageing in Kenya's Ethno-Erotic Economies," *Ethnos* vol. 80 (4): 472-496.

Mir, Farina. 2006. "Genre and Devotion in Punjabi Popular Narratives: Rethinking Cultural and Religious Syncretism," *Comparative Studies in Society and History*, vol. 48, 3 (July): 727-758.

Mittal, J.P. 2006. *History of Ancient India: From 4250 BC to 637 AD*. New Delhi: Atlantic Publishers.

Mol, Annemarie. 2002. *The Body Multiple: Ontology in Medical Practice*. Durham: Duke University Press.

Mongomerie, Thomas George. 1875. "Extracts from an Explorer's Narrative of His Journey from Pitgarh in Kumaon via Jumla to Tadum and Back Along the Kali Gandaki to British Territory," *Journal of the Royal Geographic Society of London* 45: 350-63.

Morgan, David. 1999. *Visual Piety: A History and Theory of Popular Religious Images*. Berkeley: University of California Press.

Morgan, David, ed. 2010. *Religion and Material Culture: The Matter of Belief.* London: Routledge.

Morinis, Alan. 1992. *Sacred Journeys: The Anthropology of Pilgrimage* (Contributions to the Study of Anthropology). Westport: Praeger.

Muktinath Foundation International (MFI) 2015. Website. http://www.muktinath.org/. Accessed 5 January 2015.

Munn, Nancy D. 1970. "The Transformation of Subjects into Objects in Walbin and Pitjantjatjara Myth." In *Australian Aboriginal Anthropology: Modern Studies in the Social Anthropology of the Australian Aborigines*, edited by Ronald M. Berndt. Nedlands, W.A.: University of Western Australia Press.

—. 1992. "The Cultural Anthropology of Time: A Critical Essay," *Annual Review of Anthropology* 21: 93-123.

Nandy, Ashish. 1990. "The Politics of Secularism and the Recovery of Religious Tolerance." In *Mirrors of Violence: Communities, Riots and Survivors in South Asia*, edited by Veena Das. New Delhi: Oxford University Press.

Nash, Joshua. 2012. "Re-examining Ecological Aspects of Vrindavan Pilgrimage." In *Flows of Faith: Religious Reach and Community in Asia and the Pacific*, edited by Lenore Manderson, Wendy Smith, and Matt Tomlinson. Dordrecht: Springer Science & Business Media.

Nelson, Clifford M. 1968. "Ammonites: Ammon's Horns into Cephalopods," *Journal of the Society for the Bibliography of Natural History* 5: 1-18.

Olwig, Karen Fog, and Kirsten Hastrup, eds. 1997. *Siting culture: The shifting anthropological object*. New York: Routledge.

Oppert, Gustav. 1901. "On Shaligrams: Sacred Stones of Indian Aborigines Become Emblems of the God Vishnu," *Review of the History of Religions* vol. 43: 325-332. (Sur Les Sālagrāmas: Pierres Sacrées Des Aborigènes De L'inde Devenuesem-blèmes Du Dieu Vishnou. *Revue de l'histoire des religions*).

Oppitz, Michael. 1968. *History and Social Order of the Sherpa*. (*Geschichte und Sozialordnung der Sherpa*). Innsbruck: Universitasverlag Wagner.

Ortner, Sherry B. 1989. *High Religion*. Princeton: Princeton University Press.

Östör, Ákos. 1980. *The Play of the Gods: Locality, Ideology, Structure, and Time in the Festivals of a Bengali Town*. Chicago and London: University of Chicago Press.

Otto, Rudolf. 1958. *The Idea of the Holy*. Oxford: Oxford University Press.

Page, Kevin N. 2008. "The Evolution and Geography of Jurassic Ammonoids," *Proceedings of the Geologists' Association*. vol. 119 (1): 35-57.

Palmié, Stephan. 2013. "Mixed Blessings and Sorrowful Mysteries: Second Thoughts about 'Hybridity'," *Current Anthropology* vol. 54, no. 4 (August): 463-482.

Parmentier, Richard. 1987. *The Sacred Remains: Myth, History and Polity in Belau*. Chicago: University of Chicago Press.

Patton, Laurie L. 2005. *Bringing the Gods to Mind: Mantra and Ritual in Early Indian Sacrifice*. Berkeley: University of California Press.

—, and Doniger, Wendy, eds. 1996. *Myth and Method*. Charlottesville and London: University Press of Virginia.

Paxson, Heather. 2008. "Post-Pasteurian Cultures: The Microbiopolitics of Raw-Milk Cheese in the United States," *Cultural Anthropology* 23 (1): 15-47.

Pedersen, Morten Axel. 2011. *Not Quite Shamans. Spirit Worlds and Political Lives in Northern Mongolia*. Ithaca, NY: Cornell University Press.

—. 2012. "The Task of Anthropology is to Invent Relations: For the Motion," *Critique of Anthropology* 32 (1): 59-65.

Pennington, Brian K. 2005. *Was Hinduism Invented? Britons, Indians, and the Colonial Construction of Religion*. Oxford: Oxford University Press.

Pigg, Stacy Leigh. 1996. "The Credible and the Credulous: The Questions of 'Villagers' Beliefs' in Nepal," *Cultural Anthropology* 11 (2): 160-201.

Povinelli, Elizabeth. 2006. *The Empire of Love: Towards a Theory of Intimacy, Genealogy, and Carnality*. Durham: Duke University Press.

—. 2016. *Geontologies: A Requiem to Late Liberalism*. Durham: Duke University Press.

Pye, Michael. 1991. "Reflections on the Treatment of Tradition in Comparative Perspective, with Special Reference to Ernst Troeltsch and Gerardus Van Der Leeuw." In *Syncretism in Religion: A Reader*, edited by Anita Maria Leopold and Jeppe Sinding Jensen, pp. 101-111. London: Routledge.

Raffles, Hugh. 2010. *Insectopedia*. 1st edition. New York: Pantheon Books

Rai, Ratan Kumar. 1994. *The Thakalis, Bon DKar, and Lamaist Monasteries Along the Kali Gaṇḍaki, the Ancient Salt Route in Western Nepal*. Delhi: Book Faith India.

Ramanujadas, Madhu Sudhan. 2003. *Muktichhetra Mahatmyamam*. Kathmandu: Dhaulagiri Osset Press.

Ramberg, Lucinda. 2014. *Given to the Goddess: South Indian Devadasis and the Sexuality of Religion*. Durham: Duke University Press.

Ramble, Charles. 1983. "The Founding of a Tibetan Village: The Popular Transformation of History," *Kailash* 10, no. 1: 267-290.

—. 1990. "How Buddhist are Buddhist Communities? The Construction of Tradition in Two Lamaist Villages," *Journal of the Anthropological Society of Oxford* 21, no. 2: 185-97.

—. 1992. The Archive of Baragaon. http://www.tibetanhistory.net/documents-from-mustang/baragaou/. Accessed 10 October 2016.

—. 2002. "Temporal disjunction and collectivity in Mustang, Nepal," *Current Anthropology* 43, supplement, August-October.

—. 2008a. *Tibetan Sources for a Social History of Mustang, Nepal. Volume 1: The Archive of Te*. Halle: International Institute for Tibetan and Buddhist Studies.

—. 2008b. *The Navel of the Demoness: Tibetan Buddhism and Civil Religion in the Highlands of Nepal*. Oxford: Oxford University Press.

—. 2014. "The Complexity of Tibetan Pilgrimage." In *Searching for the Dharma, Finding Salvation – Buddhist Pilgrimage in Time and Space*. Proceedings of the Workshop "Buddhist Pilgrimage in History and Present Times" at the Lumbini International Research Institute (LIRI), Lumbini, 11-13 January 2010. Edited by Christoph Cueppers and Max Deeg. Lumbini International Research Institute.

—, and C. Seeber. 1995. "Dead and Living Settlements in the Shoyul of Mustang," *Ancient Nepal* no. 138: 107-130.

—, and Michael Vinding. 1987. "The Bem-chag Village Record and the Early History of Mustang District," *Kailash* 13, nos. 1-2: 5-47.

Rapport, Nigel. 1998. "Coming Home to a Dream: A Study of the Immigrant Discourse of 'Anglo-Saxons' in Israel." In *Migrants of Identity: Perceptions of Home in a World of Movement*, edited by N. Rapport and A. Dawson. Oxford: Berg.

Regmi, Mahesh. 1970. (ed.) "Mustang Rajya," *Regmi Research Series* 2 (4): 99.

—. 1976. *Landownership in Nepal.* Berkeley: University of California Press.

—. 1978. *Thatched Huts and Stucco Palaces: Peasants and Landlords in 19th-Century Nepal.* New Delhi: Vikas.

Romano, Marco. 2014. "From petrified snakes, through giant 'foraminifers', to extinct cephalopods: the early history of ammonite studies in the Italian peninsula," *Historical Biology* 27 (2) 214-235.

Rubin, Gayle. 1975. "The Traffic in Women: Notes on the 'Political Economy' of Sex." In *Toward an Anthropology of Women*, edited by Rayna R. Reiter, pp. 157-210. New York: Monthly Review.

Sahlins, Marshall. 2013. *What Kinship Is – And Is Not.* Chicago: University of Chicago Press.

Sakai, Harutaka. 1989. "Rifting of the Gondwanaland and Uplifting of the Himalayas Recorded in Mesozoic and Tertiary Fluvial Sediments in the Nepal Himalayas." In *Sedimentary Facies in the Active Plate Margin*, edited by A. Taira and F. Masuda, pp. 723-732. Tokyo: Terra Scientific Publishing Company.

Salazar, Noel B. 2010. "Towards an Anthropology of Cultural Mobilities," *Crossings: Journal of Migration and Culture* 1 (1): 53-68.

Sanders, Clint. 1999. *Understanding dogs.* Philadelphia: Temple University Press.

Schaik, Sam van. 2011. *Tibet: A History.* New Haven: Yale University Press.

Schattschneider, Ellen. 2003. *Immortal Wishes: Labor and Transcendence on a Japanese Sacred Mountain.* Durham: Duke University Press.

Schieffelin, Edward. 1985. "Performance and the Cultural Construction of Reality," *American Ethnologist* 12, 4: 707-724.

Schneider, David Murray. 1984. *A Critique of the Study of Kinship.* Ann Arbor: University of Michigan Press.

Schuh, Dieter. 1994. "Investigations in the History of the Muktinath Valley and Adjacent Areas, Part I," *Ancient Nepal* 137: 9-91.

—. 1995. "Investigations in the History of the Muktinath Valley and Adjacent Areas, Part II," *Ancient Nepal* 138: 5-54.

Schweig, Graham M. 2007. *Dance of Divine Love: India's Classical Sacred Love Story, The Rasa Lilas of Krishna.* Delhi: Motilal Banarsidass Publishers.

—. 2004. "Krishna, the Intimate Deity." In *The Hare Krishna Movement: The Postcharismatic Fate of a Religious Transplant*, edited by Edward F. Bryant and Maria L. Ekstrand. New York: Columbia University Press.

Sharma, Prayag Raj. 1977. "Caste, Social Mobility, and Sanskritization: A Study of Nepal's Old Legal Code," *Kailash* 5, no. 4: 277-99.

Shaw, Rosalind, and Charles Stewart, eds. 1994. *Syncretism/Anti-Syncretism: The Politics of Religious Synthesis* (European Association of Social Anthropologists). London: Routledge.

Shedden, Rikardo. 2008. "Of Kalinga Rice Rituals and the Conflation of Traditions." In *Culture and development in Southeast Asia*, edited by D. Wangsgard, pp. 109-23. Bangkok: White Lotus Press.

Shukla, Deeptima. 2006. "India-Nepal Relations: Problems and Prospects," *The Indian Journal of Political Science*, vol. 67, no. 2 (April-June): 355-374.

Siddhantashastree, Rabindra Kumar. 1985. *Vaishnvaism Through the Ages*. Delhi: Munshiram Manoharlal Publishers.

Singh, Vikash. 2017. *Uprising of the Fools: Pilgrimage as Moral Protest in Contemporary India*. Stanford: Stanford University Press.

Snellgrove, David. 1958. *Buddhist Himalaya*. Oxford: Bruno Cassirer.

—. 1961. *Himalayan Pilgrimage: A Study of Tibetan Religion by a Traveller Through Western Nepal*. Oxford. Cassirer.

—. 1965. "Review of *The Sherpas of Nepal* by C. Furer-Haimendorf," *Asian Review of Art and Letters* New Series 2.1.

Spivak, Gayatri Chakravorty. 1988. "Can the Subaltern Speak?" In *Marxism and the Interpretation of Culture*, edited by Cary Nelson and Lawrence Grossberg. Urbana: University of Illinois Press.

Stallybrass, Peter, and Allon White. 2002. "The Politics and Poetics of Transgression." In *A Reader in the Anthropology of Religion*, edited by Michael Lambeck. Oxford: Blackwell Publishing.

Stewart, Charles. 2005. "Relocating Syncretism in Social Science Discourse." In *Religious Syncretism: A Reader*, edited by Anita Marie Leopold and Jeppe Sinding Jensen. Routledge: New York.

Stewart, Pamela J., and Andrew Strathern. 2015. *Landscape, Memory and History: Anthropological Perspectives* (Series: Anthropology, Culture and Society). London: Pluto Press.

Stewart, Susan. 1993. *On Longing: Narratives of the Miniature, the Gigantic, the Souvenir, the Collection*. Durham and London: Duke University Press.

Stiller, Ludwig F. 1973. *The Rise of the House of Gorkha: A Study in the Unification of Nepal, 1768-1816*. Kathmandu: Ratna Pustak Bhandar.

Stone, Linda. 1976. "Concepts of Illness and Curing in a Central Nepal Village," *Contributions to Nepalese Studies* 3 (Special Issue): 55-80.

Strathern, Marilyn. 2005. *Kinship, Law, and the Unexpected: Relatives Are Always a Surprise*. New York: Cambridge University Press.

Sweetser, Eve. 2012. "Viewpoint and Perspective in Language and Gesture, from the Ground Down." In *Viewpoint in Language: A Multimodal Perspective*, pp. 1-24. Cambridge: Cambridge University Press.

Taves, Ann. 2009. *Religious Experience Reconsidered: A Building-Block Approach to the Study of Religion and Other Special Things*. Princeton: Princeton University Press.

Taylor, Paul D. 2012. "Shaligrams and Vishnu: Ammonites in Hinduism," *Batalleria* 17: 39-44.

Thapa, Netra B. 1990. *A Short History of Nepal*. Kathmandu: Ratna Pustak Bhandar.

Thompson, Matt. 2012. "Digital Anthropology Group Is Happening Now," *Savage Minds*. Retrieved 31 January 2017.

Tilley, Christopher. 2001. "Ethnography and Material Culture." In *Handbook of Ethnography*, edited by P. Atkinson et al., pp. 258-272. London: Sage Publications.

Tilman, Harold W. 1952. *Nepal Himalaya*. Cambridge: Cambridge University Press.

Trawick, Margaret. 1990. *Notes on Love in a Tamil Family*. Berkeley and Los Angeles: University of California Press.

Tremlett, Paul-Francois. 2003. "The problem of belief: A response to Matthew Engelke," *Anthropology Today* 19 (4): 24.

Tsing, Anna Lowenhaupt. 1993. *In the Realm of the Diamond Queen*. Princeton: Princeton University Press.

—. 2005. *Friction: An Ethnography of Global Connection*. Princeton and Oxford: Princeton University Press

—. 2016. *The Mushroom at the End of the World: On the Possibility of Life in Capitalist Ruins*. Princeton: Princeton University Press.

Tucci, Giuseppe. [1953] 1982. *Journey to Mustang*. Translated by Diana Fussell. Kathmandu: Ratna Pustak Bhandar.

Turner, Victor. 1969. "Planes of Classification in a Ritual of Life and Death." In *The Ritual Process: Structure and Anti-Structure*. London and New York: Routledge.

—, and Edith Turner. 1978. *Image and Pilgrimage in Christian Culture*. New York: Columbia University Press.

Uberoi, Patricia. 1993. *Family, Kinship, and Marriage in India*. Delhi: Oxford University Press.

Urry, John. 2000. *Sociology Beyond Societies: Mobilities for the Twenty-First Century*. London: Psychology Press.

—. 2002. "Mobility and Proximity," *Sociology* 36 (2): 255-73.

Van der Geer, Alexandra, Michael Dermitzakis, and John De Vos. 2008. "Fossil Folklore from India: The Siwalik Hills and the Mahabharata," *Folkore* 119: 71-92.

Viveiros de Castro, Eduardo. 2009. *Métaphysiques cannibales: Lignes d'anthropologie post-structurale* (Cannibalistic Metaphysics: Lines of Post-structural Anthropology). Paris: Presses universitaires de France.

—. 1998. "Cosmological deixis and Amerindian perspectivism," *Journal of the Royal Anthropological Institute* 4 (3): 469-88.

Vroom, Hendrik M. 2005. "Syncretism and Dialogue: A Philosophical Analysis." In *Religious Syncretism: A Reader*, edited by Anita Marie Leopold and Jeppe Sinding Jensen. Routledge: New York.

Vygotsky, Lev Semyonovich. (1933) 1976. "Play and its Role in the Mental Development of the Child." Translated by C. Mulholland, in *Play – Its Role in Development and Evolution*, edited by Jerome Brunner, Alison Jolly, and Kathy Sylva, pp. 537-554. New York: Basic Books.

Walter, Damian. 2003. "Among Spirits and Deities: Diverse Shamanisms in the Nepal Himalayas." *CSQ* 27.2 (Summer) Shamanisms and Survival. Available online at: http://www.culturalsurvival.org/publications/cultural-survival-quarterly/nepal/among-spirits-and-dietiesdiverse-shamanisms-nepal-him. Accessed 5 October 2016.

Walters, Holly. 2016. "Playing God: Participant Frameworks in the Ras Lilas of Krishna," *Journal of Religion and Popular Culture* vol. 28, 2-3 (Fall).

Waterhouse, E. 1940. *Religion in Essence and Manifestation*. By G. Van Der Leeuw. Translated by J. E. Turner. (London: George Allen & Unwin.1938, pp. 709).

Weiner, Annette. 1992. *Inalienable Possessions: The Paradox of Keeping-While-Giving*. Berkeley: University of California Press.

Weston, Kath. 1997. *Families We Choose: Lesbians, Gays, Kinship*. Rev. ed. New York: Columbia University Press.

Whitehead, Amy. 2013. *Religious Statues and Personhood: Testing the Role of Materiality*. London: Bloomsbury Academic Publishers.

Williams, Raymond. 1977. *Marxism and Literature*. Oxford: Oxford University Press.

Winnicott, Donald W. 1971. *Playing and Reality*. London: Routledge.

Wulff, Donna M. 1984. *Drama as a Mode of Religious Realization: The Vidagdhamadhava of Rupa Gosvami*. Chico, CA: Scholars Press.

—. 1982. *The Divine Consort: Radha and the Goddesses of India*, Berkeley: University of California Press.

—, and Hawley, John Stratton. 1996. *Devi: Goddesses of India*. Berkeley and Los Angeles: University of California Press.

Yamba, C. Bawa. 1995. *Permanent Pilgrims: The idea of pilgrimage in the lives of West African Muslims*. London: Edinburgh University Press for the International African Institute.

Zelliot, Eleanor. 2010. "The Early Voices of Untouchables." In *From Stigma to Assertion: Untouchability, Identity and Politics in Early and Modern India*, edited by Aktor, Michael and Robert Deliège. Copenhagen: Museum Tusculanum Press.

Index

For Product Safety Concerns and Information please contact our EU representative GPSR@taylorandfrancis.com
Taylor & Francis Verlag GmbH, Kaufingerstraße 24, 80331 München, Germany

www.ingramcontent.com/pod-product-compliance
Ingram Content Group UK Ltd.
Pitfield, Milton Keynes, MK11 3LW, UK
UKHW020015200926
475702UK00014B/1890
9781041185994